D1566270

CIRCUITS AND SYSTEMS FOR WIRELESS COMMUNICATIONS

Circuits and Systems for Wireless Communications

Edited by

Markus Helfenstein

and

George S. Moschytz

Swiss Federal Institute of Technology, Zurich

KLUWER ACADEMIC PUBLISHERS

BOSTON / DORDRECHT / LONDON

A C.I.P. Catalogue record for this book is available from the Library of Congress.

ISBN 0-7923-7722-2

Published by Kluwer Academic Publishers,
P.O. Box 17, 3300 AA Dordrecht, The Netherlands.

Sold and distributed in North, Central and South America
by Kluwer Academic Publishers,
101 Philip Drive, Norwell, MA 02061, U.S.A.

In all other countries, sold and distributed
by Kluwer Academic Publishers,
P.O. Box 322, 3300 AH Dordrecht, The Netherlands.

Printed on acid-free paper

Printed in the Netherlands.

Contents

Preface

This book contains revised contributions by the speakers of the 1st IEEE Workshop on Wireless-Communication Circuits and Systems, held in Lucerne, Switzerland, from June 22–24, 1998. The aim of the workshop was to apply the vast expertise of the CAS Society in the area of circuit and system design to the rapidly growing field of wireless communications. The workshop combined presentations by invited experts from academia and industry with panel and informal discussions. The following topics were covered:

- RF System Integration (single-chip systems, CMOS RF circuits),
- RF Front-End Circuits (CMOS RF oscillators, broadband design techniques),
- Wideband Conversion for Software Radio (A/D conversion issues, wideband sub-sampling, low-spurious A/D conversion),
- Process Technologies for Future RF Systems (Si, SiGe, GaAs, CMOS, packaging technologies),
- DSP for Wireless Communications (DSP algorithms, fixed-point systems, DSP for baseband applications),
- Blind Channel Equalization (adaptive interference suppression, design techniques, channel estimation).

The workshop was a great success, with over 130 participants from 19 countries, from the U.S. to Europe and Asia, including a large contingent of participants from industry (60 %). Feedback from the participants showed that the carefully selected combination of tutorial-like lectures with lectures on specialized and advanced topics was a feature of the workshop that was particularly appreciated. Due to the relatively strong involvement of industry — both in the form of lecturers and listeners — a high level of discussion was attained in both panel sessions and informal gatherings.

By a stroke of luck, Philips Semiconductors, Zurich, Switzerland, celebrated their 50th anniversary in 1998. With the well-deserved spirit of celebration and generosity that this anniversary triggered at Philips, and the goodwill and encouragement of the Board of Governors and Excom members of the CAS Society, the finances were guaranteed at an early stage. For this the organizers were very grateful.

The format of this workshop and book has strongly been influenced by the AACD workshop series organized by J. H. Huijsing, R. J. van de Plassche and W. Sansen. The editors greatly appreciate the inspiration provided by this series.

It is a pleasure to acknowledge the speakers and authors for making available their expertise. Our sincerest thanks go also to D. Arnold, M. Goldenberg, D. Lím, F. Lustenberger, H. Mathis, and H. P. Schmid for helping in the preparation of the workshop and book. Moreover, D. Lím, H. Mathis and H. P. Schmid were strongly involved in the technical editing of this book, which the editors very gratefully acknowledge.

Zurich, September 1999
Markus Helfenstein and George S. Moschytz

I RF System Integration

1 RF SYSTEM INTEGRATION

Chris Toumazou

Dept. Elect. & Electron. Eng.
Imperial College
Exhibition Rd.
London SW7 2BT, U.K.

1.1 INTRODUCTION TO THE FOLLOWING PAPERS

The idea for this part of the book arose from a need to convey the intricacies of RF system integration in today's wireless information system arena. The proliferation of portable communication devices has created a high demand for small and inexpensive transceivers with low power consumption. The Radio-Frequency (RF) and wireless-communications market has suddenly expanded to unimaginable dimensions. Sources predict that mobile telephony subscriptions will increase to over 350 Million in the year 2000.

While this trend continues and challenges still exist, the RF practice at present is one which requires the so-called "green fingers" of design. System integration from the interconnection of sub-blocks at the front-end to the interconnection of metal layers within the IC is all highly interactive. However, what has now become apparent at the low-GHz frequency band is that as feature sizes of silicon devices shrink, there is a more natural move towards VLSI at high frequencies where lumped circuit design is beginning to find a new home. As we start integrating more of the board-level components onto a single chip, the board layout will then also form a major part of the "lumped" circuit.

Part I has brought together key engineers from industry and academia to shed light on performance demands, board level design, and sub-micron silicon CMOS solutions, taking the reader through realistic design scenarios for RF system integration.

Devices and systems, such as pagers, cellular and cordless phones, cable modems, mobile faxes, PDAs (Personal Digital Appliances), wireless LANs, and RF identification tags are rapidly penetrating all aspects of our lives, evolving from luxury items to indispensable tools. Semiconductor and system companies, small and large, analogue and digital, have seen the statistics and are striving to capture their own market share by introducing various RF products. Today's pocket phones contain more than one million transistors, with only a small fraction operating in the RF range and the rest performing low-frequency "base-band" analogue and digital signal processing. However, the RF section is still the design bottleneck of the entire system.

In contrast to other types of analogue and mixed-signal circuits, RF systems demand not only a good understanding of integrated circuits, but also of many areas that are not directly related to integrated circuits.

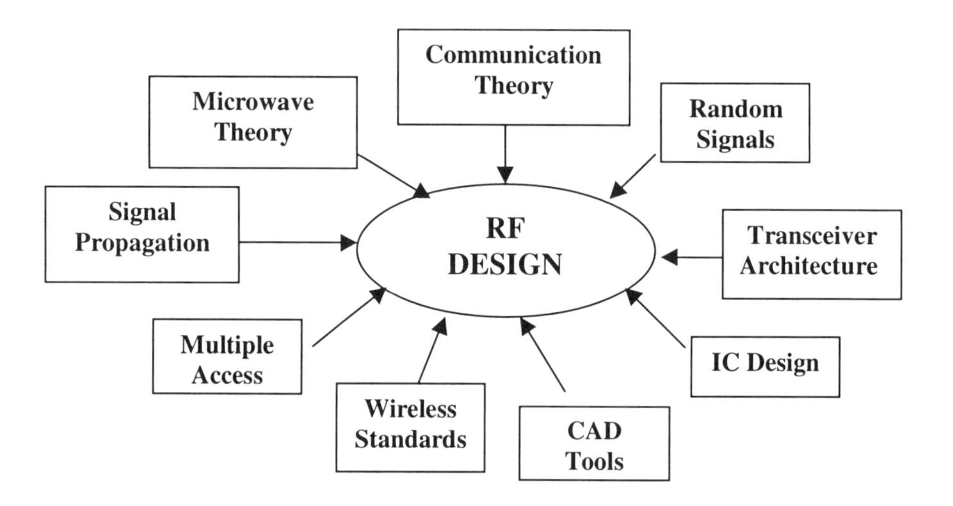

Figure 1.1 RF design disciplines.

Most of the areas shown in Fig. 1.1 have been studied extensively for more than two decades, making it difficult for an IC designer to acquire the necessary knowledge in a reasonable amount of time. Traditional wireless system design has thus been carried out at somewhat disjointed levels of abstraction: communication theorists create the modulation scheme and base-band signal processing; RF system experts plan the transceiver architecture; IC designers develop each of the building blocks; and manufacturers "glue" the ICs and other external components together. In fact, architectures are often planned according to the available off-the-shelf components, and ICs are designed to serve as many architectures as possible, leading to a great deal of redundancy at both system and circuit levels. This results in higher levels of power consumption and generally lower performance.

Most recently, as the industry moves toward higher integration and lower cost, RF and wireless design increasingly demands more "concurrent engineering," thereby requiring IC designers from both industry and academia to combine forces and to have a sufficient and integrated knowledge of all the disciplines [1]. RF circuits must process analogue signals with a wide dynamic range at high frequencies. It is interesting to note that the signals must be treated as analogue even if the modulation is digital or the amplitude carries no information.

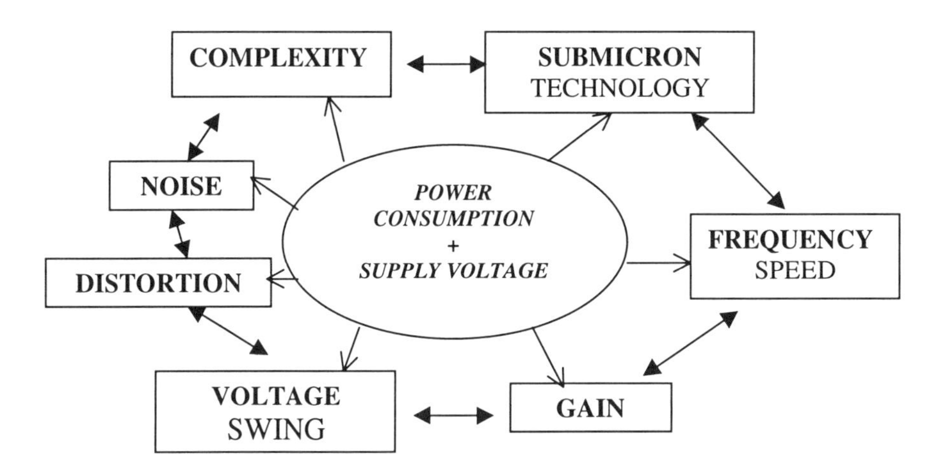

Figure 1.2 RF design hexagon.

The trade-offs involved in the design of such circuits can be summarised in the "RF design hexagon" shown in Fig. 1.2. While any of the seven parameters trade to some extent, all these parameters are severely constrained by the core parameters, namely the power consumption and supply voltage. It is important to recognise that, while digital circuits directly benefit from advances in IC technologies, RF circuits do not benefit as much. This issue is exacerbated by the fact that RF circuits often require external components—for example, inductors—that are difficult to bring onto the chip even in modern IC processes. RF design techniques are thus becoming highly sensitive to device physics, and so analog characterisation of digital VLSI technology is of primary concern. One of the major challenges is implementing RF circuits on ICs instead of PCBs, offering advantages including lower production cost, high functionality, small physical size, high reliability, and low power requirements. It now becomes very necessary to achieve better co-ordination between the "system design" activity and the "RF circuit design" activity.

A few years ago Gallium-Arsenide (GaAs) technology was the primary-choice semiconductor for implementing RF ICs due to its low noise figure, higher gain and higher output power. Advances in sub-micron silicon CMOS, however, have made it possible to achieve higher levels of RF system integration at lower cost than with GaAs, predominantly for low-GHz-band wireless applications [2]. The other benefits of CMOS RF are the greater manufacturability and minimised power requirements to drive off-chip loads. While integrated silicon BJT transceivers are still more desirable for today's products, CMOS RF solutions are looking very promising, with the realistic prospect of a single-chip transceiver in a plastic package. Furthermore, newer device technologies such as Silicon Germanium are maturing rapidly and offer the high mobility necessary for today's RF wireless products. This array of competing technologies offers system designers more creative opportunity, and the best wireless transceiver solutions may well emerge from system design evolving together with architecture, circuits, antennas, and power allocation plans. In the future, base-band signal processing will inevitably make up for imperfections in the front end (e.g. software radio).

Part I begins with a section by Gordon Aspin from TTP Communications, a company with vast experience in RF system integration for cellular products. The section describes the realistic design of a part of a fully integrated transceiver IC from Hitachi which satisfies multi-band GSM RF specifications. Some of the subtleties of board-level integration are presented, coupled with a design approach which attempts to make practical GSM handset design a more straightforward task. Emphasis is placed upon the importance of understanding total system-level requirements when designing a chip, and upon how board level design influences low-level requirements.

In the next section, Peter Mole from Nortel Semiconductors gives an overview of system integration on a chip. Peter takes us through a number of practical RF design issues and then discusses general problems that radio systems must overcome to achieve acceptable performance. The section overviews practical concerns for both receiver and transmitter and how different radio architectures can be utilised to overcome some of the problems. The section concludes with a number of practical design issues for integrating radio circuitry in silicon technology.

The final two sections concentrate on the design of fully integrated transceiver chips in sub-micron and deep sub-micron Silicon CMOS technologies. Michiel Steyaert from the Katholieke Universiteit of Leuven introduces us to the arena of using deep sub-micron CMOS to create single-chip transceiver blocks and components such as LNAs, VCOs , up-converters , synthesisers etc. to satisfy cellular performance specifications above 1 GHz. The section discusses all the bottlenecks and challenges of RF CMOS using plain deep sub-micron devices for integration within systems such as DECT, GSM, and DCS 1800.

Finally, Qiuting Huang et. al. from the Integrated Systems Laboratory at ETH Zurich presents a practical high-performance GSM transceiver front-end in a 0.25 μm CMOS process. This section concludes Part I by taking the reader through a practical RF system integration example. The work demonstrates that excellent RF performance is feasible with 0.25 μm CMOS, even in terms of the requirements of the super-heterodyne architecture. Design for low noise and low power for GSM handsets has been given particular attention.

In conclusion, Part I will give the reader a practical evaluation of state- of-the-art RF system design and integration for GHz wireless communications. The chapters in Part I encompass the failures, successes, and most of all the realistic RF challenges to enable total integration of portable future wireless information systems.

References

[1] B. Razavi, "Challenges in Portable Transceiver Design", *Circuits and Devices Magazine*, IEEE 1996..

[2] K. T. Lin, *Private Communication*, Imperial College 1999.

2 RF SYSTEM BOARD LEVEL INTEGRATION FOR MOBILE PHONES

Gordon J. Aspin

TTP Communications Ltd.
Melbourn Science Park
Royston, SG8 6EE, U.K.

Abstract: This paper highlights the importance of system design choices in the development of RF chip sets for cellular mobile phones. By way of example, the design of the BRIGHT family of RF chip sets for GSM is described. Performance results are presented based on these chip sets.

2.1 INTRODUCTION

Traditionally, the RF system represents one of the most difficult challenges for companies developing digital cellular phones, particularly for companies with limited development experience. It is now possible to buy more or less complete baseband-chip-set solutions together with associated software from a number of vendors. However, although off-the-shelf radio-chip-set solutions are available, they nevertheless require a significant level of design expertise on the part of the handset designer in order to realise a manufacturable product which meets the necessary performance requirements. This paper describes the results of some of our work to meet the challenge of realising an RF chip set which makes designing a GSM mobile phone a straightforward task.

This work, the development of the BRIGHT[1] RF chip set family, has been carried out in collaboration with Hitachi Semiconductor of Japan.

The initial devices were targeted at single-band products — the main GSM market at that time. Even more important commercially, however, has been the emergence of dual-band 900/1800 capability as a key market requirement. Fortunately, the BRIGHT architecture is particularly well suited to multi-band operation, and the latest-generation devices, BRIGHT2, are designed to support this mode.

The objective of the development has been to achieve the maximum level of integration in the radio system, compatible with the objectives of lowest total system cost, lowest part count, and ease of design and manufacture, within the available IC process technology. The approach is a systems approach with the focus upon achieving a device design in which the radio requirements of the whole product are most simply and elegantly met.

2.2 DESIGN APPROACH

What makes RF IC design particularly interesting is that the silicon represents only a small part of the total RF system but can make a very large difference to how easy or difficult it is to design the rest of the system. It is therefore vital to understand the total system requirements when designing the chip. It is also important to be able to validate the design at both device and system level. This is achieved by means of a System Evaluation Board — essentially a complete handset — on which all the system performance parameters can be measured and confirmed (Fig. 2.1). Validation of the system is often one of the most expensive parts of the development.

This radio System Evaluation Board is designed to be compatible with Baseband Evaluation Boards used for baseband chip set and protocol software development work. The two boards can be plugged together to emulate complete handset operation. As well as testing in our own laboratories we have even been able to take such a system through "Type Approval" at a GSM test house to verify radio system performance.

2.3 KEY GSM SYSTEM SPECS

As with all modern radio standards, there are many specification points for GSM that need to be met in order for the radio to conform to the standard. However, a small number of these specification points turn out to be critical to the radio architecture and design.

2.3.1 Transmitter phase error

As a digital phase modulation system, GSM controls how closely the transmitted phase follows the ideal modulated phase trajectory. It does this by specifying the phase error in terms of an RMS value (5°) and a peak value (20°), across the useful part of the burst. The phase error is a very critical parameter for the transmitter, because so many things can contribute to it, including digital modulator phase error, synthesiser settling

[1]BiCMOS Radio IC for GSM by Hitachi and TTPCom

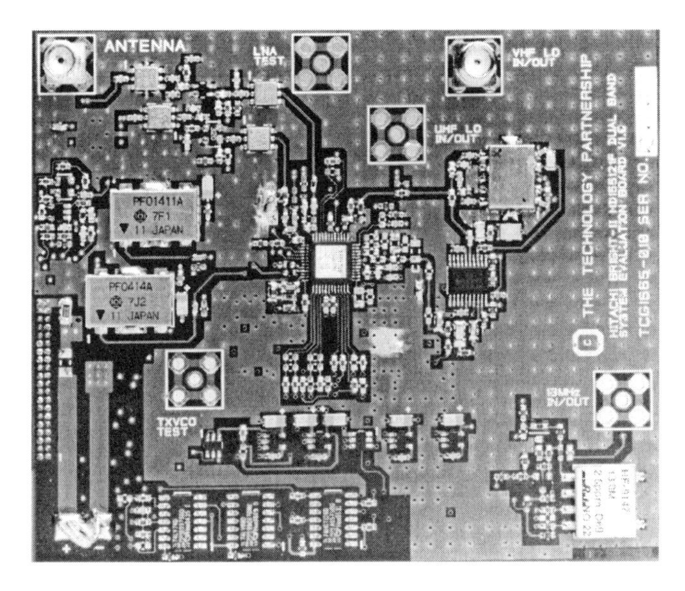

Figure 2.1 BRIGHT 2 system evaluation board.

time, switching transients, I/Q gain and phase imbalance, up-converter phase noise and spurious modulation, and power amplifier AM–to–PM conversion, etc.

In putting together a radio system design, budgets are allocated to each of these parameters. Some, however, can be difficult to quantify theoretically (e.g. switching transients) and may only become apparent once the design is realised in hardware. The task of the system designer is to identify a system architecture in which uncontrollable and unquantifiable effects are minimised.

A classic example of a problematic architecture is the direct-up-converter transmitter, in which a baseband I/Q modulator is mixed up to the final frequency in a single stage. As a result, the up-converter's local oscillator runs also at the final frequency. Maintaining sufficient isolation between the modulated power amplifier output and the unmodulated low-power local oscillator is very difficult, particularly within a miniature handset.

2.3.2 *Transmitter modulation spectrum (Fig. 2.2)*

Like the phase error, the modulation spectrum is affected by many factors, including digital modulator spectrum, I/Q gain and phase imbalance, up-converter phase noise and spurious modulation, power amplifier AM–to–PM conversion, transmitter noise floor, etc. A good system design will minimise these effects inherently in the design.

2.3.3 *Transmitter noise in the receiver band (Fig. 2.3)*

To avoid interference between handsets in close proximity, GSM limits the amount of spurious radiation emitted from the antenna into the receive band. In the non-extended band, this is limited to −79 dBm in a 100kHz bandwidth, or −129 dBm/Hz. Reflected

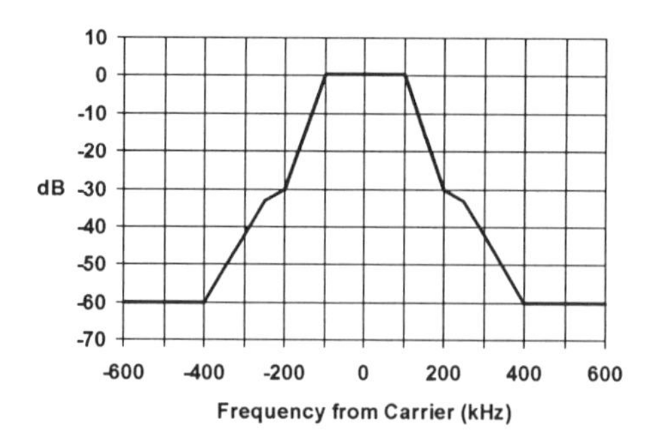

Figure 2.2 Transmit spectrum mask.

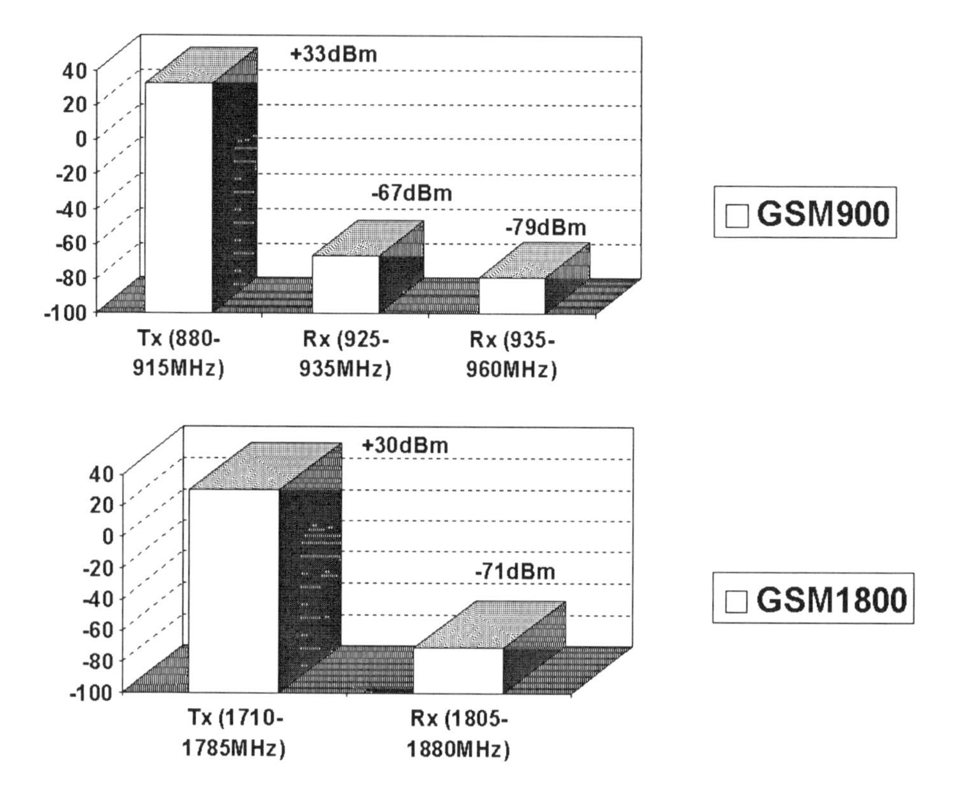

Figure 2.3 Transmit noise in receive bands.

back to the input of the power amplifier this typically corresponds to a noise level of −162 dBm/Hz or 12 dB above the ideal thermal noise floor. In practice, modulators cannot achieve this level of performance, and typically around 20 dB or more of filtering is required in a duplexer — which is a relatively bulky and expensive component.

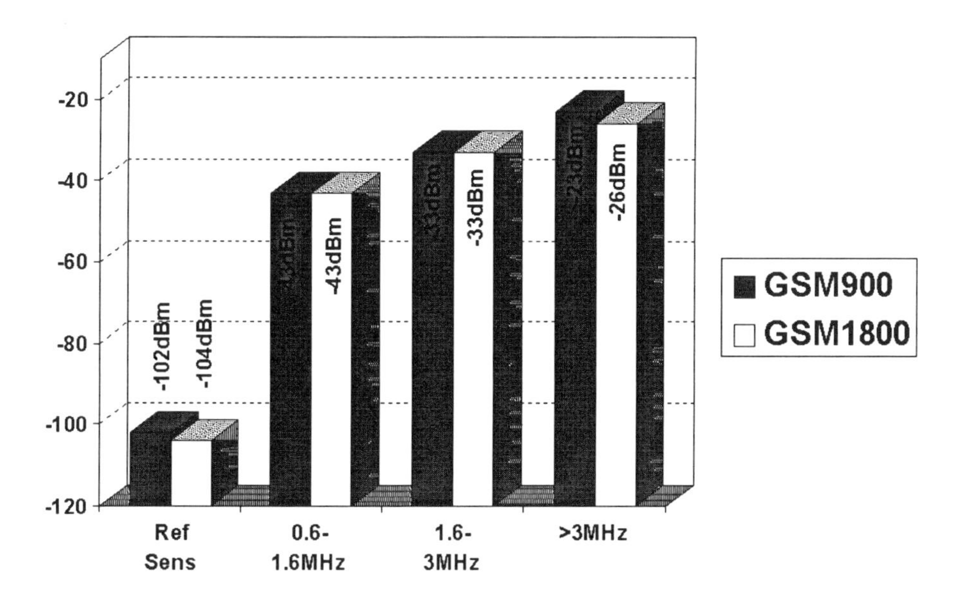

Figure 2.4 Receiver sensitivity & blocking requirements.

In the extended GSM900 band, the specification is relaxed by around 12 dB to avoid even more severe duplexer filtering requirements. In the 1800 MHz bands, the specifications are comparable to the E-GSM requirements scaled by the frequency.

2.3.4 Receiver blocking vs. sensitivity (Fig. 2.4)

By themselves, neither the receiver blocking nor the receiver sensitivity requirements is intrinsically difficult to meet: it is achieving both requirements simultaneously within a low-power, low-cost handset that is the design challenge. The sensitivity requirements translate into system noise figures of about 10.2 dB at 900 MHz and 12.2 dB at 1800 MHz for a typical GSM equaliser realisation. It is good practice to design for a nominal performance at least 4 dB better than these figures to give adequate production margin. The most challenging blocking requirement is the 3 MHz blocking figure which is almost 80 dB up on the wanted signal (74 dB for GSM1800). Translated into voltages on chip, it is necessary to receive a signal at a level of around 2 μV, without distortion, in the presence of a blocking signal of around 20 mV. This degree of linearity must be maintained by the receiver front end, i.e. the LNA, and particularly the mixer, before the blocking signal amplitude can be reduced in the first IF filter.

2.4 ARCHITECTURE CHOICES

The requirements highlighted in the previous section are reflected into various architectural choices in the design of the BRIGHT devices. On the transmitter, the key choice

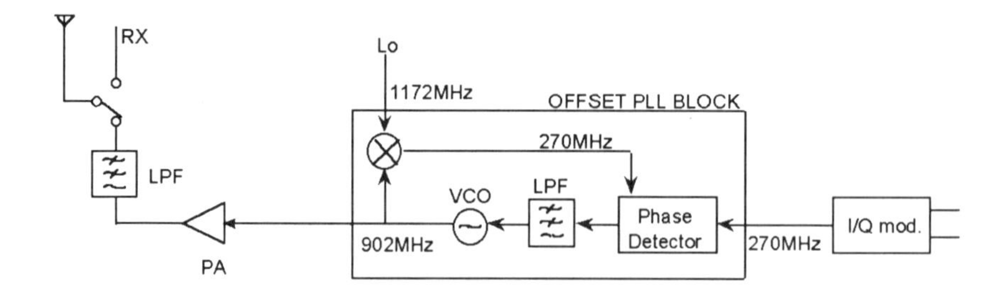

Figure 2.5 Offset phase locked loop transmitter.

is the use of an offset phase-locked loop. This approach has substantial benefits for systems like GSM which employ constant envelope modulation.

As a result of the strict phase and modulation accuracy requirements of GSM, it is only really feasible to perform transmit carrier modulation in the digital domain. This must then be converted into an analogue signal and translated to the required transmit frequency. In a conventional approach, the frequency translation is performed in one or more up-conversion stages. One of the problems of trying to do this operation in a single stage, when the mixer's local oscillator runs at the same frequency as the transmitted signal, is that of cross-coupling between the modulated high-level transmitter power output and the unmodulated low-level VCO. A solution to this problem is to use multiple conversions, but each conversion stage introduces spurious conversion products which must be filtered out. A better approach is the one illustrated in Fig. 2.5. In this scheme, the required signal modulation is impressed upon a VCO by means of a phase locked loop. The modulated VCO output is mixed down to a suitable intermediate frequency and compared with the signal from the digital modulator, which has been up-converted to the same IF. The loop comparison frequency is not critical and is chosen to provide a suitable frequency plan. Channelisation is supported by stepping the local oscillator as normal.

The bandwidth of the loop has a significant effect upon overall transmitter performance. The low-pass filter in the offset PLL must be chosen to optimise the following parameters:

- Minimum loop settling time (following frequency steps).
- Good phase tracking (to give minimum phase error).
- Minimum in-band spurii (to meet modulation spectrum requirements).
- Suppression of wide-band modulator noise (to meet transmit noise in receive band requirements).

Typically, a loop bandwidth of around 1 MHz is found to be optimum. Using a basic offset PLL design of the form shown has clear benefits:

- The VCO is inherently a constant-envelope device, hence no spurious amplitude modulation occurs in the signal driving the PA, and therefore we find no spurious PM generated in the PA through AM–PM conversion effects.

- The noise floor of the VCO is sufficiently low so that it can meet the receiver-band noise-floor requirements without further filtering. The duplex filter can be replaced by a simple low-pass harmonic filter and a transmit/receive switch, with considerable cost and space savings.

- Removing the duplex filter reduces the loss in the transmit path by around 1 dB. This implies up to 25% longer talk time for the handset.

- Removing the duplex filter removes a major cause of ripple in the transmit band. This allows the handset manufacturer to make use of the margin in the transmit level specifications to operate the handset closer to the minimum level, thus giving further battery life improvements.

In the receiver, the architectural choice is mainly about the number of conversion stages and the frequency plan. A two-IF approach has been chosen for BRIGHT, with a 225 MHz SAW filter defining the first IF. The use of a relatively high first IF means that image frequencies and other spurious responses from the first mixer cause no particular problems. The majority of receiver gain occurs at the second IF of 45 MHz, and it is here also that the AGC is applied. This approach minimises power consumption in the receiver. In our reference baseband solution (the Analog Devices GSM baseband chip sets), the channel filtering occurs digitally as part of the analogue–to–digital converter — another example of system design choices. Thus, at 45 MHz, only a relatively wide-band LC filter is required to provide protection against blocking of subsequent stages. The architecture of the complete 900 MHz BRIGHT device is shown in Fig. 2.6. The main UHF local oscillator runs at 1150 to 1185 MHz and drives both the receiver first mixer and the transmitter offset mixer. By ensuring that the receiver first IF and the transmitter offset loop comparison frequency are spaced by the Tx/Rx duplex offset (i.e. 45 MHz) the pulling range of the first local oscillator is minimised to the operating bandwidth of 35 MHz.

The loop comparison frequency is chosen to be a multiple of 45 MHz (in this case $6 \times 45\,\text{MHz} = 270\,\text{MHz}$), to give a simple scheme for generating all other required frequencies by on-chip division as shown.

In terms of integration, all the transmit and receive silicon functionality is integrated onto a single device, but the PLL synthesiser functions for the two main oscillators are on a separate device. This choice was based on an assessment of the risk factors involved versus the benefits. There is virtually no difference in printed circuit board area whether or not the synthesisers are integrated, yet there is a significant risk of noise in the digital PLL synthesiser leaking into the analogue parts of the chip. Furthermore, the PLL synthesisers are established parts.

The front-end low noise amplifier is also not fully integrated within the BRIGHT device. Initial studies identified that it would be very difficult to achieve the necessary noise figure and blocking performance from an integrated amplifier. Instead it was decided to integrate the bias function for the LNA, and to use an external transistor as the amplifying device. The final die for the BRIGHT 900 MHz part is shown in Fig. 2.7. Some key process and packaging parameters are summarised in Tab. 2.1.

When the first BRIGHT devices were designed, the main market interest was in GSM900. However, support of other single-band standards, DCS1800 and PCS1900, was also of interest and was a factor in the choice of architecture. As a result, a high

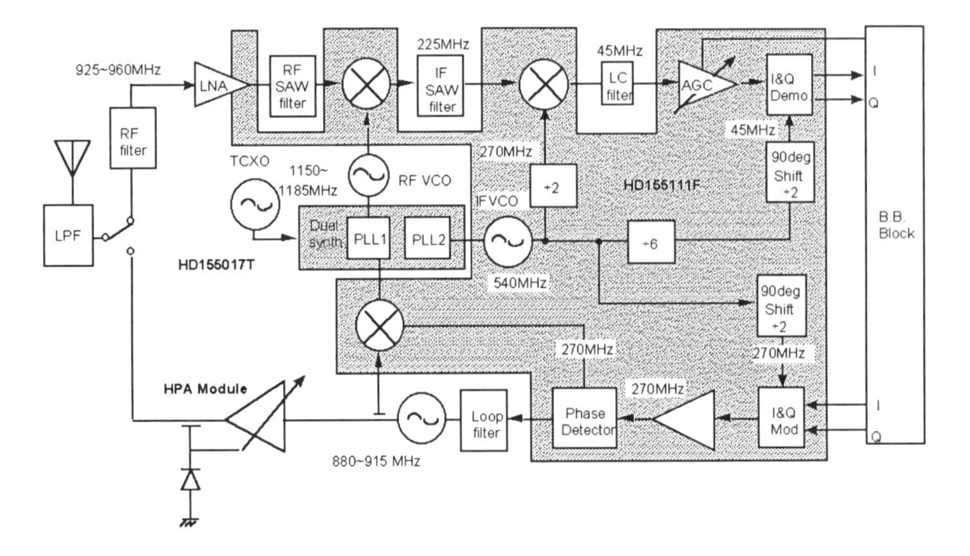

Figure 2.6 BRIGHT architecture.

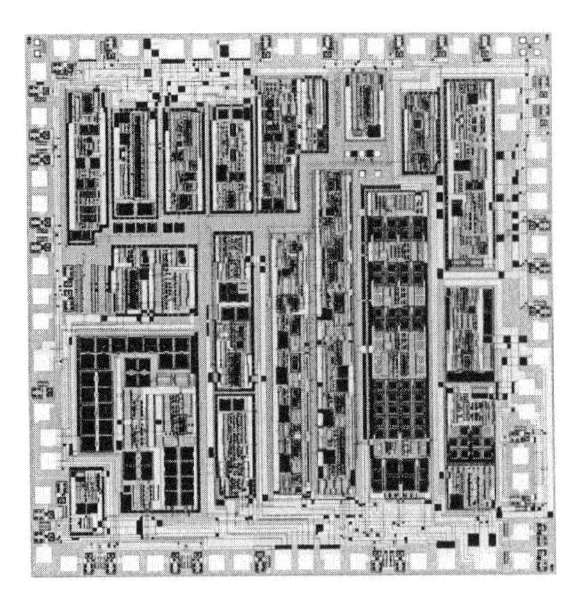

Figure 2.7 BRIGHT chip die.

frequency variant BRIGHT-HF was developed with improved high frequency mixer performance. Only towards the end of 1997 did it become apparent that dual-band would become a major market requirement, and with that came the further develop-

Table 2.1 BRIGHT process parameters.

Parameter	*Value*
Process Technology	0.6μm BiCMOS
F_t	15 GHz
$r_{bb'}$	150Ω
Supply voltage	2.7 V
Package	LQFP-48

ment of the BRIGHT2 device. The BRIGHT architecture is well suited to dual-band operation, with only small adaptations. On the receive side, the only change is the inclusion of a second mixer to support the second band: the first IF and subsequent circuitry is shared between both bands. On the transmit side, the only change is the inclusion of a different divide ratio for operation at 1800 MHz, giving an offset loop comparison frequency of 135 MHz. For minimum VCO range, the difference between transmit and receive IFs should equal the Tx/Rx duplex offset. In this case, for the 1800 MHz operation, we have a 90 MHz IF separation, compared to a 75 MHz duplex offset — thus giving around 20% excess pulling requirement over the theoretical minimum.

All other areas of the architecture are essentially unchanged. For market reasons, however, a digital AGC is included in BRIGHT2, compared to the analogue AGC scheme used in the original BRIGHT. It is found that the digital AGC gives better performance than the analogue AGC, because the on-chip DAC can be better matched to the characteristics of the AGC amplifier.

2.5 RESULTS

2.5.1 Transmitter phase error

Fig. 2.9 plots the measured peak phase error as a function of channel number in the GSM band. The RMS phase error is very consistent at around 3°, peaking up to just under 4° in one place. The truly remarkable factor is the independence of the phase error with transmit power level. This is testimony to the real benefits of the offset phase locked loop approach, where some of the sources of phase error, which cause major problems with other architectures, are completely removed by design. Similar results have also been reported to us for single-band GSM900 and DCS1800 phones in production, based on the single-band BRIGHT variants. The peak phase error is also well within the specification of 20°. Comparable results are obtained at 1800 MHz (see Fig. 2.10).

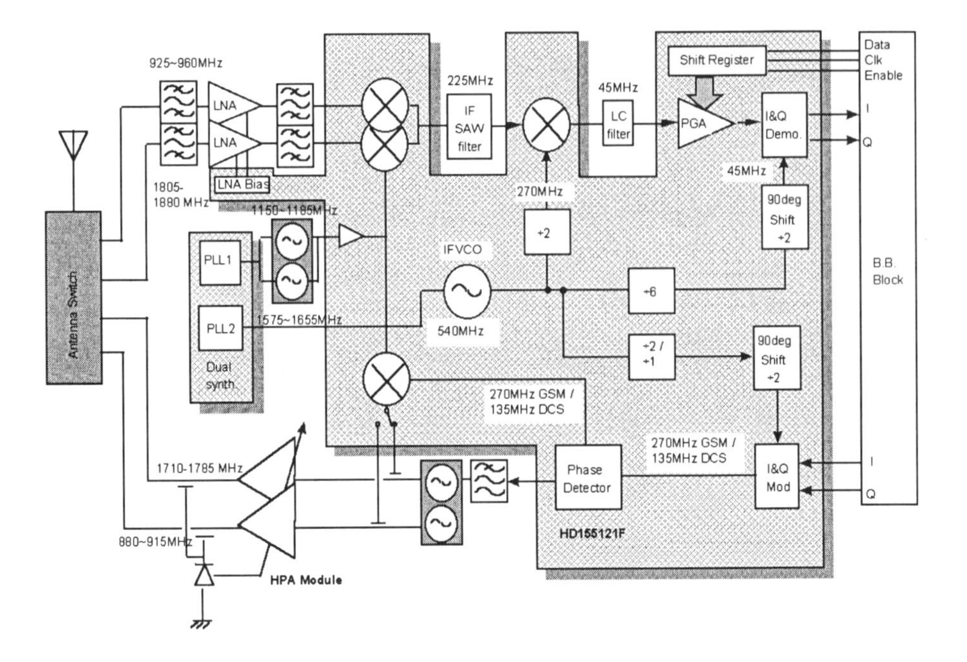

Figure 2.8 BRIGHT 2 architecture.

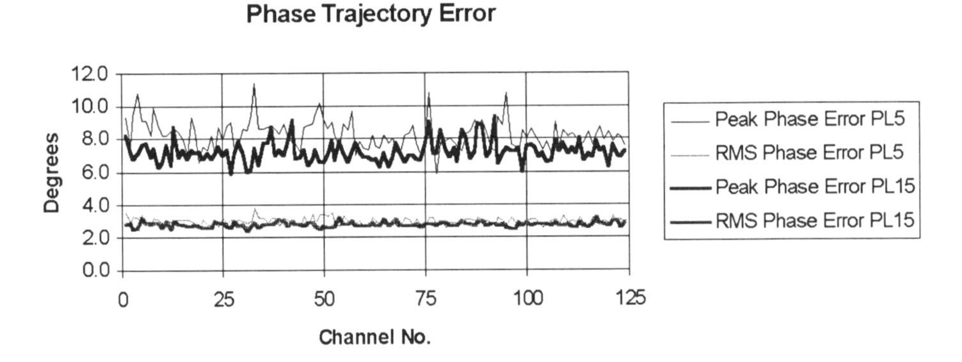

Figure 2.9 GSM900 phase error.

2.5.2 Transmitter modulation spectrum

Fig. 2.11 illustrates the typical modulation spectra achieved with the BRIGHT2 design at 900 MHz. The spectrum falls comfortably within the specification mask. Similar results are obtained at 1800 MHz.

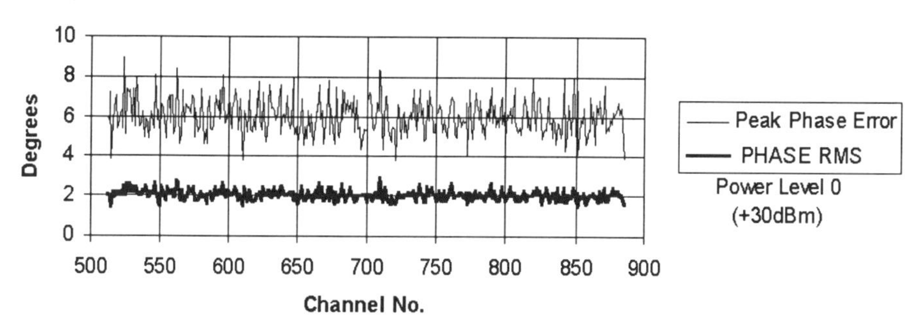

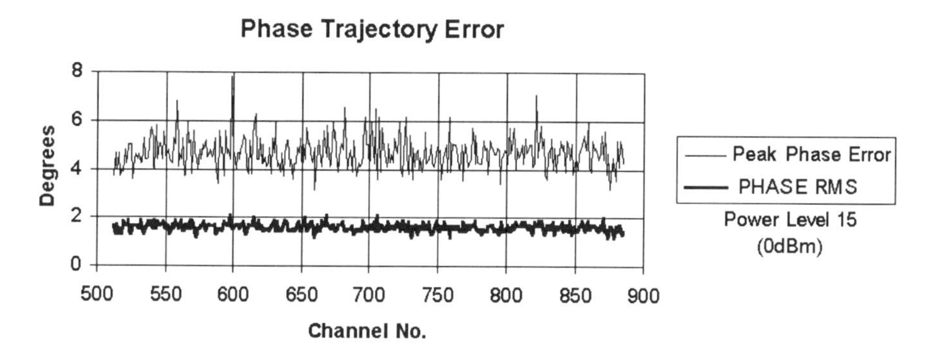

Figure 2.10 GSM1800 phase error.

2.5.3 *Receive sensitivity*

Fig. 2.12 shows the measured receive sensitivity in both bands plotted as a function of channel number. For GSM900, the specification is −102 dBm, for GSM1800, the specification is −100 dBm. A minimum margin of 4 dB is achieved on these figures, with 5–7 dB typical margin. This provides sufficient margin for production and temperature tolerances.

The underlying sensitivity variation displayed in the graphs is due to ripple in the front-end SAW filter. There are two well-known "deaf" channels in GSM900 receivers, channels 5 and 70, at the 72nd and 73rd harmonic of the GSM reference frequency of 13 MHz. These are clearly visible in the figure. The other deaf channel is thought to be due to a local GSM base station transmitting on this channel in the vicinity of the laboratory.

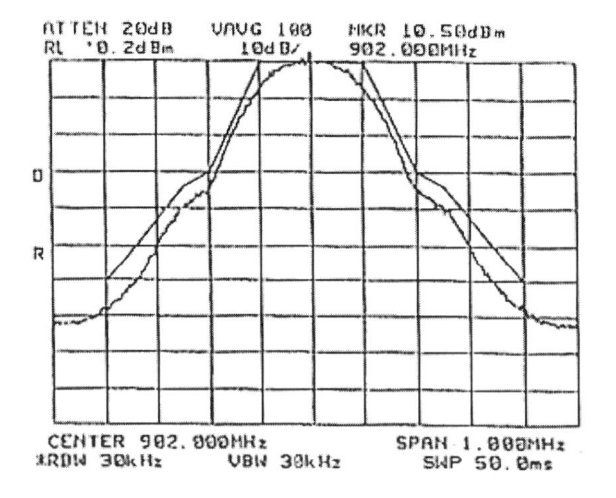

Figure 2.11 Transmitter modulation spectrum.

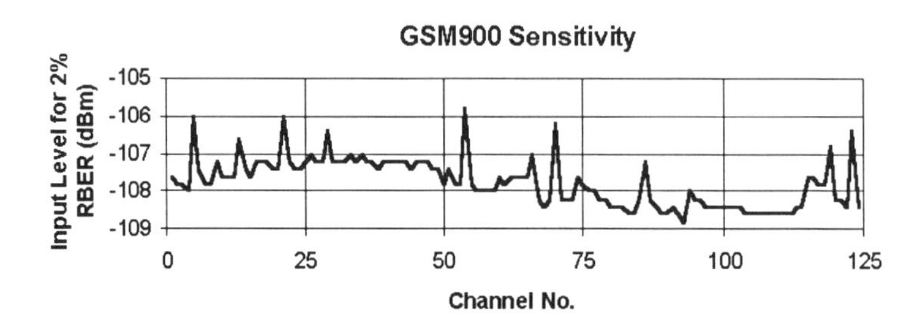

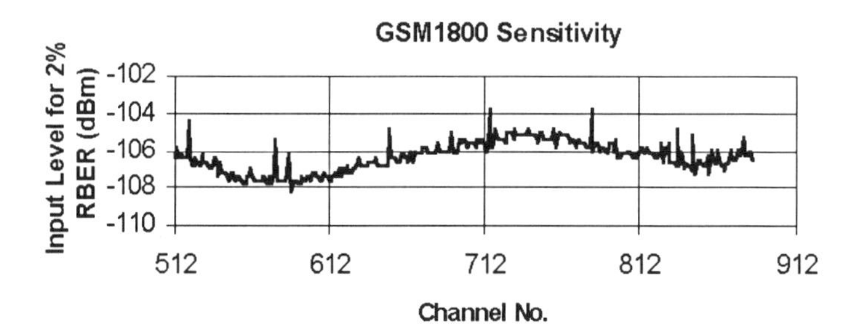

Figure 2.12 Receive sensitivity.

2.5.4 *Blocking performance*

Fig. 2.13 shows the blocking performance for BRIGHT2 in the 900 and 1800 MHz bands. The specification points are 2% BER and −23 dBm for GSM900, −26 dBm for GSM1800. Around 2–3 dB margin is thus achieved in each band. The blocking performance of BRIGHT2 in the GSM900 band is further illustrated in Fig. 2.14. A wanted signal is applied, 3 dB above the reference sensitivity level, and then the level of blocking signal which causes the receiver BER to degrade to the nominal 2% level, is measured. The wanted signal is on channel 61, i.e. at 947.2 MHz.

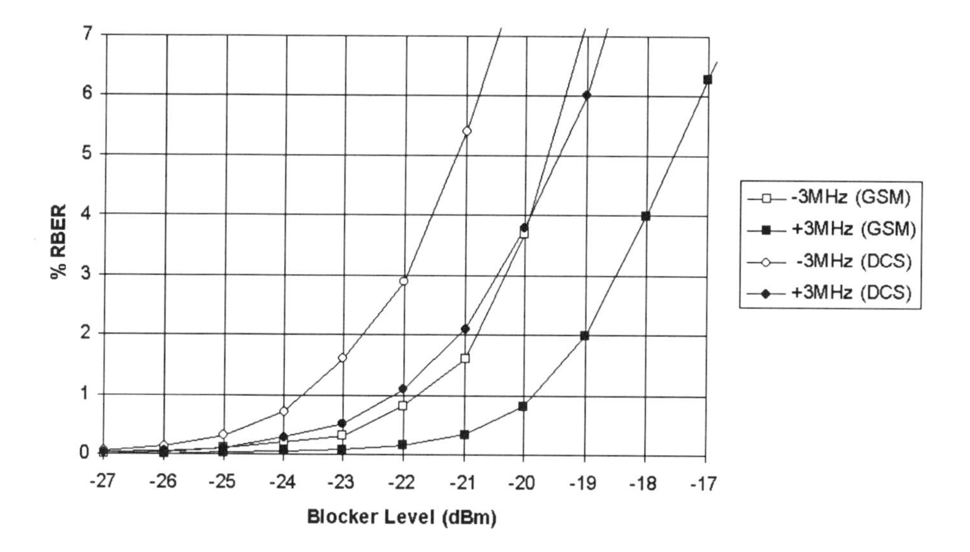

Figure 2.13 Receiver blocking performance.

The GSM requirements are as follows:

- −43 dBm at ±0.6 MHz, i.e. < 946.6 MHz & > 947.8 MHz
- −33 dBm at ±1.6 MHz, i.e. < 945.6 MHz & > 948.8 MHz
- −23 dBm at ±3 MHz, i.e. < 944.2 MHz & > 950.2 MHz

The most difficult requirement is the 3 MHz blocker, and for this, a minimum of 1.5 dB margin is demonstrated. Since these measurements were made, it was identified that the first mixer was not particularly well-matched (hence the ripple), and subsequently better performance with more than 3 dB margin has been demonstrated.

2.6 FUTURE OPTIONS

Looking to the future, there are a number of possible options. Firstly, the levels of integration can be increased, for example by including on chip the VCOs or the baseband interface. The challenge with putting VCOs on chip is to realise circuits of sufficiently high Q to meet the phase noise requirements; also low-loss varactors are not normally realisable in a BiCMOS IC process. Nevertheless, there is much interest in this area.

Including the baseband interface within the RF chip is more straightforward in principle. Whether this makes commercial sense is more questionable — integration

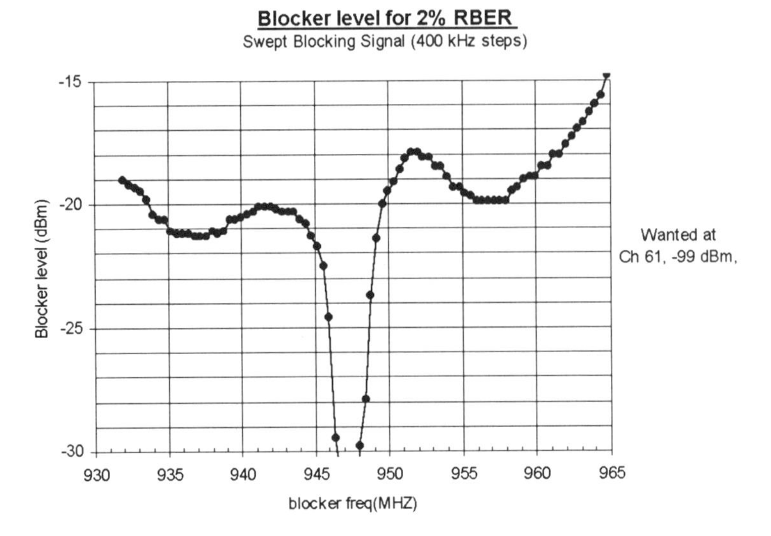

Figure 2.14 Receiver swept blocking measurements.

does not help reduce the total number of pins in the package very much, and the area for a 64-pin QFP is actually greater than for a 48-pin QFP plus a 20-pin SSOP. Also, because the baseband interface may be realised in smaller geometries than the RF device, the total power consumption may be increased by integration.

A more fruitful line is perhaps to consider greater functionality. In this regard, triple-band phones may be a market requirement soon, a requirement which may be supported as a further extension of the BRIGHT architecture.

In conclusion, it is clear that radio design remains one of the most challenging and exciting areas in this industry. Whatever happens in wireless communications over the next decade, elegant design at both circuit and system level will continue to be a major factor in the success of products, as it has been to date with BRIGHT.

3 INTEGRATION OF RF SYSTEMS ON A CHIP

Peter J. Mole

Nortel Networks
London Road
Harlow, Essex, CM17 9NA, U.K.

3.1 RF ISSUES

Before considering the details of implementation of an RF system on a chip, it is worth spending some time discussing the problems that any radio system must overcome if to achieve an acceptable performance. The important aspect to remember is that the radio system never operates in isolation. It is not sufficient to think only of the transmitter, the link and the receiver, though many problems lie in this simple chain alone. It is also important to realise that the transmitter can interfere with other links, and a receiver may be unduly sensitive to unwanted, but entirely legitimate, signals.

3.1.1 Receiver concerns

In any system, a band is defined in which the receiver may receive signals. The receiver must be able to reject signals outside the band without loss of performance. These signals may be very large (e.g. TV transmissions) and quite capable of overloading a sensitive receiver. If the receiver is overloaded, amplifier compression will reduce the receiver gain and hence the ability to detect weak signals, or the non-linearities in the receiver that are excited by the overload will allow unwanted signals to intermodulate. This may result in a distortion product falling onto the wanted signal and effectively masking it. It is therefore imperative that these signals are heavily filtered before the input to the receiver.

The receiver must respond to any signal in the system band, however, it must also reject unwanted in-band signals without suffering from overload. Such unwanted signals may be significantly larger than the wanted; consider the situation when you are trying to make a cellular phone link to your service provider's remote base-station but, unfortunately, you are physically adjacent to a second service provider's base-station. The system specification provides guidelines to the levels of signals which can exist. Fig. 3.1 shows the relative levels for signals in the DECT cordless phone system. In this system, where communication channels are allocated to both time and frequency slots, the frequency channels are separated by 1.728 MHz. A weak signal, (defined in the DECT specification as −73 dBm) must be receivable in the presence of a signal 13 dB stronger in the neighbouring channel and 34 dB stronger in the next neighbouring channel. To achieve this rejection of neighbouring interferers, a tunable filter is required. In practice this is achieved by mixing the signal to a fixed intermediate frequency and then filtering. The specifications reflect the fact that it is more difficult to filter the neighbouring channel than more distant ones. Until the signals can be filtered, however, the radio must be designed so that the interferers do not create overload. Unlike out-of-band blocking signals, in-band interferers cannot be filtered prior to the sensitive input of the receiver.

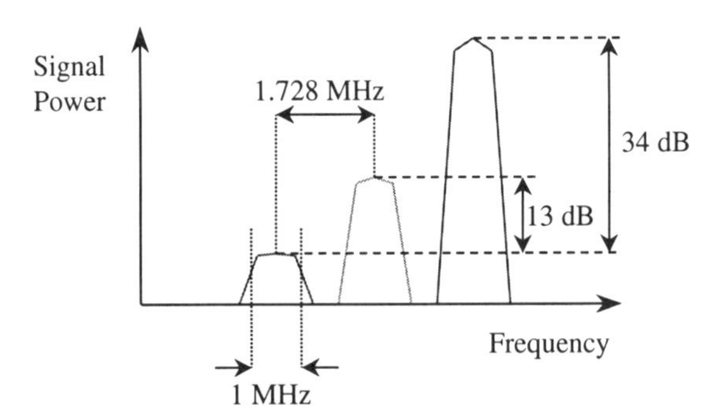

Figure 3.1 Adjacent channel interferers for DECT system.

The frequency translation that is required in a receiver to move the radio frequency input to the required intermediate frequency for filtering is achieved with a mixer. A mixer effectively multiplies the RF input signal with a chosen local oscillator (LO) signal. Thus we can ideally represent the RF input by $\sin(\omega_r t)$ and the local oscillator by $\cos(\omega_L t)$. Thus the output can be written as:

$$\sin\omega_r t \cdot \cos\omega_L t = 0.5(\sin(\omega_r + \omega_L)t + \sin(\omega_r - \omega_L)t)\,.$$

From this we can see immediately that an unwanted output is produced at the sum frequency, but this is easily filtered because of the frequency difference. A more subtle defect is that two input frequencies ($\omega_L \pm \omega_{\mathrm{IF}}$) can produce signals at the intermediate frequency (ω_{IF}). Thus there are two frequencies we are sensitive to, the wanted and the image frequency. It is therefore essential that any signal present at the image frequency is filtered out in front of the mixer.

In practice, to achieve efficient mixing, it is normal that the LO signal is close to a square wave. Thus the LO signal will contain the third harmonic of the LO (about 9 dB lower in signal amplitude) which is also multiplied by the input signal. This immediately means that signals at $(3\omega_L \pm \omega_{\mathrm{IF}})$ will also appear at the output as an intermediate-frequency signal. Other harmonics will behave similarly. Again input signals at these frequencies must be removed prior to mixing.

Another defect of the LO signal is that it suffers from phase noise (or jitter). This means the LO is not a pure tone but a spectrum of signals centred on the desired frequency. Thus the mixer will respond to signals close to the wanted signal with a sensitivity falling off with the phase-noise spectrum. This problem is known as reciprocal mixing. Thus, to ensure that the receiver can reject neighbouring channels, it is important to provide an LO signal with low phase noise. These issues are illustrated in Fig. 3.2.

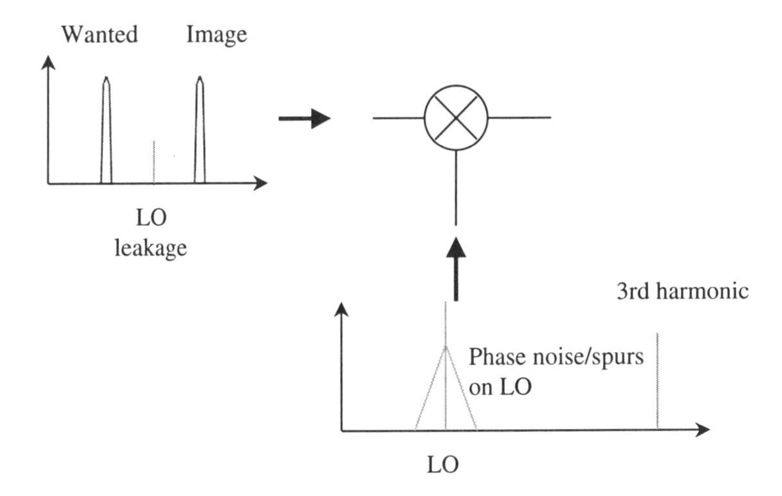

Figure 3.2 Non-idealities in a mixer.

3.1.2 Transmitter concerns

The transmitter must efficiently produce enough output power to ensure that the system has sufficient range whilst ensuring that power is not transmitted into neighbouring channels. It is important to understand how power can be transmitted at unwanted frequencies. Firstly, harmonics will be generated in an efficient power amplifier. These can usually be filtered out before they reach the antenna, but it must be remembered that any filter will also attenuate the wanted output, hence reducing the overall efficiency and increasing the current consumption from the battery. Non-linearity in the power amplifier will spread the spectrum of the signal into neighbouring bands. This behaviour is known as spectral regrowth. It is worse with modulation schemes that are not constant amplitude. Intermodulation is usually at its worst in the power amplifier itself, but intermodulation throughout the transmitter must be considered.

Broadband noise in the transmitter either due to noise in analogue circuit elements or due to quantisation noise if the signal is generated from a D-to-A converter must be considered. The latter, in particular, will require careful attention to band filtering to ensure low spurious emissions. A low-phase-noise local oscillator is also required to keep the noise emissions to acceptable levels.

In a mechanism similar to phase noise, spurious signals present on the local oscillator will also give rise to spurious emissions on the output signal. Spurious signals can arise from the frequency synthesis process or from unintentional couplings to the VCO. On single-chip systems, the identification of these couplings can be particularly difficult, but attention must be paid to single-ended CMOS input signals which couple strongly to the substrate and also to high-power outputs. An example of the former is given in [1]. These effects are illustrated in Fig. 3.3.

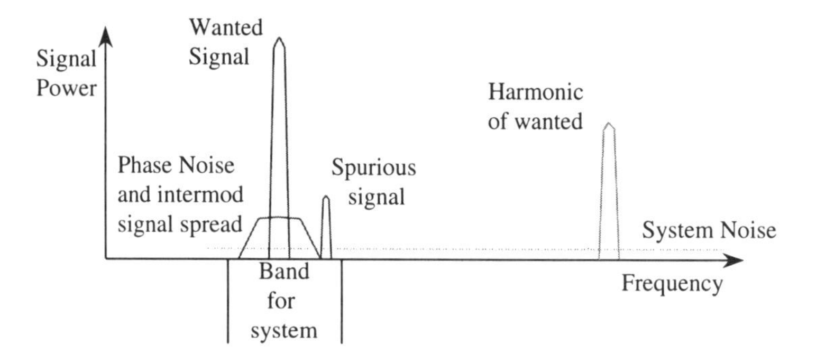

Figure 3.3 Non-idealities in a transmitter.

3.2 RADIO ARCHITECTURES

With the issues discussed in the previous section, the advantages of different architectures can be discussed both in terms of performance and suitability for integration.

3.2.1 Receiver architectures

The double-superheterodyne architecture (Fig. 3.4). This is probably the most commonly employed architecture in current wireless systems. The out-of-band blocking signals are reduced by an RF bandpass filter placed immediately after the antenna. The signal is then amplified by an LNA, which must have a sufficiently low noise to allow detection of weak signals but must also have the dynamic range to handle in-band interferers. The bandpass filter is usually insufficient to reduce signals at the image frequency to the system noise level, and so a second image filter is inserted prior to mixing. To ensure that the image is sufficiently far away from the wanted signal to allow effective filtering, a relatively high first intermediate frequency must be chosen (for 1 to 2 GHz RF systems an IF of 100–200 MHz is common).

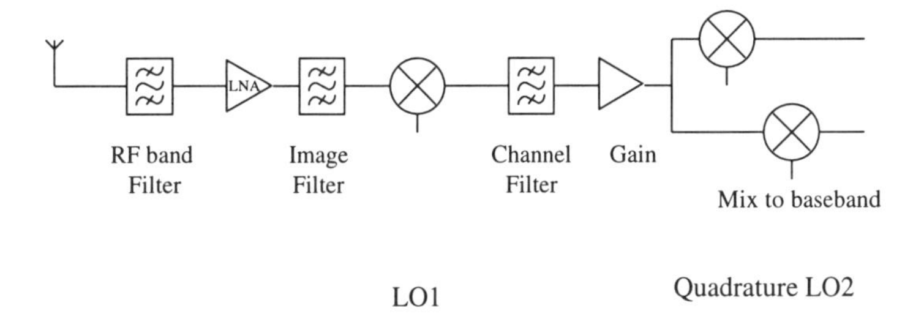

Figure 3.4 The double-superheterodyne architecture.

The mixer must still handle the complete dynamic range of the in-band signal. After the mixer, a SAW filter can be used to achieve the channel filtering. At these frequencies the SAW filter is small but usually has a large in-band loss when complete channel filtering is to be achieved. The output drive of the mixer must therefore boost the signal level to allow for this loss.

Once the interfering channels have been attenuated, the signal can be boosted to a high level (it can be limited if a constant-amplitude modulation scheme is used). The signal is then reduced to baseband frequency for demodulation. It is of course possible to split the channel filtering between the two intermediate frequencies. This will require a greater dynamic range in the second mixer.

This architecture requires the synthesis of two local oscillators, and their frequencies must be chosen so that spurious responses from the radio are kept to a minimum. This aspect of frequency planning, which will not be discussed in more detail here, is a well-understood design process which requires considerable care and experience.

This design requires several external filters and therefore does not lend itself to easy integration as the pin count increases. Moreover, the filters are usually single ended—though this is not essential—and hence achieving isolation between pins becomes an issue. In particular, the channel filter will often need to provide 50 dB of attenuation at key frequencies, thus implying that greater isolation must be achieved between the pins and with respect to signal ground if the filter response is not to be degraded. The image filter can be eliminated if an image-rejecting mixer is used. This will prevent the need to come off chip after the LNA and makes an LNA plus image-reject mixer a useful integrated building block.

The direct-conversion architecture (Fig. 3.5). The direct-conversion receiver, because of its simplicity, appears to offer the best opportunity for integrated systems. Some examples of its use in wireless systems do exist today, but it is not as simple in practice.

Once again, an RF bandpass filter is placed at the input. The LNA's output is passed into the mixer. The LNA must handle the same dynamic range as for the superheterodyne architecture and it must have enough gain to lift weak signals above the noise of the mixer. The mixer however, now converts directly to baseband. Thus the signal is its own image, and channel filtering can now be carried out by low-pass

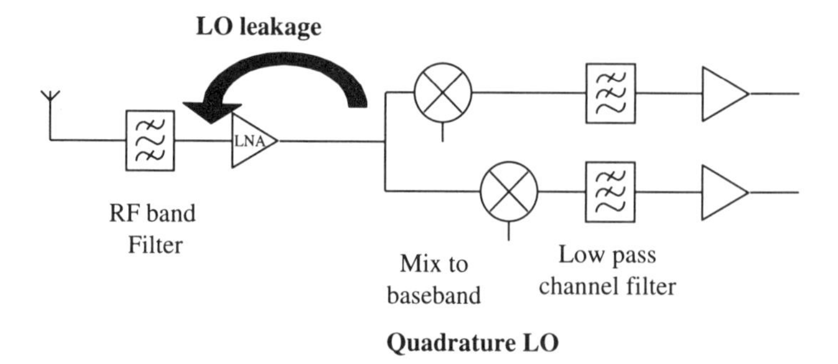

Figure 3.5 The direct-conversion architecture.

baseband filters. Only one local-oscillator frequency needs to be synthesised, and frequency planning is straightforward.

To achieve a precise conversion to baseband, it is important that good quadrature can be achieved at the local-oscillator frequency. Unlike the superheterodyne architecture, this quadrature now needs to be achieved at the RF frequency, which poses some problems that are generally soluble.

The main issue is associated with DC signals which are generated by defects in the mixer. These signals are implicitly in-band, it is therefore difficult, and sometimes impossible, to filter them from the wanted signal. It is necessary to keep them sufficiently below the signal. Unfortunately, the amplification that can be applied to the signal before mixing is limited by the level of in-band interferers which must not overload the mixer.

The DC offsets are caused by a collection of effects. Imbalance in the mixers will lead to a DC output. This is generally a constant quantity and could be cancelled with suitable circuitry. Any leakage of the local oscillator to the input of the receiver will result in a DC signal being generated. If this leakage is via radiation coupling into the antenna, then this may vary with the local environment. Finally, non-linearities in the mixer may cause signals to be generated at DC from other interferers. The latter mechanisms can be time varying, and any offset cancellation needs to be able to respond to time variation. It is only when special precautions are taken to cancel DC offsets that direct-conversion architectures can be used. More details of the direct-conversion issues are given in [2].

Digitisation at IF (Fig. 3.6). This architecture is still a research activity. After the first mixer the full dynamic range of the in-band signals may still exist. This full range needs to be captured by the A-to-D converter. A bandpass sigma-delta converter offers potential for a very flexible system as it does not place high demands on the anti-alias filter. With such a system, the final channel filtering can be made programmable in the DSP filter to allow a single radio architecture to adapt itself to a range of systems. At present the complexity and power consumption in the A-to-D converters provide a significant challenge. As an alternative to bandpass sigma-delta converters, sub-sampled or pipeline converters may provide a viable alternative, but the dynamic range which can be accommodated will be limited by the number of available bits and the linearity of the converter. It should be noted that a first IF of about 100 MHz will be required to enable adequate image rejection to be obtained at the radio frequency.

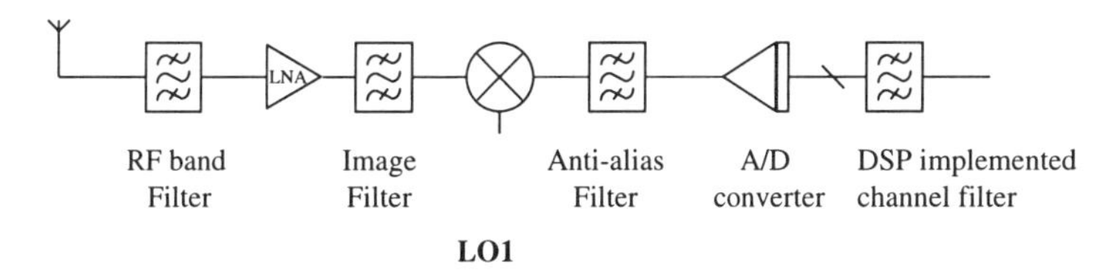

Figure 3.6 Architecture for digitalisation at IF.

If a very-wide-bandwidth A-to-D converter can be used, this system also offers the potential to capture all the activity in the RF band from one receiver. There would be no need to tune to individual channels; this could be done by tuning the digital filter. This would enable radio base-stations to operate on many channels with one radio unit. The number of simultaneous channels that can be received would be limited by the number of DSP filters present.

3.2.2 *Transmitter architectures*

Direct up-conversion (Fig. 3.7). Direct up-conversion offers a very flexible means of implementing a general modulation. The baseband I and Q signals contain the phase and amplitude modulation information and can be generated digitally. They are converted to analogue signals using D-to-A converters. The major imperfections of the D-to-A converters are quantisation noise and replications of the wanted signal at multiples of the D-to-A clocking frequency. To prevent spurious emissions at unacceptable levels, the signal must be band filtered. The filtering is most easily performed while the signal is at baseband.

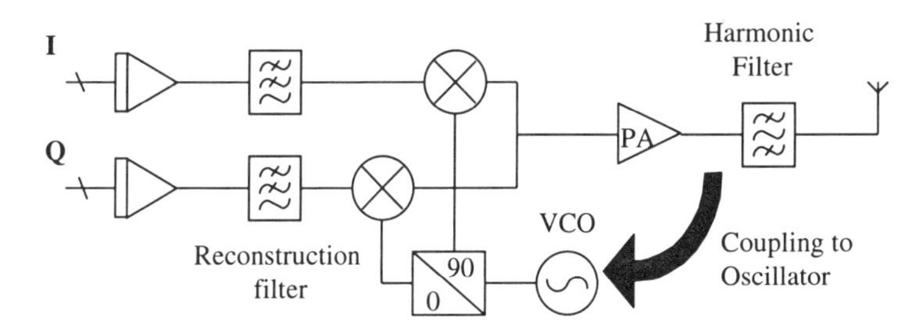

Figure 3.7 Direct up-conversion architecture.

Whilst the signals are at baseband, any non-linearities will create distortion products which will fall in neighbouring channels. It is therefore important that these stages are highly linear as it will be impossible to filter such products once they have been converted to RF.

To convert the signal to RF, a quadrature up-mix is required. The quality of the quadrature will determine the accuracy of the modulation. Any DC offsets will appear as a leakage of the local oscillator through to the output. Once there it will appear as a modulation error. These up-mixer defects must be kept small (about −40 dBc) in order to obtain a high-integrity communication link. Integration and highly balanced circuitry offer this possibility.

The VCO will be very sensitive to pulling from the output signal as it is operating in the centre of the output band. Whether this architecture is implemented in an integrated or a discrete form, unwanted coupling which causes pulling of the VCO will require careful consideration. In integrated form, however, evaluation of the interaction between on-chip components is poorly developed. The design uncertainties in this architecture need to be balanced carefully against its undoubted simplicity when considering integration.

Offset up-conversion (Fig. 3.8). In this architecture, the first up-mix is the quadrature mix and it is to a low first IF. The lower local-oscillator frequency means that a good quadrature local oscillator is more easily built and local oscillator feed-through is less of an issue. As for the direct-conversion architecture, quantisation noise and replications of the signal must be filtered out, though with this architecture the harmonic filter can share this task. The harmonic filter also has the job of removing all the unwanted mixer products from the first mix to ensure they do not appear as unwanted emissions at the output.

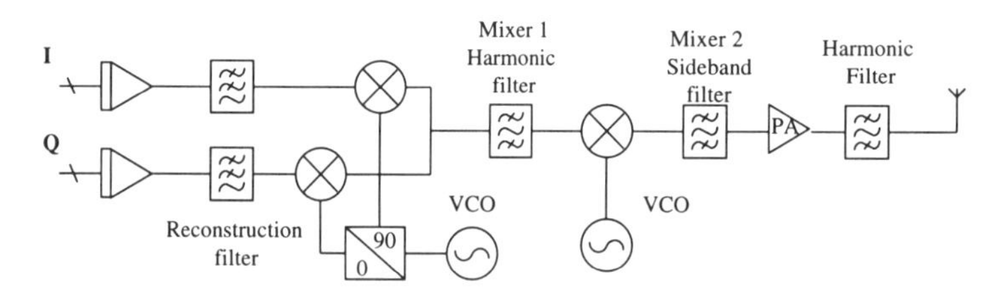

Figure 3.8 Offset up-conversion architecture.

The second mixer needs to be a good single-sideband mixer, as the unwanted sideband will be offset from the wanted RF signal by twice the first intermediate frequency and will cause a significant interference problem if not well suppressed. The final mixer can use a quadrature structure to suppress the unwanted sideband, but even with balanced on-chip components some degree of filtering will be required to keep the unwanted image emission low. To achieve this, the unwanted sideband must lie significantly outside the system band, and this will determine the lowest practical limit for the first intermediate frequency.

The selection of the first and second intermediate frequencies must ensure that unwanted mixer products never result in unacceptably high emissions; this problem is very similar to that of ensuring lack of spurious responses in the double-superheterodyne receiver. It does however enable an architecture to be designed where the VCOs are operating well away from the power output frequency and are not susceptible to pulling. The need for external filtering will make integration of this type of architecture more difficult, but if an external PA is used anyway, this constraint is not severe. The double up-mixer with external harmonic filter is a very useful integrated block.

Loop-locked and modulated VCO (Fig. 3.9). In this architecture the concept is to reduce the need for signal-path filtering by generating a pure modulated signal from a VCO. With this concept, since the modulation is effectively applied to the VCO input, amplitude modulation of the signal is not possible. However for phase/frequency modulation schemes this is an attractive option.

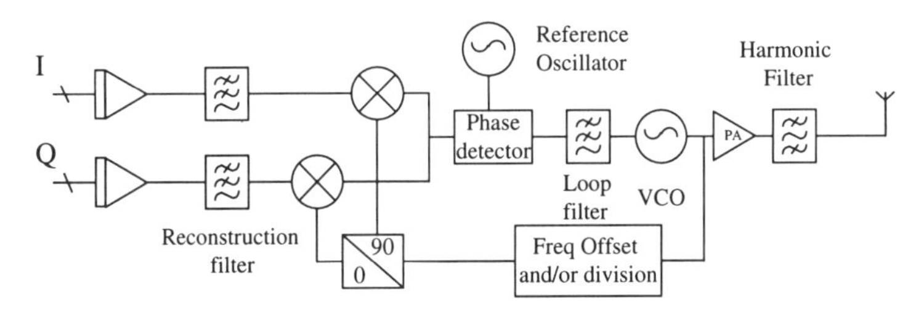

Figure 3.9 Loop-locked and modulated VCO architecture.

To ensure that the modulation is precisely controlled about a desired centre frequency, the modulation is applied in a feedback loop. The aim of the loop is to ensure that phase modulation sampled at the VCO output is in exact anti-phase with the phase at the I and Q baseband inputs. This will then result in a constant-phase signal at the phase detector. Using the high-gain phase detector output to drive the VCO, the resulting loop ensures that this condition is closely maintained. The spurii that can enter the VCO are limited by the bandwidth of the lowpass loop filter, enabling simple control of spurious signals. Because the signal at the PA is very pure, this simplifies the filtering after the PA too. Here, only PA generated harmonics need be removed, which requires a lower-loss output filter and hence improved power efficiency.

The absolute frequency is determined by the reference oscillator and the frequency offset/division block. These need to be carefully planned. Offsetting the frequency requires down-mixing, and it is important to filter unwanted products, as for the superheterodyne receiver. The frequency planning will however ensure that the VCOs should not be susceptible to pulling, and the main VCO is not susceptible as it is phase coherent with the PA. It is not clear, however, whether for a given frequency requirement, this architecture can produce a design with low-frequency external filter which is easily integrable.

3.2.3 Architectures overview

The architectures under consideration are still very similar to those used in radios built from discrete components. SAW filters still play a significant role as they provide good filter characteristics with a very wide dynamic range whilst consuming no power themselves (though we should not overlook the power required to compensate for their loss). Architectures which depend on many SAW filters place limits on the integrability due to the need to get signals on and off chip, whilst maintaining good signal and ground isolation.

SAWs are also fixed-performance devices, thus if the radio is required to operate across systems, several filter devices which can be switched into the path will be required. Architectures which will allow the programming of the filters will be very attractive in the future for such multi-mode systems. Architectures where the signal is digitised with a wide dynamic range and is filtered and demodulated using the power and low cost of data processing implemented in a small-dimension CMOS process look attractive for this purpose. However, sufficiently high performance and sufficiently low power consumption has not been achieved in the analogue to digital conversion.

Implementation in integrated form offers significant advantages for matched and balanced circuitry. Thus architectures which exploit this, for example image-rejecting mixer based circuits, will evolve as prime candidates for integrated radios. However, in any integrated radio, we need to be able to guarantee isolation between the sections. Currently, isolation is achieved by PCB and screened-compartment design. The move to integration will require that isolation is achieved on chip. At present, design skills in this area are low. The choice of architecture to minimise isolation requirements, and the development of techniques to calculate the isolation across the chip are going to be key factors, if the level of integration of radios is to increase.

3.3 SOME DESIGN ISSUES

In this section, some of the design issues for integrating radio circuitry in silicon technology are introduced.

3.3.1 Power supply and ground coupling

The essential problem here is that RF currents flowing in the power-supply and ground leads can drop significant potentials across track and bond inductances. Moreover, capacitances can resonate with the bond/track inductances causing the impedance seen by the circuitry to become high at certain frequencies. These frequencies must be separated from key circuit frequencies. Supply return currents may cause potentials to be dropped on signal grounds. This can give rise to feedback to the input, poor isolation and even instability. Fig. 3.10 shows the situation for power supply decoupling.

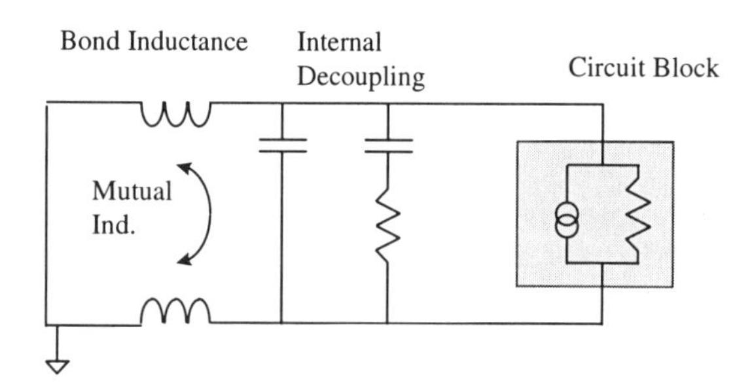

Figure 3.10 Power supply decoupling.

In this circuit, some on-chip capacitance has been added to decouple the power supply. If we assume the power supply is perfectly decoupled close to the chip, then the bond inductance—modified by any mutual coupling between the supply and ground lines—will resonate with the internal decoupling capacitance. If the on-chip circuitry takes a significant current at this resonant frequency (represented by the current source) then a large power supply ripple will occur. This ripple is limited by the loss in the circuitry itself and the loss in the decoupling capacitors. Simulation can be used to estimate the magnitude of this ripple and also to decide on the values of the capacitance to use (to put the resonance in an innocuous range) and how much damping to provide if the circuit does not provide sufficient loss.

As general design principles:

- Separate power supplies for blocks which require isolation.
- Keep ground and power supply pins close to exploit mutual inductance.
- Add on-chip decoupling, choose resonance away from circuit frequencies.
- Damp resonance if circuit does not provide sufficient loss.
- Try and separate signal grounds and ground power returns if possible.

3.3.2 Substrate coupling

As a general rule, as interconnection and circuitry scales down, cross-talk capacitance decreases. This occurs because the length of interconnect required decreases as the circuit compacts with scaling, whilst the cross-talk capacitance per unit length varies relatively little. This implies that to make higher speed circuitry, all dimensions will need to be reduced, and integration is a natural route forward.

However, with integration, there is also the possibility that signals will couple through the substrate. With silicon technology, where the substrate is conducting, it is possible for the signal to be conducted through the substrate. Fig. 3.11 illustrates this.

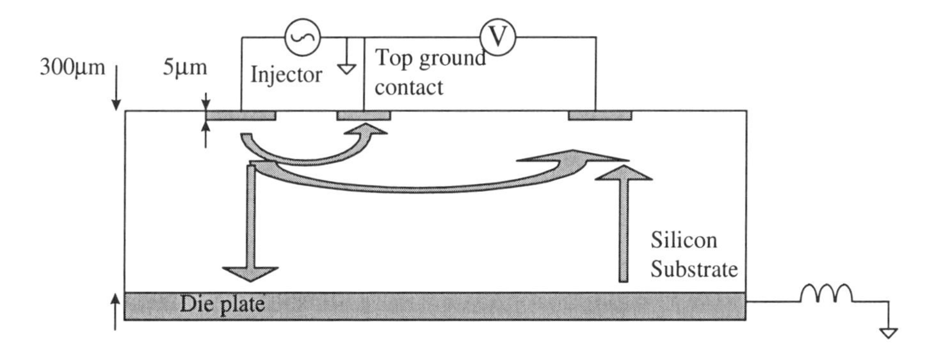

Figure 3.11 Coupling through the substrate.

Devices sitting in the top few microns of the silicon can inject charge capacitively into the substrate. Even on a silicon-on-insulator (SOI) technology this will happen, but a good SOI technology will have a low coupling capacitance. The major coupling will be to nearby devices or local contacts. Some of the charge will be injected more deeply and couple to distant circuit elements. This will be a source of interference. Finally, some charge will return via the rear die plate. The die plate potential will vary in response to the charge and the return path impedance (which will include capacitive coupling to other parts of the circuit and the impedance to ground of the die plate). If the die plate does not have a low ground impedance then it can couple the injected signal effectively across the whole chip.

These effects are being studied now. Means of simulation are being developed (for example, see [3]). As a general rule, balanced signals will always help injection, and local grounded substrate contacts are beneficial.

3.3.3 On-chip oscillators

Oscillators on chip always need careful consideration. Although the maintaining amplifier can be easily accommodated on chip, there is an advantage to keeping the whole oscillator on chip. This however requires an on-chip resonator. The phase-noise performance is directly dependent on the resonator's quality factor. Spiral inductors can be manufactured on chip, but their quality factor is limited by the resistance of the metal used to wind the inductor (losses in the substrate are usually a smaller effect). Measurements of the phase noise achieved using a 2 μm-thick Al/Si/Cu metallisation show that whilst the phase noise required for DECT is attainable, the requirements of a system with much narrower bands such as GSM is rather too demanding.

The second issue to be considered with on-chip oscillators is pulling. The oscillator can be considered as a highly shaped and amplified noise source. Essentially, at the resonant frequency, the feedback circuit provides just enough gain to amplify the natural circuit noise to the level required, so this gain is controlled by the limiting mechanism. Away from the natural resonant frequency, the loop gain falls away; the higher the Q the more rapidly the gain falls off. This leads to two defect mechanisms: if a large signal close to resonance couples into the resonator, it will get amplified until it causes the limiter to reduce the gain. The noise content at the output will fall away, and the oscillator will have locked to the interferer. If an interferer is coupled in at a much lower level, then it will not trigger the limiting action; the shaped noise will still exist, but the spurious tone will interact with the oscillations and modulate the frequency. A side-tone will appear at the interferer frequency and a second one on the other side of the main tone—characteristic of FM modulation. Now the sensitivity to such an interferer depends on how close to the main tone the interferer is. If it is at the resonant frequency itself, the coupling only needs to be at circuit noise levels, the level required increases as the coupling tone is offset in frequency. The higher the resonator's Q the higher the level of the interfering signal that can be tolerated. Since very small signals may pull an oscillator, the choice of architecture is critical. The aim, if an on-chip oscillator is to be included, will always be to ensure that spurious signals which could pull or lock the VCO are not present.

References

[1] Jeff Durec, "An Integrated Silicon Bipolar Receiver Subsystem for 900 MHz ISM Band Application", *JSSCC*, vol. 33, Sept. 1998, pp. 1352–1372.

[2] Behzad Razavi, "Design Considerations for Direct Conversion Receivers", *IEEE Transaction on Circuits and Systems-II*, Vol. 44, June 1997, pp. 428–435.

[3] Ranjit Gharpurey and Robert G. Meyer, "Modeling and Analysis of Substrate Coupling in Integrated Circuits", *IEEE JSSC*, vol. 31, Mar. 1996, pp. 344–353.

4 TOWARDS THE FULL INTEGRATION OF WIRELESS FRONT-END CIRCUITS IN DEEP-SUBMICRON TECHNOLOGIES

Michiel Steyaert

K.U. Leuven, ESAT-MICAS
Kardinaal Mercierlaan 94
B–3001 Heverlee, Belgium

Abstract: Research into the potentials of CMOS technologies for RF applications has been growing enormously in the past few years. The trend towards deep sub-micron technologies allows operation frequencies of CMOS circuits above 1 GHz, which opens the way to integrated CMOS RF circuits. Several research groups have developed high performance down-converters, low phase noise voltage-controlled oscillators and dual-modulus pre-scalers in standard CMOS technologies. Research has already resulted in fully integrated receivers and VCO circuits with no external components, neither tuning nor trimming. Further research on low noise amplifiers, up-converters and synthesisers has recently resulted in fully integrated CMOS RF transceivers for DCS-1800 applications.

4.1 INTRODUCTION

A few years ago, the world of wireless communications and its applications started to grow rapidly. The driving force for this was the introduction of digital coding and digital signal processing in wireless communications. This digital revolution is driven by the development of high performance, low cost CMOS technologies which allow the integration of an enormous number of digital functions on a single die. This in turn allows the use of sophisticated modulation schemes, complex demodulation algorithms and high-quality error detection and correction systems, resulting in high-performance, lossless digital communication channels. Low cost and a low power consumption are the driving forces, and they make the analog front-end the bottle neck in future RF designs. Both low cost and low power are closely linked to the trend towards full integration. An ever higher level of integration renders significant space, cost and power reductions. Many different techniques to obtain a higher degree of integration for receivers, transmitters and synthesisers have been presented over the past years [1–3].

Parallel to the trend to further integration, there is the trend to the integration of RF circuitry in CMOS technologies. The mainstream use of CMOS technologies is the integration of digital circuitry. If possible, using CMOS technologies to integrate high-performance analog circuits has many benefits. The technology is cheap if used without any special adaptations towards analog design. Plain CMOS has the additional advantage that the performance gap between devices in BiCMOS, nMOS devices in deep sub-micron CMOS, and even NMOS devices in the same BiCMOS process is becoming smaller and smaller due to the much higher investments into the development of CMOS rather than bipolar technologies. Nowadays, NMOS devices have even higher transition frequencies (f_t) than NPN devices.

Although some research has been done in the past on the design of RF circuits in CMOS technologies [4], close attention has been given to it only in the past few years [5]. Today, several research groups at universities and in industry are researching this topic [2, 3, 6, 7]. Since bipolar devices are inherently better than CMOS devices, RF CMOS is seen by some people as a possibility for only low performance systems with reduced specifications (like ISM) [8]. Some say that the CMOS processes need adaptations, like substrate etching under inductors. Others feel, however, that the benefits of RF CMOS can be much larger, and that it will be possible to use plain deep sub-micron CMOS for the full integration of transceivers for high performance applications like GSM, DECT and DCS 1800 [2, 3].

In this chapter, some trends, limitations and problems in technologies for high frequency design are analysed. Second, the down-converter topologies and implementation problems are reviewed. Third, the design and trends towards fully integrated low phase-noise PLL circuits are discussed. Finally, the design of fully integrated up-converters is addressed.

4.2 TECHNOLOGY

Due to the never ending progress in technology down-scaling and the requirement to achieve a higher degree of integration for DSP circuits, sub-micron technologies are nowadays considered standard CMOS technologies. The trend is even towards deep sub-micron technologies, e.g. transistor lengths of 0.1 μm and below. Transistors whose f_t is near 100 GHz have recently been built in 0.1 μm technologies [9, 10].

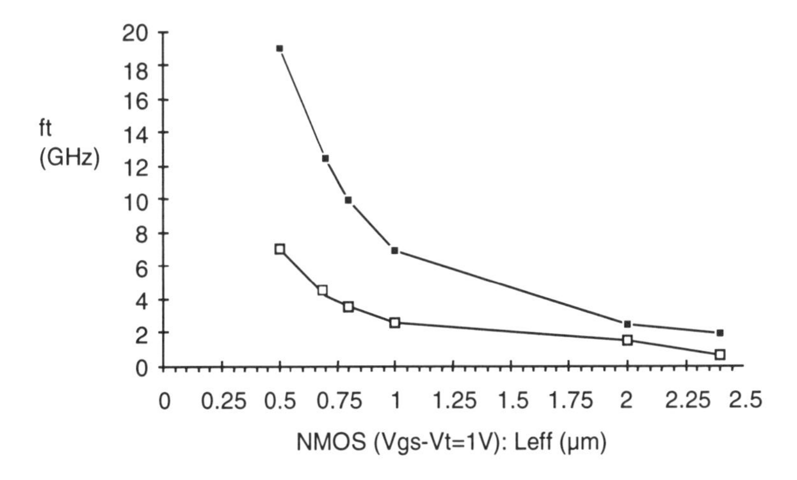

Figure 4.1 Comparison of f_t (■) and f_{max} (□).

However, the speed increase of deep sub-micron technologies is reduced by the parasitic capacitance of the transistor, meaning the gate-drain overlap capacitances and drain-bulk junction capacitances. This can clearly be seen in Fig. 4.1 in the comparison for different technologies of the f_t and the f_{max}, defined as the 3 dB point of a diode-connected transistor [11]. The f_{max} is more important, because it reflects the speed limitation of a transistor in a practical configuration. As can be seen, the f_t rapidly increases, but for real circuit designs (f_{max}) the speed improvement is only moderate.

Finally, recent integrations of CMOS RF circuits [5] have made it clear that the limiting factor will not be the technology, but the packaging. Since the RF signals have to come off the chip sooner or later, and since the RF antenna signal has to get into the chip, any parasitic PCB, packaging or bond-wire capacitance in combination with the ESD protection network and packaging pin capacitances will strongly affect and degrade the RF signal.

4.3 FULLY INTEGRATED CMOS DOWN-CONVERTERS

The heterodyne or IF receiver is the best known and most frequently used receiver topology. In the IF receiver, the wanted signal is down-converted to a relatively high intermediate frequency. A high-quality passive bandpass filter is used to prevent a mirror signal to be folded upon the wanted signal on the IF frequency. Very high performance can be achieved with the IF receiver topology, especially when several IF stages are used. The main problem of the IF receiver is the poor degree of integration that can be achieved because every stage requires going off-chip and using a discrete bandpass filter. This is both costly (the cost of the discrete filters and the high pin-count of the receiver chip) and power consuming (often the discrete filters have to be driven by a 50 Ω signal source). Moreover, in CMOS RF circuit design input and output is becoming a serious problem already in the lower GHz frequency range.

The homodyne or zero-IF receiver has been introduced as an alternative for the IF receiver. It can achieve a much higher degree of integration. The zero-IF receiver uses a direct, quadrature down-conversion of the wanted signal to the baseband. The wanted signal is its own mirror signal, and sufficient mirror signal suppression can therefore be achieved even with a limited quadrature accuracy. Theoretically, there is thus no discrete high-frequency bandpass filter required in the zero-IF receiver, allowing the realisation of a fully integrated receiver, especially if the down-conversion is performed in a single stage (e.g. directly from 900 MHz to the baseband) [6]. The problem of the zero-IF receiver is its poor performance compared to IF-receivers. The zero-IF receiver is intrinsically very sensitive to parasitic baseband signals like DC-offset voltages and crosstalk products caused by RF and LO self-mixing. These drawbacks have kept the zero-IF receiver from being used on a large scale in new wireless applications. The use of the zero-IF receiver has therefore been limited to low-performance applications like pagers and ISM, in which the coding can be scrambled so that a high-pass filter can be inserted to avoid the DC offset problems. Another application is the use as a second stage in a combined IF/zero-IF receiver topology [12, 13]. It has, however, been shown that with the use of dynamic non-linear DC-correction algorithms implemented in the DSP, the zero-IF topology can be used for high performance applications like GSM and DECT [1, 14].

In recent years, new receiver topologies like the low-IF receiver [2, 15] were introduced for use in high-performance applications. The low-IF receiver performs a down-conversion from the antenna frequency directly down to, as the name already indicates, a low IF (i.e. in the range of a few 100 kHz) [2]. Down-conversion is done in quadrature and the mirror signal suppression is performed at low frequency, after down-conversion, in the DSP. The low-IF receiver is thus closely related with the zero-IF receiver. It can be fully integrated (it does not require an HF mirror signal suppression filter) and uses single stage direct down-conversion. The difference is that the low-IF receiver does not use baseband operation, resulting in a total immunity to parasitic baseband signals, resolving in this way the main disadvantage of the zero-IF receiver. The drawback is that the mirror signal is different from the wanted signal in the low-IF receiver topology, but by carefully choosing the IF frequency, an adjacent channel with low signal levels can be selected for which the typical mirror signal suppression (i.e. a 3° phase accuracy) is sufficient.

Fig. 4.2 shows the block diagram of a fully integrated quadrature down-converter realised in a 0.7 μm CMOS process [2]. The proposed down-converter does not require any external components, nor does it require any tuning or trimming. It uses a newly developed double-quadrature structure having a very high quadrature accuracy (less than 0.3° in a very large passband). The topology used for the down-converter is based on NMOS transistors in the linear region [2, 5]. In combination with capacitors on virtual ground, only a low-frequency opamp is required. Using the MOS transistors in the linear region results in very high linearity for both the RF and the LO input. This is reflected in a very high third order intercept point (IP3) for the mixers, which is over +45 dBm. The advantages of such a high linearity on both inputs are that the mixer can handle a very high third order intermodulation-free dynamic range IMFDR3, resulting in no need for any kind of HF filtering.

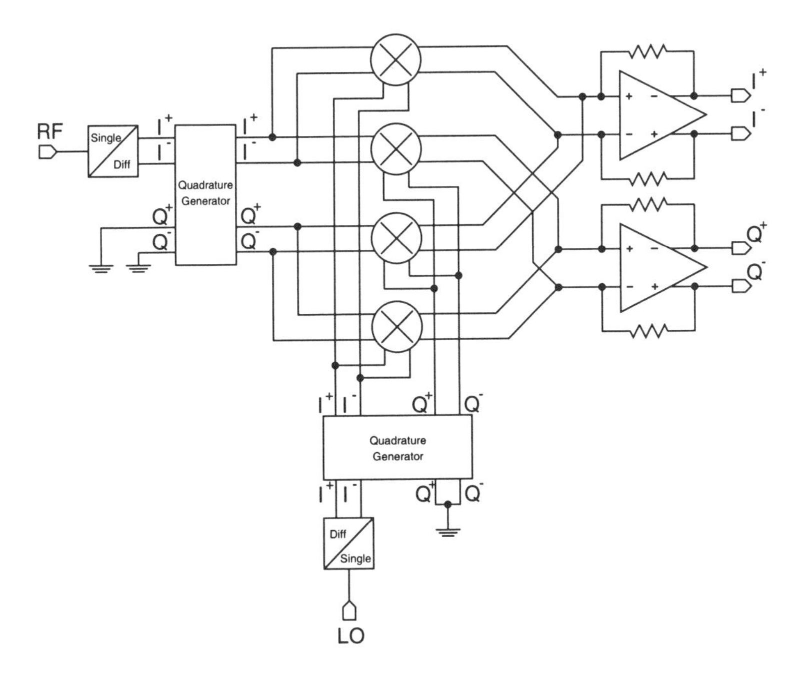

Figure 4.2 A double-quadrature down-conversion mixer.

4.4 THE SYNTHESISER

The local oscillator is responsible for the correct frequency selection in the up- and down-converters. Since the frequency spectrum in modern wireless communication systems must be used as efficiently as possible, channels are placed together very closely. The signal level of the desired receive channel can be very small, whereas adjacent channels can have very large power levels. Therefore the phase noise specifications for the LO signal are very hard to meet.

Meanwhile, mobile communication means low power consumption, low cost and low weight. This implies that a completely integrated synthesiser is desirable, where "integrated" means a standard CMOS technology without any external components or processing steps. Usually, the LO is realised as a phase-locked loop. The very hard specs are reflected in the design of the Voltage-Controlled Oscillator (VCO) and the Dual-Modulus Pre-scaler (DMP).

The phase noise of a VCO ($L\{\Delta\omega\}$) is one of the most important specifications for telecommunication applications. If there is sideband phase noise at an offset $\Delta\omega$ equal to the channel spacing of the system, the down-converter will superimpose the adjacent channel onto the signal in the receive channel. As a result, this will reduce the signal to noise ratio (SNR) of the wanted signal.

For the realisation of a gigahertz VCO in a sub-micron CMOS technology, two options exist: ring oscillators or oscillators based on the resonance frequency of an LC-tank. The inductor in this LC-tank can be implemented as an active inductor or a passive one. It has been shown that for ring oscillators as well as active LC-oscillators, the phase noise is inversely related to the power consumption [16].

$$\text{Ring oscillator [10]:} \qquad L\{\Delta\omega\} \sim kT \cdot R\left(\frac{\omega}{\Delta\omega}\right)^2 \text{ with } g_m = \frac{1}{R}$$

$$\text{Active-LC [7]:} \qquad L\{\Delta\omega\} \sim \frac{kT}{2\omega C}\left(\frac{\omega}{\Delta\omega}\right)^2 \text{ with } g_m = 2\omega C \tag{4.1}$$

Therefore, the only viable way to obtain a low-power, low-phase-noise VCO is an LC-oscillator with a passive inductor. In this case, the phase noise changes proportionally with the power consumption:

$$\text{Passive-LC [7]:} \qquad L\{\Delta\omega\} \sim kT \cdot R\left(\frac{\omega}{\Delta\omega}\right)^2 \text{ with } g_m = R \cdot (\omega C)^2 \tag{4.2}$$

As could be expected, the limitation in this oscillator is the integrated passive inductor. Equation (4.2) shows that for low phase noise, the resistance R (i.e. the equivalent series resistance in the LC-loop) must be as small as possible. Since the resistance R will be dominated by the contribution of the inductor's series resistance, the inductor design is critical. Three solutions exist:

1. Spiral inductors on a silicon substrate usually suffer from high losses in this substrate, which limit the obtainable Q-factor. Recently, techniques have been developed to etch this substrate away underneath the spiral coil in a post-processing step [17]. However, since this requires an extra etching step after normal processing of the ICs, this technique is not allowed for mass production.

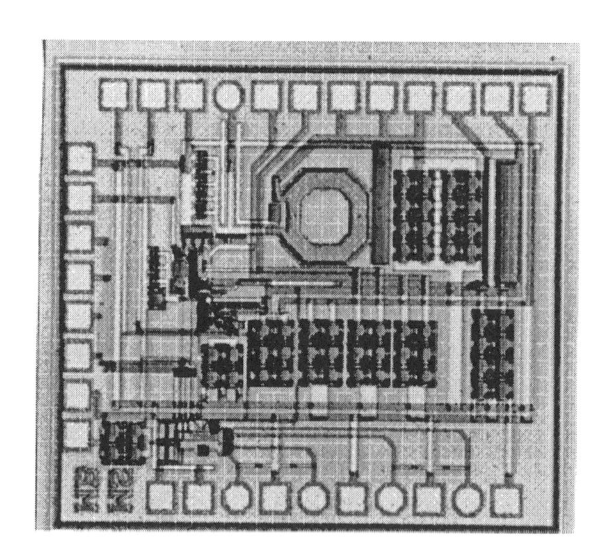

Figure 4.3 A 0.4 μm CMOS up-converter.

2. For extremely low phase-noise requirements, the concept of bondwire inductors has been investigated. Since a bondwire has a parasitic inductance of approximately

1 nH/mm and a very low series resistance, very-high-Q inductors can be created. Bondwires are always available in IC technology and can therefore be regarded as being standard CMOS components. Two inductors, formed by four bondwires, can be combined in an enhanced LC-tank to allow a noise/power trade-off [16]. A micro-photograph of the VCO is shown in Fig. 4.3 [18]. The measured phase noise is as low as -115 dBc/Hz at an offset frequency of 200 kHz from the 1.8-GHz carrier. The power consumption is only 8 mA at 3 V supply. Although chip-to-chip bonds are used in mass commercial products, they are not characterised on yield performance for mass production. Therefore, the industry is reluctant to accept this solution.

3. The most elegant solution is the use of a spiral coil on a standard silicon substrate without any modifications. Bipolar implementations do not suffer from substrate losses, because they usually have a high-ohmic substrate [19]. Most sub-micron CMOS technologies use a highly doped substrate and have therefore large induced currents in the substrate which are responsible for the high losses [20]. The effects present in these low-ohmic substrates can be investigated with finite-element simulations. This analysis leads to an optimised coil design, which is used in the spiral-inductor LC-oscillator shown in Fig 4.4. Only two metal layers are available, and the substrate is highly doped. With a power consumption as low as 6 mW, a phase noise of -116 dBc/Hz at 600 kHz offset from the 1.8-GHz carrier was obtained [21].

4.5 RF CMOS UP-CONVERTERS

Until now, mainly CMOS down-conversion mixers have been reported in the open literature. It is only recently that CMOS converters are presented and results are demonstrated [22]. In classical bipolar transceiver implementations, the up- and down-converter mixers typically use the same four-quadrant topology. There are, however, some fundamental differences between up- and down-converters, which can be exploited to derive optimal dedicated mixer topologies. In a down-conversion topology, the two input signals are at a high frequency (e.g. 900 MHz for GSM systems) and the output signal is a low-frequency signal of at most a few MHz for low-IF or zero-IF receiver systems. This extra degree of freedom has been used in the design of a very successful down-converter-only CMOS mixer topology [5].

For up-conversion mixers, the situation is totally different. The high-frequency local oscillator (LO) and the low-frequency baseband (BB) input signal are multiplied to form a high-frequency output signal. All further signal processing has to be performed at high frequencies, which is very difficult and power consuming when using current sub-micron CMOS technologies. Furthermore, all unwanted signals like the intermodulation products and LO leakage have to be limited to a level below e.g. -30 dB of the signal level.

Many published CMOS mixer topologies are based on the traditional variable-transconductance multiplier with cross-coupled differential modulator stages. Since the operation of the classical bipolar cross-coupled differential modulator stages is based on the translinear behaviour of the bipolar transistor, the MOS counterpart can only be effectively used in the modulator or switching mode. Large LO signals have to be used to drive the gates, resulting in a huge LO feed-through. This is a problem

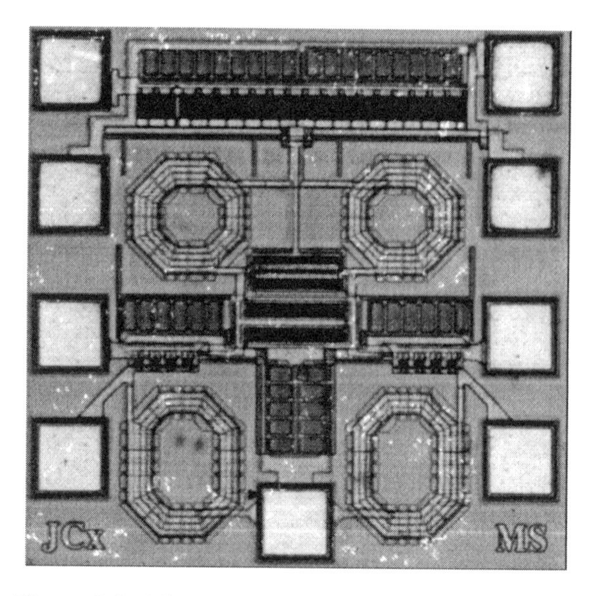

Figure 4.4 Micro-photograph of the hollow-spiral VCO.

of CMOS down-converters as well; e.g. in [23] the output signal level is -23 dBm with a LO feed-through signal of -30 dBm, which represents a suppression of only 7 dB. This gives rise to very severe problems in direct up-conversion topologies. The problems above can be overcome in CMOS by linearly modulating the current of a MOS mixing transistor biased in its linear region. For a gate voltage of $V_1 + v_{\text{in}_1}$, a drain voltage of $V_2 + v_{\text{in}_2}/2$ and a source voltage of $V_2 - v_{\text{in}_2}/2$, the current through the transistors is given by:

$$i_{DS} = \beta \cdot (v_{\text{in}_1} \cdot v_{\text{in}_2}) + \beta \cdot (V_1 - V_T - V_2) \cdot v_{\text{in}_2} \tag{4.3}$$

When the LO signal is connected to the gate, and the baseband signal to v_{in_2}, the current contains frequency components around the LO due to the first term, and components of the baseband signal due to the second term in equation (4.3), respectively. Based on this principle, a 1 GHz up-converter has been presented in a standard CMOS technology [22] (see Fig. 4.3).

All unwanted measured signals are below -30 dBc. For example, the mixer of [12] is implemented in a 0.7 μm CMOS technology which achieves an $f_{\max}$ of only 6 GHz for a gate over-drive of 1 V or a gm/I ratio of only 2. Typical bipolar technologies used for the implementation of 900 MHz fully integrated transceivers have cut-off frequencies of over 20 GHz [12]. Due to the low gm/I of present sub-micron technologies suitable for high frequency operation, the power consumption of CMOS pre-amplifiers will be up to 20 times higher than of bipolar ones. However, thanks to the rapid down-scaling of CMOS technologies, the present CMOS building block realizations show that full CMOS transmitters with an acceptable power consumption are feasible in very deep sub-micron CMOS, but some considerable research has still to be performed.

4.6 FULLY INTEGRATED CMOS TRANSCEIVERS

Combining all the techniques described above has recently resulted in the development of single-chip CMOS transceiver circuits [24]. In Fig. 4.5, a micro-photograph of the 0.25 μm CMOS transceiver is presented. The chip does not require a single external component, nor does it require any tuning or trimming. The objective of this design is to develop a complete system for wireless communications at 1.8 GHz that can be built with a minimum of surrounding components: only an antenna, a duplexer, a power amplifier and a baseband signal processing chip are needed. The high level of integration is achieved by using a low-IF topology for reception, a direct quadrature up-conversion topology for transmission, and an oscillator with on-chip integrated inductor. The presented chip has been designed for the DCS-1800 system, but the broadband nature of the LNA, the down-converter, the up-converter and the output amplifier makes the presented techniques equally suitable for use at other frequencies, e.g. for use in a DCS-1900 or a DECT system. The presented circuit consumes 240 mW from a 2.5 V supply and occupies a die area of 8.6 mm^2.

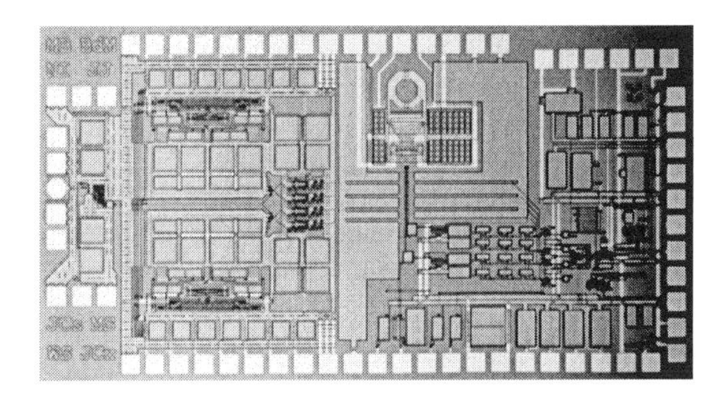

Figure 4.5 A 0.25 μm CMOS transceiver.

4.7 CONCLUSIONS

The trend towards deep sub-micron technologies has resulted in the exploration of the possible use of CMOS technologies for the design of RF circuits by several research groups. Especially the development of new receiver topologies, such as low-IF topologies, in combination with highly linear down-converters, has opened the way to fully integrated down-converters with no external filters or components. The trends towards deep sub-micron technologies will allow to achieve those goals as long as the short-channel effects will not limit the performance concerning linearity and intermodulation problems.

Fully integrated CMOS VCO circuits having high performance, low phase noise and low power drain were shown in this paper. In a first step, the use of post processing techniques has enabled the use of standard CMOS technologies, using bondwires as inductors.

Today, low phase noise performance can be achieved with optimised integrated spiral inductors in standard CMOS technologies even without any post-processing, tuning, trimming or external components. This opens the way towards fully integrated receiver circuits.

Recently, CMOS up-converters with moderate output power were presented in the open literature. Again, thanks to the trends towards deep sub-micron, this opens the way towards fully integrated transceiver circuits in standard CMOS technologies.

References

[1] J. Sevenhans, A. Vanwelsenaers, J. Wenin and J. Baro. "An integrated Si bipolar transceiver for a zero IF 900 MHz GSM digital mobile radio front-end of a hand portable phone," *Proc. CICC*, pp. 7.7.1–7.7.4, May 1991.

[2] J. Crols and M. Steyaert. "A Single-Chip 900 MHz CMOS Receiver Front-end with a High Performance Low-IF Topology," *IEEE J. of Solid-State Circuits*, vol. 30, no. 12, pp. 1483–1492, Dec. 1995.

[3] P. R. Gray and R. G. Meyer. "Future Directions in Silicon ICs for RF Personal Communications," *Proc. CICC*, May 1995.

[4] Bang-Sup Song. "CMOS RF Circuits for Data Communications Applications," *IEEE JSSC*, SC-21, no.2, pp. 310–317, April 1986.

[5] J. Crols and M. Steyaert. "A 1.5 GHz Highly Linear CMOS Downconversion Mixer," *IEEE JSSC*, SC-30, no.7, pp. 736–742, July 1995.

[6] D. Rabaey and J. Sevenhans. "The challenges for analog circuit design in Mobile Radio VLSI Chips," *Proc. of the AACD workshop*, vol. 2, pp. 225–236, Leuven, March 1993.

[7] F. Brianti *et al.* "High integration CMOS RF Transceivers," *Proc. of the AACD workshop*, Lausanne, April 1996.

[8] C. H. Hull, R. R. Chu and J. L. Tham. "A Direct-Conversion Receiver for 900 MHz (ISM Band) Spread-Spectrum Digital Cordless Telephone," *Proc. ISSCC*, pp. 344–345, San Francisco, Feb. 1996.

[9] R. Yan *et al.* "High performance 0.1 micron room temperature Si mosfets," *Digest of technical papers, 1992 Symposium on VLSI technology,* 2–4 June 1992.

[10] J. Chen *et al.* "A high speed SOI technology with 12 ps/18 ps gate delay operation at 1.5 V," *Proceedings of IEEE International Electron Devices Meeting*, SF, CA, 13–16 Dec.1992.

[11] M. Steyaert and W. Sansen. "Opamp Design towards Maximum Gain-Bandwidth," *Proc. of the AACD workshop*, pp. 63–85, Delft, March 1993.

[12] T. Stetzler, I. Post, J. Havens and M. Koyama. "A 2.7 V to 4.5 V Single-Chip GSM Transceiver RF Integrated Circuit," *Proc. ISSCC*, pp. 150–151, San Francisco, Feb. 1995.

[13] C. Marshall *et al.* "A 2.7 V GSM Transceiver ICs with On-Chip Filtering," *Proc. ISSCC*, pp. 148–149, Feb. 1995.

[14] J. Sevenhans *et al.* "An Analog Radio front-end Chip Set for a 1.9 GHz Mobile Radio Telephone Application," *Proc. ISSCC*, pp. 44–45, San Francisco, Feb. 1994.

[15] M. Steyaert *et al.* "RF CMOS Design, Some untold pitfalls" *Proc. of the AACD workshop*, Lausanne, April 1996.

[16] J. Craninckx and M. Steyaert. "Low-Noise Voltage Controlled Oscillators Using Enhanced LC-tanks," *IEEE Trans. on Circuits and Systems - II: Analog and Digital Signal Processing*, vol. 42, no. 12, pp. 794–804, Dec. 1995.

[17] A. Rofourgan, J. Rael, M. Rofourgan, and A. Abidi. "A 900-MHz CMOS LC-Oscillator with Quadrature Outputs," *Proc. ISSCC*, pp. 392–393, Feb. 1996.

[18] J. Craninckx and M. Steyaert. "A 1.8-GHz Low-Phase-Noise Voltage-Controlled Oscillator with Prescaler," *IEEE Journal of Solid-State Circuits*, vol. 30, no. 12, pp. 1474–1482, Dec. 1995.

[19] N. M. Nguyen and R. G. Meyer. "A 1.8-GHz Monoithic LC Voltage- Controlled Oscillator," *IEEE Journal of Solid-State Circuits*, vol. 27, no. 3, pp. 444–450, March 1992.

[20] J. Crols, P. Kinget, J. Craninckx and M. Steyaert. "An Analytical Model of Planar Inductors on Lowly Doped Silicon Substrates for High Frequency Analog Design up to 3 GHz," *Proc. VLSI Circuits Symposium*, June 1996.

[21] J. Craninckx and M. Steyaert. "A 1.8-GHz Low-Phase-Noise Spiral-LC CMOS VCO," *Proc. VLSI Symposium*, June 1996.

[22] M. Borremans and M. Steyaert. "A 2 V, Low Distortion, 1 GHz CMOS Upconversion Mixer," *Proc. CICC*, session 24.3, pp. 517–520, May 1997.

[23] A. N. Karanicolas. "A 2.7 V 900 MHz CMOS LNA and Mixer," *Proc. ISSCC*, pp. 50–51, San Francisco, Feb. 1996.

[24] M. Steyaert *et al.* "A Single Chip CMOS Transceiver for DCS1800 Wireless Communications," *Proc. IEEE-ISSCC'98*, Feb. 1998

5 GSM TRANSCEIVER FRONT-END CIRCUITS IN 0.25 μm CMOS*

Qiuting Huang, Paolo Orsatti, and Francesco Piazza

Integrated Systems Laboratory
Swiss Federal Institute of Technology (ETH) Zurich
ETH-Zentrum, CH – 8092 Zurich, Switzerland

Abstract: So far, CMOS has been shown to be capable of operating at RF frequencies, although the inadequacies of the device-level performance often have to be circumvented by innovations at the architectural level that tend to shift the burden to the circuit building blocks operating at lower frequencies. The RF front-end circuits presented in this paper show that excellent RF performance is feasible with 0.25 μm CMOS, even in terms of the requirements of the tried-and-true superheterodyne architecture. Design for low noise and low current consumption targeted for GSM handsets has been given particular attention in this paper. Low-noise amplifiers with sub-2 dB noise figures, a double-balanced mixer with 12.6 dB SSB NF, as well as sub-25 mA current consumption for the RF front-end (complete receiver), are among the main achievements.

5.1 INTRODUCTION

The quest for miniaturized cellular telephone handsets has resulted in the development of commercial transceiver ICs of increasing integration levels in the last decade. The more traditional RF front-ends comprising a single LNA or LNA-mixer combination, though they still have a strong presence in the market, are expected to slowly give way to the more highly integrated front-ends that combine the RF mixer, or even the LNA, with the IF-strip and baseband demodulators in the receiver [1–5]. The dominant technology for commercial RF front-ends has so far been advanced bipolar junction transistor (BJT) technology.

The last five years have also seen CMOS, traditionally confined to the digital and baseband part of radio transceivers, make successful in-roads into the RF sections. At the research level, highly integrated CMOS RF transceivers have been reported for wireless LAN and cordless telephones [6, 7], that are highly innovative in terms of architecture design. For cellular applications, where weight, size, and stand-by power requirements of the handset are much more demanding and the radio network environment more complex (less restricted in terms of possible interfering signals) and hostile, CMOS implementations of RF front-ends have still fallen far short of BJT performance, especially in terms of current consumption [6, 8]. Compared to the typical current consumption of 50 mA in highly integrated BJT receivers, CMOS receivers published so far typically consume 100 mA, which makes them an unattractive alternative to BJT solutions because of the corresponding drop in stand-by time or increase in battery size and weight.

In this work, we report a low-power CMOS RF front-end which is part of our on-going effort to develop a complete RF transceiver in deep-submicron CMOS, and which not only meets the more stringent requirements of cellular telephony, but is also competitive to BJT implementations in terms of current consumption. Instead of taking on simultaneously the difficulties of the development of both a novel architecture and a low power implementation of RF circuits using CMOS, we have opted for the path of first gaining experience in the latter. The receiver architecture for our RF front-end is therefore the well-established and widely used superheterodyne receiver. Since more than 70% of the power in a typical receiver is consumed by the building blocks that operate at RF frequencies, our initial work has concentrated on the LNA and mixer for the receiver, and the preamplifier for the transmitter, which we report in this paper. The IF-strip plus the baseband circuits will form another chip that will be the subject of another report [9]. The ultimate goal of this work is to combine the complete receiver and transmitter into a single IC that meets at least the performance required for small mobile stations (MS).

The paper is organized as follows. Section 5.2 provides a general description of the transceiver chip architecture and the required performance. Design considerations of the LNA and mixer, including the many trade-offs involved, are presented in Sections 5.3 and 5.4, before the transmitter preamp design is described in Section 5.5. Sections 5.6 and 5.7 conclude the paper with discussions of the achieved power consumption and measured performance, respectively.

5.2 TRANSCEIVER ARCHITECTURE AND RELEVANT GSM SPECIFICATIONS [10]

Receiver planning typically consists of trade-offs, such as the number of IFs, the exact frequency of each IF, and the distribution of gain, linearity, and NF to each block in the chain, etc. In addition to the minimum performance for type approval, many other important parameters such as size, complexity, power consumption and cost must be carefully taken into consideration.

In our case, the partners responsible for system design have specified the single-IF superhet architecture shown by the block diagram in Fig. 5.1. The relatively high IF of 71 MHz was primarily dictated by image-rejection requirements and the availability of low-cost commercial filters. Realizing high gain at such a high frequency had raised concerns over the power consumption and stability of the IF AGC. Those concerns have been addressed when an earlier test chip [11] we implemented showed that with a good design, a relatively low-power IF amplifier is feasible at 71 MHz. Since complete channel filtering at the 71 MHz IF is not possible without using high-quality filters, whose price is not justified for this application, intermodulation performance of the IF amp is critical, adding further complexity to its design. To limit these problems and achieve a feasible design, 32 dB of the total gain was allocated to baseband, and the maximum gain of the IF amplifier has been limited to 60 dB. Eighty decibels of AGC range is implemented in the IF amp which is digitally programmable from −20 to +60 dB in 2 dB steps.

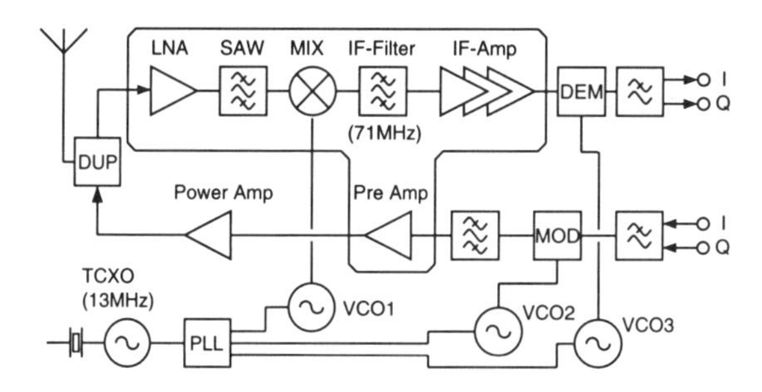

Figure 5.1 Transceiver block diagram—single-IF superhet receiver, direct up-conversion transmitter.

The large dynamic range (−102 dBm to −15 dBm) of the input signal as specified in GSM standards, for which a BER $< 10^{-3}$ must be maintained, requires at least 87 dB of AGC range for the receiver to avoid saturating the baseband A/D converters, in addition to the necessary linearity. Allowing for a 10–15 dB margin, the receiver needs to be designed with a 100 dB AGC range. Since the IF amplifier is programmable over an 80 dB range, the additional 20 dB are implemented in the LNA with a bypass switch.

The gains of the LNA and the RF mixer require another trade-off in receiver planning. The main parameters to be taken into consideration are sensitivity, intermodulation and blocking performance, as well as power consumption. Using high gain in the LNA helps to reduce the noise figure by compensating the insertion loss of the interstage filter and scaling down the noise contribution of the mixer. That is, however, at the expense of higher power consumption and risk of early overloading of both the LNA output stage and of the mixer.

The blocking signal levels that may overload the LNA are depicted in Fig. 5.2, as prescribed by GSM testing for type approval. At 20 MHz away from the MS receive bands, 0 dBm can be expected at the receiver front-end. After 25–30 dB attenuation by a typical RF filter or duplexer, the blocking signal at the output of a 15 dB LNA can still be as high as −10 dBm. For 5–10 mA output current, the −1 dB compression point can not be expected to be much higher than −5 dBm in a 50 Ω environment at the output of the LNA. If we expect the LNA to operate with blocking signals some 6 dB below its compression point, then its gain must be limited to 15 dB. Since 3–4 dB loss is expected to come from the interstage filter after the LNA, the noise power of the mixer is only divided by 16 before it is added to the input noise of the LNA. If the overall NF specification is tight, both the LNA and the mixer must also have low NF.

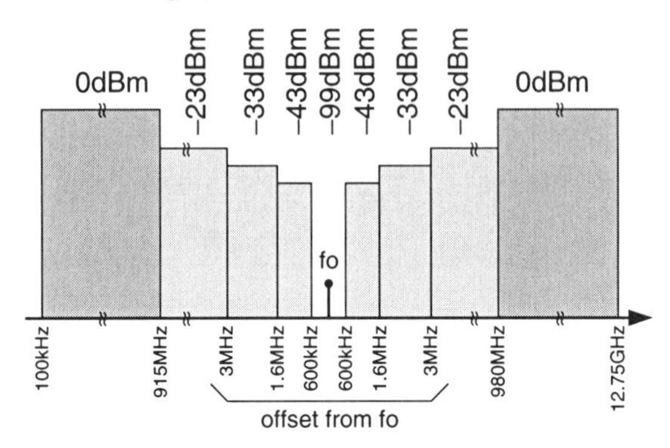

Figure 5.2 Template of blocking signal levels specified by the GSM standard.

In digital wireless communications, the noise figure of the receiver is a particularly critical parameter, because, at the boundary between very low and very high bit-error rate, the transition can be effected by just 1–2 dB of signal–to–noise ratio. A low noise figure enables satisfactory reception even for low levels of received signal power which, in turn, allows lower-power transmission from the base station and lower levels of interference in the radio environment in general.

For GSM type approval, a reference sensitivity of −102 dBm is specified for small mobile stations. This means that for static additive white Gaussian noise (AWGN) channel conditions, a BER $< 10^{-4}$ must be achieved when the input signal power P_i is −102 dBm. For the GMSK modulation scheme used in GSM, an SNR of about 9 dB is required to reach such a BER, assuming an optimum maximum-likelihood sequence estimation (MLSE) receiver. If a sub-optimal MSK-type detector is used, an SNR of 10 dB is required [12]. The simplicity of the MSK-type detector is of interest here as it makes the latter well suited for low power implementation, while offering essentially the same performance as MLSE, especially under conditions more realistic than a static AWGN channel. Bearing in mind that the (antenna) source noise power density at a temperature of 290 K is −174 dBm/Hz, a noise power P_n of −121 dBm results for a 200 kHz channel. The required NF after the antenna can therefore be calculated in dB as $\mathrm{NF} = P_i - P_n - \mathrm{SNR} = 9\,\mathrm{dB}$.

In order to attenuate blocking interferers an RF filter, or often a duplexer, is required between the antenna and the LNA. Its passband loss, however, translates directly into a serious NF degradation for the receiver. A typical duplexer [13], like the one which we intend to use, has an insertion loss of 3.2 dB for the Rx path (1.2 dB for the TX path). This leaves an overall NF of less than 6 dB at the LNA input.

A typical RX image-reject interstage filter has an insertion loss of 3–4 dB (3.8 dB in [13]), which degrades the NF at the mixer input by the same amount. As a result, even if the NFs of the LNA and mixer are as low as 2 dB and 12 dB (single sideband, SSB) respectively, the overall NF will still be 7.6 dB, leaving only a small margin for gain tolerances, insertion loss of the IF filter, noise contribution from the IF amplifier and the imperfect rejection of the LNA's noise at the image frequency. To prevent excessive NF contributions from the IF-strip, a gain of 10 dB is specified for the mixer. If we restrict the NF of the IF-strip (including the IF filter's insertion loss) to 15 dB, the overall NF is just under 8 dB. The 15 dB limit can be met, for example, by an IF AGC with a NF of 8 dB [11] combined with an IF filter having 5.1 dB insertion loss [13], or a cheaper IF filter with 11.3 dB loss [13] could be tolerated with a better AGC design with a 4 dB NF, which is more difficult. Should the gain of the LNA drop to 14 dB due to component variations, the overall NF becomes 8.5 dB, which is still within specifications. The noise figures are therefore specified as 2 dB and 12 dB SSB for the LNA and the mixer respectively, both demanding values even for BJT or GaAs technologies.

To save the interstage filter and eliminate the adverse effect of its insertion loss on NF, image-reject mixers can be used as an alternative. By operating two mixers in quadrature the image is usually attenuated by 30–35 dB, which is sufficient to prevent noise at the image frequency from contributing to the total NF excessively. Having two mixers operating at 900 MHz and a phase shifter in both the local oscillator and IF paths, however, results in power consumption roughly 3–4 times higher than that of one conventional mixer, which was considered unacceptable for the present design.

Compared to the receiver, the transmitter requirements are relatively simple. The most difficult one is perhaps meeting the −36 dBm spurious signal emission limit.

Without any IF or image frequency, a direct up-conversion transmitter is less likely to generate spurious signals. The filtering requirements to eliminate leakage are therefore greatly relaxed. Power consumption, one of the disadvantages of direct conversion, is not as critical here, since it will be dominated by the power amplifier (≈ 5 W average input power). The direct up-conversion architecture, shown in Fig. 5.1, is therefore preferred.

Since the local oscillator will be integrated on the same chip as the transmitter, oscillator pulling may be an issue. Using an offset local oscillator will solve this problem, at the expense of an extra external filter.

The integrated part of the transmitter consists of a vector modulator driving a preamp via an external interstage SAW filter [13]. In the present front-end, only the transmitter preamp has been integrated to test the ability of CMOS to deliver the 2 mW of power which are typically required by commercial power amplifiers. Thanks to the constant envelope modulation, linearity is not too critical, and the preamp can be driven at the compression point for better efficiency. The nominal gain of the preamp is 23 dB.

5.3 LOW-NOISE AMPLIFIER

In addition to the gain and noise figure, as discussed in the previous section, the requirements put on the LNA include 50 Ω matching at both the input and output, small-signal linearity (IP3) and the large-signal linearity, which is described by the −1 dB compression point. In addition, these performance parameters must be relatively insensitive to the variations of process parameters and values of passive components such as the matching inductor and capacitors [14].

5.3.1 Input stage

The most important factor that determines our choice of the LNA configuration is the achievable noise figure. Although noise figures of 3 dB and better are usually quite easy to achieve given sufficient current, integrated LNAs do not often have NFs below 2 dB even in GaAs and BJT technologies. This apparent barrier can be partially explained by the definition of the noise figure itself, which results in a law of diminishing returns.

Since the input-referred noise of a MOS transistor is given by $4\gamma kT/g_m$, with $\gamma = 2/3$, the noise factor of a common gate (C-G) amplifier is given by $1+\gamma$. The minimum NF achievable by a C-G amplifier is therefore 2.2 dB. With the additional noise contributions of pads, the substrate and the output stage, the achievable NF is closer to 3 dB, which is too high for the present application.

For a common source configuration, as shown in Fig. 5.3, the real part of the input impedance required for 50 Ω matching is generated by inductive degeneration [14–16]. The matching network formed by L_1 and L_2 and the gate capacitance C_{gs} gives a voltage gain between the LNA input (marked RF) and the gate–source of transistor M1 which is equal to the quality factor Q of the matching network at resonance.

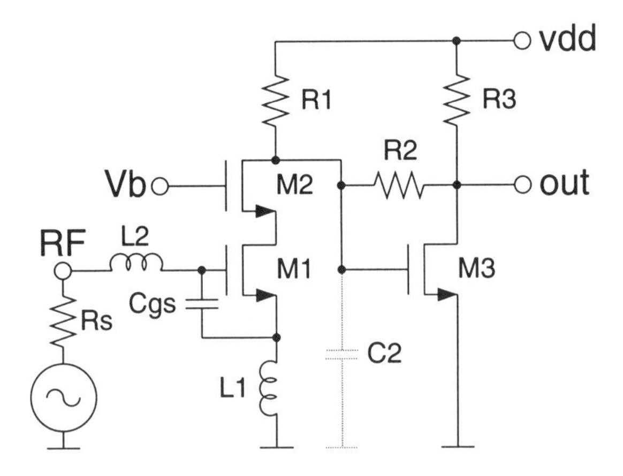

Figure 5.3 Common-source LNA with transimpedance output stage.

$$Q = \frac{1}{g_m \omega_0 L_1} \tag{5.1}$$

If Q is greater than one, this gain may help reduce the contribution of the thermal noise of M1's channel, I_n^c, and of those noise sources from the LNA's output stage which contribute to the overall LNA noise figure. For example, the NF contribution of

I_n^c alone is given under matching conditions by

$$\mathrm{NF} = 1 + \gamma \frac{\omega_0 L_1}{Q \cdot R_s} = 1 + \gamma \frac{1}{Q^2 \cdot g_m R_s}, \tag{5.2}$$

where $\gamma = 2/3$ for long-channel devices. Even if the transconductance is set to only 20 mS as in the case of the C-G configuration, the voltage gain of Q can still reduce the noise contribution from I_n^c to very low levels. Indeed, this well-defined noise contribution reduces to insignificant levels, even with only moderate values of Q and g_m. The NF of common source LNAs, typically 1.5 dB to 3 dB, (even if integrated in GaAs and BJT technologies) must be dominated by other less well-defined noise sources. This latter category of noise sources includes the contribution of substrate resistance through capacitances under bonding pads, gate-induced noise current [15] and the back-gating effect of the substrate resistance under the MOS transistor [17].

A more important reason to use $Q > 1$ to achieve some voltage gain is to reduce the current level required for a given overall transconductance of the input stage. Under matching conditions, and neglecting the parasitic capacitances associated with input pads, the ratio between the output current of the input stage and the input voltage of the LNA is exactly the inverse of L_1's impedance, independent of the transconductance g_m of M1.

$$|G_m| = \frac{1}{\omega_0 L_1} \tag{5.3}$$

The overall voltage gain of the LNA is then given by the product of G_m and the load (trans)impedance of the output stage. The latter impedance is constrained by the output matching requirement to about 50 Ω. For a LNA gain of 15 dB, $G_m \approx 120$ mS, so that at 1 GHz the value of L_1 lies around 1 nH. This happens to be the value of a typical bond-wire, so that no board-level inductor is necessary to implement L_1.

Once this overall G_m or bond-wire impedance is determined, the transconductance g_m of M1 is related to Q by Eq. (5.1). If we use the simple formula

$$g_m = \sqrt{2\mu C'_{ox} W \cdot I / L}$$

to estimate the current consumption, and bear in mind one of the conditions for impedance matching,

$$g_m L_1 = C_{gs} R_s \tag{5.4}$$

then we see that the bias current I is inversely proportional to Q. This is important for a CMOS LNA design, because without this voltage gain (i.e. of Q were unity), 10 mA would be needed for a G_m of 120 mS even at a gate length of 0.25 μm. While a high Q matching network allows the current consumption of the LNA input stage to be low, the level of reactance associated with the network inductors and capacitor becomes high. Although the nominal capacitive and inductive reactance should cancel each other if designed for 50 Ω matching, the same 10% deviation from the nominal values results in higher residual (uncanceled) reactance for higher Q. The variability of the input reflection coefficient S11 is therefore also worse. One of the serious effects of the increase in S11 and deviation from 50 Ω input resistance is the decrease of the RF filter's performance [14].

Measurements of a commercial duplexer [13] in Fig. 5.4 show that $|S11| = -6$ dB can both increase the filter's pass-band loss by 0.5 dB, causing a corresponding increase in the overall noise figure, and at the same time degrade the stop-band attenuation by as much as 6 dB at 20 MHz away from the passband, at frequencies where 0 dBm blocking signals are expected.

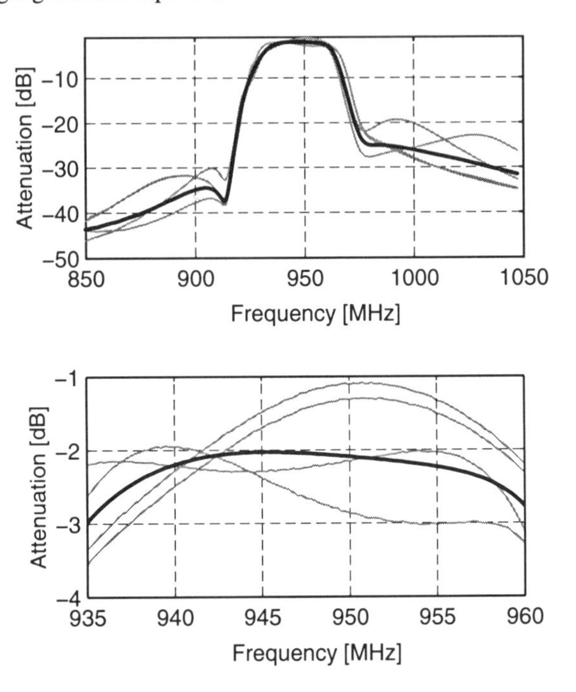

Figure 5.4 Measured duplexer characteristics. The thick curve is with exact 50 Ω matching. The other four were measured with imperfect S11: $|S11| = -6$ dB, phase of S11 = 45°, 135°, 225° and 315°. a) Overall transfer function, b) close-up of passband.

In our design, a fairly low quality factor of 2.5 is used for the input matching network, so that for the 10–20% tolerance expected of L_1, L_2 and C_{gs}, the variability of S11 is still very low. On the other hand, the bias current of the first stage is reduced to 4 mA, which is sufficiently low in the context of the target current consumption of 25–50 mA for the complete RF receiver.

Without the cascode transistor M2, the influence of M1's Miller capacitance can add quite significantly to that of the pad capacitance and cause the design to deviate from the ideal situations described by Eqs. (5.1)–(5.4). The bias current of M1, for example, increases from 4 mA to 6 mA. The cascode transistor, on the other hand, may contribute some of its own noise to the amplifier, because M1's output impedance is low due to the very short channel length and high current.

Having scaled down the contribution of the thermal noise of M1's channel by the Q of the matching network, the design of the LNA input stage for low noise figure must then concentrate on parasitic noise sources that are not as well-defined as I_n^c. Characterizing the transistors for NF before using them for the LNA, as is often done in traditional microwave LNA designs, is beyond our resources, because this would entail fabricating arrays of transistors of anticipated range of dimensions and bias currents, with anticipated layout. Since such devices are not designed to match the 50 Ω test environment, 'de-embedding' the influence of pads and other test fixtures and measuring the intrinsic noise figure at 1 GHz is a rather large undertaking. The minimal noise figure measurements made by the CMOS process provider typically come from transistors which have a specific W/L ratio and layout and are biased at unrealistically high currents (40–50 mA), and therefore do not provide us with much information either. In the absence of accurate RF models of MOS transistors backed by reliable experimental verification, the design for low NF of our LNAs had to combine general care in noise minimization with some specific, controlled experiments that were conceived to extract some data on the relative importance of certain parasitic noise sources in the overall noise figure.

One such source is the induced gate current [18, 19]. This noise source is important at higher microwave frequencies. Recently, a detailed analysis has been published [15] which not only points out that induced gate noise current is also important for CMOS at the low GHz frequency range, but also proposed ways of optimizing the LNA design parameters, such as the matching network's Q, to achieve minimum noise figure. Unfortunately, the predicted optimal NF was not confirmed out by the experimental results in [15] and so the subject is still wide open.

Another important source of noise in CMOS LNAs which was recently reported is the backgating effect of the resistive substrate under the transistor channel [17]. In the case of the 0.25 μm CMOS we used, the substrate resistivity has a fairly high value of 5 Ω-cm, which could have an even higher influence. The experimental data given in [17], however, reported a NF difference of 1 dB between a layout with sufficient substrate contacts to ground and a layout with fewer contacts, at a high level of 6–7 dB overall noise figure. The important question for us was whether there will be much difference for LNAs which already have low noise figures.

In the initial stage of our development, we had succeeded in realizing a 0.25 μm CMOS LNA, which we refer to as LNA-1, for which care was taken in pad design, fingered gate layout, etc., so that the measured noise figure was as low as 2 dB. The substrate around the gate of M1 was already well grounded by rows of contacts surrounding the periphery of the 600 μm gate formed by eighty 7.5 μm by 0.25 μm fingers, as shown in Fig. 5.5a. The input stage of LNA-1 did not use the cascode transistor M2, so that the bias current was 6 mA and the actual matching network Q was close to unity due the influence of Miller capacitance.

To investigate the influence of substrate resistance on achievable NF, we integrated a second LNA, LNA-2, that is nominally identical to as LNA-1, except that the layout of the gate of M1 was changed to allow more substrate contacts to be placed, as shown in Fig. 5.5b. The original LNA-1 was integrated again in the same run as a reference to avoid confusing NF variation from run to run with the NF improvement due to difference in layout.

Realizing that M1's Miller capacitance is causing a significant deviation of the matching network gain from that given in (5.1), which results in 6 mA instead of 4 mA in the input stage, we also introduced the cascode transistor M2 at this stage. In

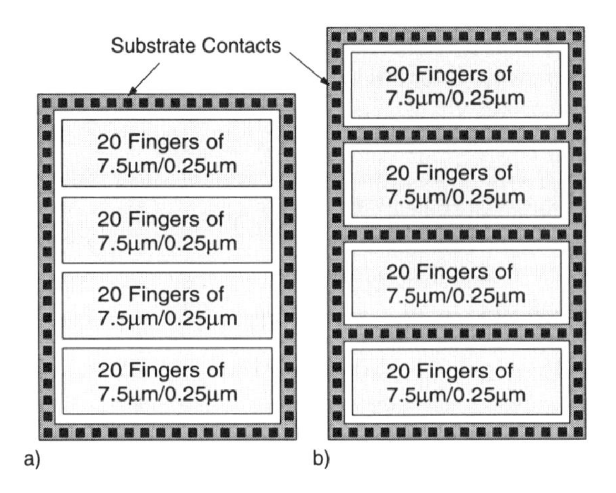

Figure 5.5 Layout of transistor M1: a) LNA-1, b) LNA-2.

doing so, the effective matching network Q increases from 1 to 2 which allows the bias current to be reduced to 4 mA. This circuit has been implemented as LNA-3 in the same run as LNA-1 and LNA-2. Although the change in Q as a result of going from a non-cascode to a cascode configuration is not exactly the same as designing for different matching network Qs for the same LNA configuration, we believe that if the matching network Q indeed has as strong an influence on the LNA's NF as [15] seems to suggest, then we should see some evidence of this in comparisons of LNA-2 and LNA-3. The gates of M1 in these two amplifiers have exactly the same layout. All three amplifiers should have noise figures less than or equal to 2 dB.

5.3.2 *Output stage*

The overall transconductance G_m of the input stage is on the order of 100–200 mS. An effective load impedance between 50 and 100 Ω is therefore needed to realize 15 dB gain. In today's CMOS technologies, however, the relatively high level of parasitic capacitance (C_2) tends to limit the achievable cut-off frequency. To make the load impedance broadband, shunt feedback in the output stage transimpedance amplifier in Fig. 5.3 is one of the most commonly used methods [1, 20]. The loop gain of the shunt feedback, approximately given by $g_{m3}R_L$, scales down the input resistance of the output stage so that the parasitic capacitance C_2 does not cut off the amplifier's frequency response as early as without feedback. The impedance R_L is the parallel combination of the output impedance of M3, the bias resistance R_3 and the 50 Ω input impedance of the interstage filter.

If the loop gain is sufficiently large, then the transimpedance of the output stage in Fig. 5.3 is approximately given by the feedback resistance R_2, and the overall LNA gain is simply $G_m R_2$. In practice, the limits of M3's transconductance g_{m3} due to current, IP3 and gate capacitance considerations imply that only moderate loop gain is achievable with an R_L that is dominated by the 50 Ω load. The LNA gain must be designed on the basis of the more accurate design equation:

$$A_{VE} = \frac{1}{\omega_0 L_1}\left[\frac{R_2+R_L}{1+g_{m3}R_L}||R_1||Z_{d12}\right]\frac{R_L}{R_L+R_2}(1+g_{m3}R_2)\,, \tag{5.5}$$

where Z_{d12} is the output impedance of M1 or the M1-M2 cascode.

The higher parasitics associated with PMOS transistors make them unsuitable for biasing purposes and resistors R_1 and R_3 are used instead. The lack of voltage headroom combined with high bias currents constrain the values of R_1 and R_3 to a couple of hundred ohms. In addition to (5.5), the choice of R_1, R_2, R_3 as well as g_{m3} are constrained by two other important requirements: realizing a 50 Ω output resistance and providing a sufficiently high −1 dB compression point (CP). The CP is limited by the maximum output voltage for a given bias current I_3,

$$V_o^{\max} = I_3 \cdot Z_3 = I_3 \cdot [(R_2+R_1||Z_{d12})||R_L] = I_3 \cdot [(R_2+R_1||Z_{d12})||R_3||Z_{d3}||R_{50}]\,, \tag{5.6}$$

where Z_3 describes the impedance seen by the drain terminal of M3. Due to the 50 Ω input impedance of the interstage filter, R_{50}, $|Z_3|$ is less than 50 Ω. This means that for a CP requirement of 0 dBm (300 mVp) the bias current I_3 must be no less than 6 mA. In practice R_1 and R_3 can not be very high because of both the voltage headroom and the need to implement the 50 Ω output impedance for the LNA. In the worst case, $|Z_3|$ can be as low as 25 Ω, and twice as much current is needed. The constraint on R_1, R_2 and R_3 in terms of 50 Ω output impedance can be expressed as

$$Z_0 = \left(\frac{R_2+R_1||Z_{d12}}{1+g_{m3}(R_1||Z_{d12})}\right)||R_3||Z_{d3} = 50\,\Omega\,, \tag{5.7}$$

where it can be seen that in the absence of M3's transconductance g_{m3}, the 50 Ω output impedance would have to be implemented completely by R_1, R_2 and R_3, in which case $|Z_3|$ in (5.6) would be as low as 25 Ω. The design of the output stage is therefore primarily a trade-off of resistance values to satisfy (5.5)–(5.7) and minimize the required output current, while ensuring that M3's gate overdrive is sufficiently high to maintain a high IP3. The latter requirement sets a limit to M3's gate width.

For LNA-1, a higher CP was sought, which led to higher output current. In the absence of the cascode, the current of the input stage is also higher, resulting in lower values of R_1, R_2 and R_3. This leads to an output bias current of 12 mA which gave us a measured output CP of −4 dBm. This output stage design has been left unchanged for LNA-2 and LNA-3, so that only one thing is different between the two successive designs. Further optimization of the output stage, which took advantage of the lower bias current of the input stage, allows the levels of R_1, R_2 and R_3 to be somewhat higher than before, so that only half as much bias current is needed for the output stage, with only a slight drop in output compression point. The increase of resistance levels is also expected to result in lower equivalent noise referred to the drain of M2, so that the overall noise figure will be slightly lower. This design is referred to as LNA-4. The differences between the four designs and the key parameters of each design are summarized in Tab. 5.1.

Table 5.1 Summary of the implemented LNAs and their key design parameters.

	LNA-1	*LNA-2*	*LNA-3*	*LNA-4*
I(M1) [mA]	6	6	4	4
I(M3) [mA]	12	12	12	6
M1 [μm]	600	600	600	600
M2 [μm]	–	–	600	600
M3 [μm]	100	100	100	300
R1 [Ω]	235	235	400	400
R2 [Ω]	120	120	120	120
R3 [Ω]	105	105	105	255
Sub. contact	Fig. 5.5a	Fig. 5.5b	Fig. 5.5b	Fig. 5.5b

5.4 SINGLE AND DOUBLE-BALANCED MIXER

The chief requirements of an RF mixer in a superhet receiver such as ours are 50 Ω matching to the interstage filter, low noise figure, high IP3, and a moderate gain to reduce the noise contribution of subsequent stages to the receiver front-end. To satisfy all of the requirements, single-balanced (SBM) or double-balanced (DBM) mixers based on the Gilbert cell are practically left as the only choice, as evidenced by the fact that they are found in virtually every BJT RF front-end IC.

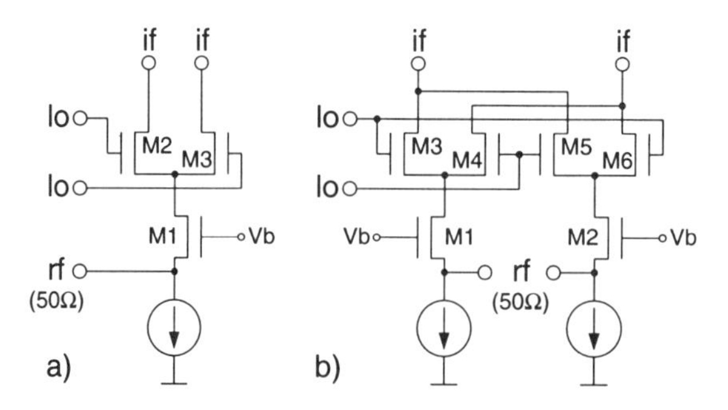

Figure 5.6 Schematic diagram of a) single-balanced mixer, b) double-balanced mixer.

In a single or double-balanced mixer (c.f. Fig. 5.6) the transconductor (g_{mi}) implemented by the (input) RF transistor(s) is followed by commutating switches that perform the function of frequency translation. The effective conversion transconductance of the mixer is g_{mi}/π for the single balanced configuration and $2g_{mi}/\pi$ for the double-balanced configuration, assuming ideal switching. Thus even in the ideal case the switches introduce a loss of $1/\pi$ (−9.9 dB) for the SBM case and $2/\pi$ (−3.9 dB) for the DBM case. With sinusoidal LO signals the switching is much less than ideal, and the loss is more likely −12 dB for the SBM case and −6 dB for the DBM case. Because of this loss, the influence of noise current introduced by the switching transistors (during the time when each pair of switches remain in the on-state simultaneously, for example), exceeds the noise of the RF input transistor(s). Combined with the inherent 3 dB NF degradation due to aliasing of the noise in the image sideband, the noise figure is therefore usually rather high for Gilbert mixers, 15 dB, for example, for BJT implementations.

Because the expected NF is high for mixers, the input transistors' contribution of 2 dB or 3 dB to the overall NF is therefore not very critical. This justifies the use of the common gate configuration for the input stage, which can be designed to provide a resistive 50 Ω input matching to the filter without any external components. The g_m of the input transistors is therefore 20 mS for the SBM case and 40 mS for the DBM case.

In cellular applications, neighboring users often generate strong interfering signals in channels adjacent to one's own. Since such interferers cannot be filtered by the RF or interstage filters and will be amplified by the LNA, the small-signal linearity (IP3) of the mixer must be sufficiently high to prevent intermodulation from degrading the bit-error rate. The type-approval requirement for GSM stipulates that the BER for reference sensitivity must be met in the presence of two −49 dBm interferers at 800 kHz and 1600 kHz away from the desired signal, respectively, when the latter's power is −99 dBm. The mixer's IP3 requirement is directly linked to the combined receiver front-end gain, A_{VF}, before the mixer. If we assume again that 10 dB SNR is required to achieve the desired BER, then the required IP3 (in dBm) is bounded by the following inequality:

$$-99 + A_{VF} - 3 \cdot (-49 + A_{VF}) - 9 + 2 \cdot \text{IP3} \geq 10 \tag{5.8}$$

$$\text{or} \qquad \text{IP3} \geq A_{VF} - 14.5\,. \tag{5.9}$$

Considering that within the passband of each filter the gain of the signal could be different by 1 dB from that of the interferers due to passband ripple, the combined gain $A_V F$ could be 2–3 dB worse than the nominal value. The latter, assuming −3 dB passband loss for each filter and 15 dB for the LNA, is 9 dB. Thus the required IP3, assuming worst-case gain combinations for the signal and interference, should be −3 dBm to −2 dBm. Taking a 3 dB margin, the required IP3 for a small mobile station is in the order of 0 dBm. For a standard mobile station, the desired signal is 2 dB weaker (−101 dBm) and the interferences are 4 dB (−45 dBm) stronger, so that the IP3 requirement can be as high as 7 dBm.

In BJT implementations, such IP3s can only be achieved with some linearization technique such as emitter degeneration. In a CMOS mixer, as long as the transistor has sufficient overdrive so that it stays in strong inversion, achieving more than 0 dBm IP3 is not a problem [14]. For a given input transconductance as required by impedance matching, however, it imposes a minimum-current requirement. In our designs, 4 mA was required for the single-balanced mixer and 6 mA for the double-balanced mixer, so that the transistor is biased halfway between weak and strong inversion.

Although the single-balanced mixer consumes less power, it provides less conversion transconductance, lower linearity and higher noise figure. The local oscillator signal is not suppressed properly by the mixer, so that it feeds through to the output. In a superhet receiver, however, this LO feedthrough is suppressed by the IF filter. Since a typical 71 MHz IF filter requires a matching load resistor on each port to its 330 Ω characteristic impedance, the RF mixer sees a 165 Ω load at its output. Even in the ideal case the conversion transconductance is 6.4 mS for the SBM and 12.7 mS for the DBM, which is not enough to provide 10 dB conversion gain. A simple LC network is therefore required to transform the load impedance to the desired level in any case, and the 6 dB difference between the conversion g_ms of single and double-balanced mixers is not critical as far as gain is concerned.

Because of the higher (by 6 dB) loss due to the switches, the SBM is expected to have a higher NF than the DBM. Assuming that the noise current of the switches dominates the mixer noise figure, and taking into consideration that the DBM has twice as many switches which therefore generate twice the noise power, the input-referred noise figure of SBM is expected to be about 3 dB higher than the DBM.

Although we favor the DBM for its expected higher performance, we have also implemented the SBM to check if the difference in NF is indeed as big as expected to justify the 50% higher current of the DBM.

The design of the switches in the mixer is primarily a trade-off between having high W/L ratios to reduce the overlapping period in which both switches in each pair remain on, having low thermal noise current during this period, and limiting the pole introduced by the switches into the RF path outside the period, for a given current. In our design, the final switch sizes are similar to those of the input RF transistors, 300 μm by 0.25 μm .

5.5 THE TRANSMITTER PREAMPLIFIER

Most modern GSM power amplifiers (PA) are constructed in a hybrid technology, using bipolar or power-MOS transistors. They require 2 mW of driving power and a supply voltage of 4.8 V (sometimes 6 V). The mandatory power control is implemented in the PA, and the output power can be varied by adjusting the DC voltage applied to a control pin. The peak power of a class-4 PA is as high as 2 W. Although its average transmitted power can be much lower because the average distance between the mobile and the base station is much less than the maximum radius of the cells in the network, the average power consumption of a mobile station is still high, judging by the low talk time advertised for current commercial handsets. Given that power consumption is dominated by the PA, the power consumption of the preamp and modulator is not as critical as in the receiver case. However, attention is paid to low-power design in the preamp so that the design will also be suitable for mobile stations with lower classes of PAs.

In this work, the transmit SAW filter required to remove the harmonics of the carrier and out of band noise is placed between the up-conversion mixers and the preamplifier. Since the loss of the filter precedes the preamp, the preamp is only required to deliver 3 dBm power to the PA instead of 7.5 dBm. This will ease the problems of delivering a substantial amount of power using CMOS, including efficiency and substrate coupling.

Since the efficiency of the preamp is not as critical as the PA, a class-C approach requiring another expensive and lossy external filter before the PA is neither necessary nor does it save power. Due to the lack of good PMOS transistors, a class-B (push–pull) approach is also unavailable. If the preamp output harmonics that could degrade the PA efficiency are kept low, a class-A amplifier can drive the PA without the external filter. The primary requirements for our preamp, a two-stage class-A implementation as shown in Fig. 5.7, are thus 50 Ω matching at both the input and the output, a high gain of 23 dB, and a high power output (for CMOS) of 3 dBm with sufficient linearity.

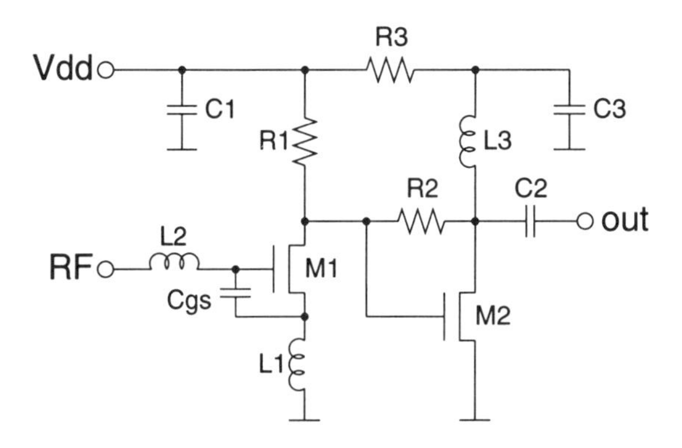

Figure 5.7 Schematic diagram of the Tx preamplifier.

The first stage, comprising M1, its biasing and the matching network, is similar to the input stage of the LNA, therefore the same design equations hold. The main difference here is the relatively high input level of -18 dBm. Since IM3 is inversely proportional to the overdrive of a MOS transistor, a relatively small transistor, biased at 11 mA, has been used for M1. Because of the relatively large overdrive of M1 that leaves little voltage headroom under a 2.5 V supply, no cascode transistor has been used in this design to ensure that M1 stays in saturation.

The second stage is a transimpedance amplifier, biased at 16 mA, as required to deliver 2 mW into the 50 Ω load with sufficient linearity. The relationship between the required output compression point and the required bias current is similar to that of the LNA. Unlike the LNA, a resistive load is ruled out by bias voltage considerations, otherwise R_3 would have to be approximately 60 Ω, which would make high gain very difficult to achieve. Instead we use inductive loading, and the RF choke L_3 resonates with the drain capacitance of M2 at 900 MHz so that the path to V_{dd} appears to be open, and the formula for the output compression point (5.6) and the output impedance (5.7) can be used with R_3 set to infinity. In fact, L_3 and the coupling capacitance C_2 could also be used as an impedance matching network at the same time, so that the impedance seen by the drain of M2 could be higher. That would allow higher power to be delivered for a given current. Due to the low supply voltage (2.5 V) of the 0.25 μm CMOS technology, this option cannot be used in our design, because impedance scaling by the matching network also scales the voltage swing of the drain of M2. Another way of achieving higher output power in traditional preamp designs is to combine such a matching network with biasing the drain of M2 at V_{dd}, which doubles the allowable voltage swing. Since the margin between the breakdown voltage and the actual supply voltage is small compared to the output swing in our 0.25 μm CMOS case, this option (which entails swinging the drain of M2 substantially above V_{dd}) was also ruled out on reliability grounds. To bias M2 in the mid-supply, resistor R_3 of approx. 60 Ω is inserted before L_3 and AC-bypassed by C_3 as shown. At DC, R_3 also helps to stabilize the bias condition through R_2.

In many bipolar transceivers [3, 5] a buffer capable of delivering 2 mW into 50 Ω has been implemented, frequently as an emitter follower biased at about 10 mA to achieve high IP3. Implementing a 20 mS (50 Ω) buffer in 0.25 μm CMOS, however, is not possible because of the large voltage swing required at the gate. Delivering 2 mW to 50 Ω corresponds to having a voltage swing of 900 mV peak–to–peak at the transistor's source and 1.8 V peat–to–peak at its gate. After adding the threshold voltage (400 mV) and some overdrive (200 mV) we have a peak voltage of 2.4 V, which is already higher than the specified minimum supply voltage. To reduce the swing, a lower impedance buffer followed by a matching network could be used, but this would result in an excessive increase in current consumption.

In our class-A implementation, the total current consumption is only 27 mA. Such a current level, required by linearity, is insignificant compared to that of the final PA.

5.6 POWER CONSUMPTION

Before describing the experimental results, it is perhaps appropriate to briefly revisit the subject of power consumption of the transceiver. Earlier we stated that for CMOS technologies to be seriously considered as a contender for RF front-ends in cellular applications and as a competitor to BJT technologies, one must *a*) prove that the complete set of performance specifications can be met and *b*) do so with a current consumption comparable to today's commercial solutions. In this section we wish to highlight the current consumption aspect.

Low power consumption is most important in the receive mode, because it defines the stand-by time. In transmit mode power consumption will be dominated by the PA. As the chief reason to purchase a mobile phone shifts from business use to private use, the toll charges serve as a natural barrier to reduce the occurrences of private users overrunning the, say, four hours of talk time possible on a single battery charge. Business and private users alike, however, may find a 40 hour or even a 72 hour stand-by time inconvenient.

The designs reported here show that by combining LNA-4 and the double-balanced mixer, the nominal supply current of the front-end chip can be as low as 16 mA. Ongoing design and experimental work on the remaining part of the receiver gives us good reasons to believe that the IF-strip plus the demodulator will consume no more than 6 mA, which shows that the front-end is indeed the dominant consumer of power (over 70%) of the receiver. Even if we add the VCO and the prescaler, which dominate the current consumption in a synthesizer (3 mA in Rx mode), the total nominal current consumption will only be 25 mA, i.e. 60 mW power consumption. This compares very favorably with existing chipsets, such as those reported in [2] (51 mA in the paper, 46 mA total in the datasheet), [3] (57 mA), and even [4] (33 mA), for example. The last reference derives its high performance from an advanced 25 GHz bipolar technology.

5.7 MEASURED RESULTS

All measurements were performed using bare chips assembled on a PCB with all the needed external SMD components. The dies were directly glued to the PCB ground plane which provides an excellent ground connection to the chip substrate. Some care was taken with the length of the bondwire to the source of M1 in each LNA and the preamplifier so that correct matching is achieved. No external trimming was performed. The supply voltage was set to its nominal value of 2.5 V. Among the chips measured, one is a combination of LNA-1, the SBM mixer and an early version of the IF-AGC amplifier [11]. Fig. 5.8 shows its micrograph. The measured parameters of the implemented LNAs, mixers, the transmit preamplifier as well as the particular receiver front-end are summarized in Tab. 5.2.

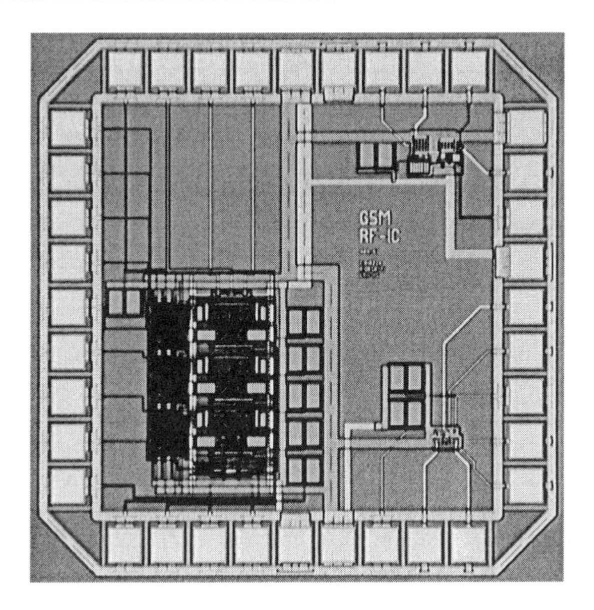

Figure 5.8 Photomicrograph of front-end IC.

Table 5.2 Summary of measured front-end circuit performances.

	LNA-1	*LNA-2*	*LNA-3*	*LNA-4*	*SBM*	*DBM*	*TX-PRE*	*RX-FE*
Gain [dB]	13.8	14.3	14.3	16.4			21.1	85 max
Conv. g_m [mS]					4.9	8.5		
SSB NF [dB]	2.0	1.74	1.97	1.6	15.9	12.6		
iIP3 [dBm]	−2.0	−2.8		−7.3		9.0		
iCP [dBm]		−17		−20	−4	−4	−14.6	
S11 [dB]	−5.6	−6.8	−4.0	−8.0	<-15	−11	−6.6	
S22 [dB]	−8	−22	−11	−12			−8.5	
S12 [dB]	<-30	<-30	<-40	<-40				
I_{dd} [mA]	18.3	18.1	16.6	10.8	3.5	6.0	27.2	25

For the low-noise amplifiers, we are particularly interested in the noise figures, as well as the influence of the various refinements on the NF in our controlled experiments. The reference amplifier LNA-1 achieves again a low NF of 2 dB, while providing the expected gain, input and output impedances and excellent linearity. The fact that a 2 dB NF has been achieved over consecutive runs shows that the design is suitable for manufacturing. With merely an increased number of substrate contacts, (of a transistor that already has many contacts,) LNA-2 shows a significant improvement in noise figure, to 1.74 dB. This demonstrates the significant contribution of back-gate resistance to the overall NF, at least for the 5 Ω-cm substrate of the technology we used. The main influence of the cascode transistor in the input stage (LNA-3) is a 30% reduction in current consumption in the first stage due to the higher Q and lower g_m needed, and a 10 dB improvement in reverse isolation. The price paid is a slight increase of the noise figure which we believe to be due to the additional contribution of M2. No rapid improvement of the NF, as predicted for the particular 0.35 μm LNA in [15] for a similar change of Q, is observed here. The last amplifier, LNA-4, achieves the best performance trade-off for our application in terms of gain, noise figure and power consumption, thanks to a better design of the output stage. The measured output compression point (−5 dBm oCP) has only decreased slightly (1 dB) from that of the other 3 LNAs despite saving half of the output stage current. A very low noise figure of 1.6 dB is measured, which is near the optimum noise match. This is the best result reported to date for an integrated CMOS implementation. In Fig. 5.9, starting with the minimum noise figure close to the center of the Smith chart, constant NF and gain circles have been measured in terms of the source impedance seen from the noise-matched LNA. Thanks to the low-Q design, these circles are big and widely separated for small increments of gain and NF. This means that variations the of matching network impedance due to component tolerances in production will not cause drastic degradation of gain and NF. Indeed, in 50 Ω power-matched conditions, the measurements of LNA-4 in Fig. 5.10 show that 15 dB gain, −8 dB S11 and 1.9 dB NF can be achieved simultaneously. Due to the effect of the bond pad, the protection diode and the PCB parasitic capacitances at both sides of L_2, the actual 50 Ω power-match conditions of the LNA are slightly different from the ideal equations (including (5.2) in this paper) normally described in recent publications [14–16].

Turning our attention to linearity, Fig. 5.11 shows the measured iCP and iIP3 of LNA-2 and LNA-4. High IP3 values far exceeding requirements have been measured, confirming the intrinsic high linearity of MOS devices. The measured compression points meet the requirements of the blocking test and are 3–4 dB below the limit set by the output current, as expected.

The measurements of the SB and DB mixers have been performed using −1.5 dBm of LO power with 50 Ω terminations at the LO ports. The measured conversion g_ms are 4.9 mS and 8.5 mS for the SB and the DB mixer, respectively. Due to imperfect switching, they are about 30% lower than the theoretical maximum, which was expected. Although the measured NF of 15.9 dB for the SB mixer is higher than we would like it to be, it is in line with most reported mixers with gain at this frequency, including BJT implementations [1,3]. The measured 12.6 dB SSB NF of the DB mixer, on the other hand, is one of the best reported to date for fully integrated mixers, and facilitates the use of RF and interstage filters with moderate specs (and hence price) in superhet GSM receivers. The S11 measurements are shown in Fig. 5.12. The common gate input guarantees excellent input matching with a broadband characteristic for both mixers. Indeed, the measured S11 is limited by the balun needed to match the DB mixer, but nevertheless stays under −10 dB at all frequencies below 1.2 GHz. The S11 degradation at higher frequencies is due to the increasing influence of parasitic capacitances. Fig. 5.13 shows that the measured iIP3 is as high as 9 dBm, and iCP

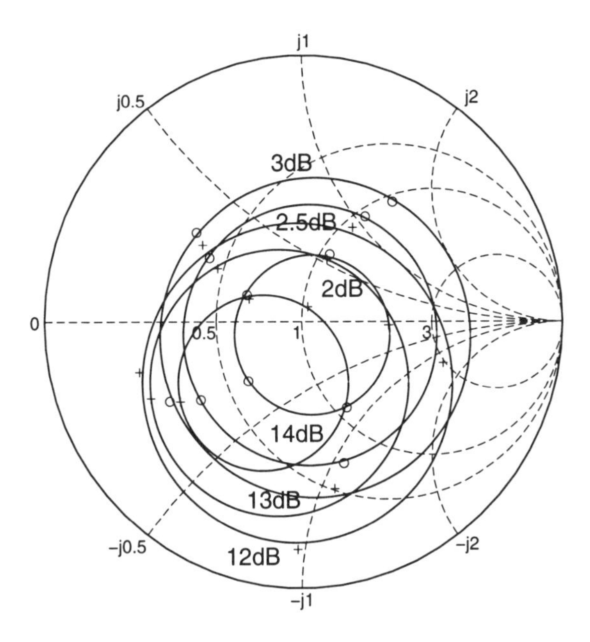

Figure 5.9 Constant NF and gain circle measurements.

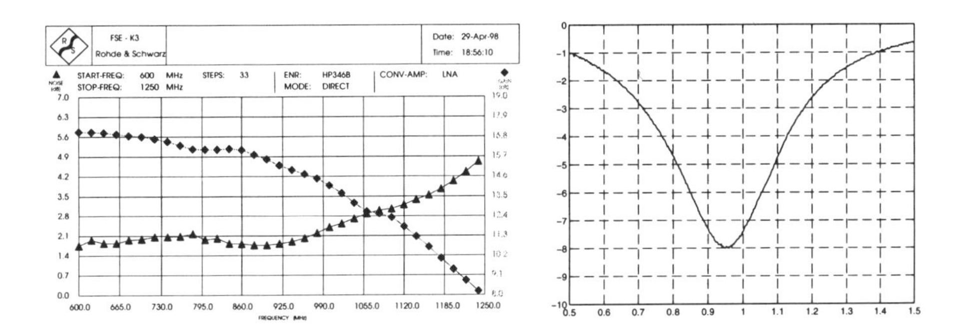

Figure 5.10 Gain, NF and S11 of impedance-matched LNA-4.

−4 dBm, for the DB mixer, which thus meets the specs outlined earlier by a comfortable margin. This has been achieved, without source degeneration, by the right choice of the g_m/I ratio of the CMOS devices, and by differential implementation.

The measurements of the transmitting preamplifier show that high gain, adequate input and output impedances as well as good linearity can be achieved even if a relatively high amount of power must be delivered. Fig. 5.14 shows the measured frequency response of the TX preamplifier. The measured iCP of the preamp is −14.6 dBm, which corresponds to +5.8 dBm oCP. This translates into more than 3.5 mW of power delivered to the 50 Ω load at compression.

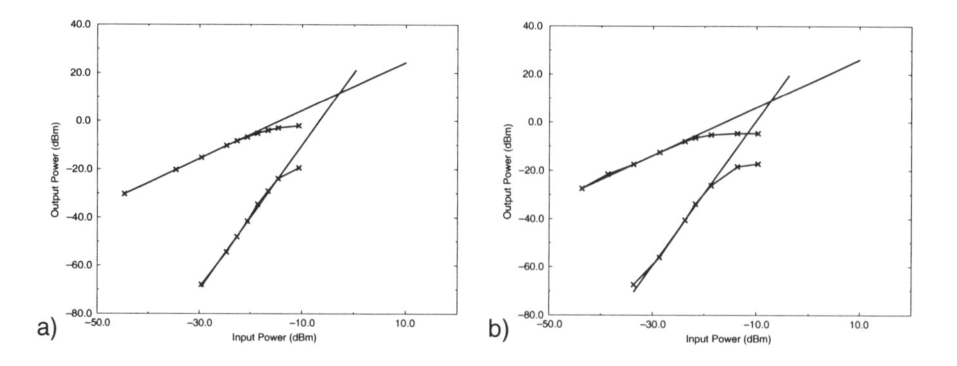

Figure 5.11 Measured, input-referred third order intercept point iIP3 and -1 dB compression point iCP for LNA-2 and LNA-4, respectively.

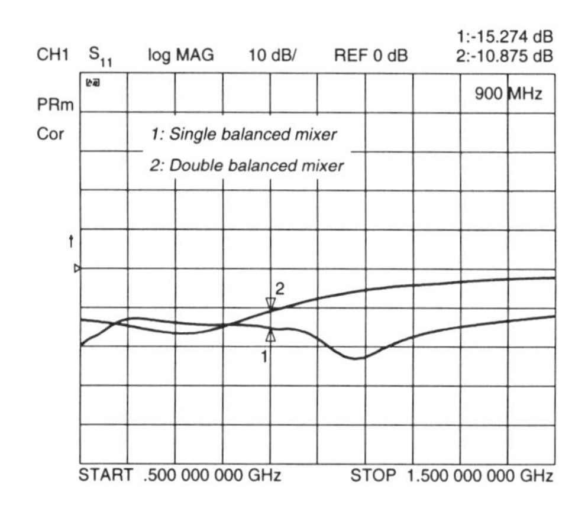

Figure 5.12 Measured S11 for single-balanced mixer and double-balanced mixer.

5.8 CONCLUSIONS

Power consumption is more important than the cost of chips in today's cellular phones. To prove that CMOS is feasible and competitive with BJTs for implementing the RF front-end of a GSM receiver, circuits operating at radio frequencies must show competitive current consumption while meeting all specs required for type approval. Type approval requirements have been discussed and translated into circuit specs where necessary, and measurements of the front-end integrated circuits reported in this paper show that they are indeed met. We have shown that current consumption lower than today's BJT implementations is indeed feasible by carefully optimizing the front-end design. Excellent noise figures are also shown to be reproducibly obtained for CMOS LNAs and mixers. Controlled experiments also confirm that backgating by substrate resistance can degrade the NF, and as many contacts as possible should be used to ground the substrate.

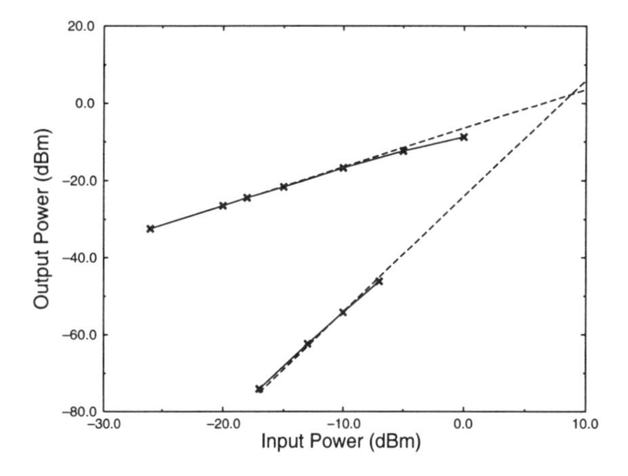

Figure 5.13 Measured, input-referred third order intercept point iIP3 and -1 dB compression point iCP, respectively, of the double-balanced mixer.

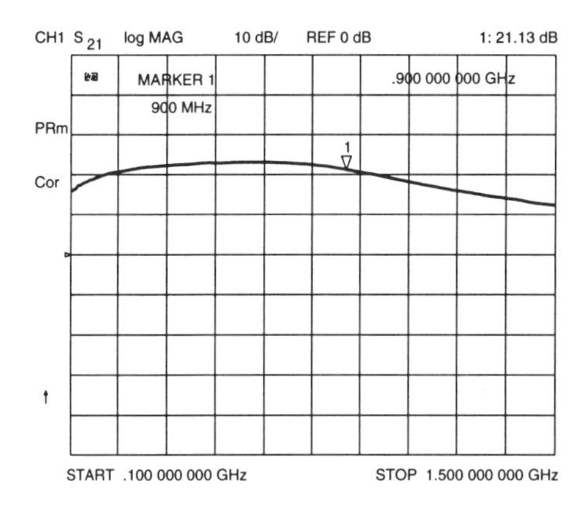

Figure 5.14 Measured frequency response of the TX preamplifier.

Acknowledgments

The contributions to this project by T. Yoshitomi and T. Morimoto of Toshiba Co. and R. Rheiner of the ETH Zurich are gratefully acknowledged.

References

[1] R. G. Meyer, W. D. Mack, "A 1-GHz BiCMOS RF Front-End IC", *IEEE J. Solid-State Circuits*, pp. 350–355, Vol. 29, No. 3, March 1994.

[2] C. Marshall et al, "A 2.7V GSM Transceiver ICs with On-Chip filtering", *ISSCC Digest of Technical Papers*, pp.148–149, Feb. 1995, San Francisco, USA.

[3] T. Stetzler et al, "A 2.7 V–4.5 V Single-Chip GSM Transceiver RF Integrated Circuit", *IEEE J. Solid-State Circuits*, pp. 1421–1429, Vol. 30, No. 12,,Dec. 1995.

[4] W. Veit et al, "A 2.7 V 800 MHz–2.1 GHz Transceiver Chipset for Mobile Radio Applications in 25 GHz f_t Si-Bipolar", *1994 Bipolar/BiCMOS Circuits and Technology Meeting*.

[5] K. Irie et al, "A 2.7 V GSM Transceiver IC", *ISSCC Digest of Technical Papers*, pp. 302–303, Feb. 1997, San Francisco, USA.

[6] A. Abidi et al, "The Future of CMOS Wireless Transceivers", *ISSCC Digest of Technical Papers*, pp. 118–119, Feb. 1997, San Francisco, USA.

[7] J. Rudell et al, "A 1.9 GHz Wide-band IF Double Conversion CMOS Integrated Receiver for Cordless Telephone Applications", *ISSCC Digest of Technical Papers*, pp. 304–305, Feb. 1997, San Francisco, USA.

[8] M. Steyaert et al, "A Single-Chip CMOS Transceiver for DCS-1800 Wireless Communications", *ISSCC Digest of Technical Papers*, pp. 48–49, Feb. 1998, San Francisco, USA.

[9] P. Orsatti et al, "A 0.25 μm CMOS Fully Integrated IF-Baseband-Strip for a Single Superheterodyne GSM Receiver", *ESSCIRC Proceedings*, pp. 64–67, Sept. 1998, The Hague, The Netherlands.

[10] GSM Standard—GSM 05.05 version 4.19.1, 11th Ed. ETSI, December 1997.

[11] F. Piazza et al, "A 2 mA/3 V 71 MHz IF Amplifier in 0.4 μm CMOS Programmable over 80 dB Range", *ISSCC Digest of Technical Papers*, pp. 78–79, Feb. 1997, San Franciso, USA.

[12] R. Steele, *Mobile Radio Communications*, IEEE Press, New York, 1992.

[13] Datasheets: duplexer, Murata DFY2R902CR947BHGF, RX interstage filter, Murata SAFC947.5MC70T, TX interstage filter, Murata SAFC902.5MA70N and IF filter, Siemens B4556 and B4568.

[14] Q. Huang et al, "The Impact of Scaling Down to Deep-Submicron on CMOS RF Circuits", *IEEE J. Solid State Circuits*, vol. 33, pp. 1023–1036, July 1998.

[15] D. Shaeffer and T. Lee, "A 1.5 V, 1.5 GHz CMOS Low Noise Amplifier", *IEEE J. Solid-State Circuits*, May 1997, pp. 745–759.

[16] A. Karanicolas, "A 2.7 V 900 MHz CMOS LNA and mixer", *ISSCC Dig. Tech. Papers*, 1996, pp. 50–51.

[17] Y. Shin and K. Bult, "An Inductorless 900 MHz RF Low-Noise Amplifier in 0.9 μm CMOS", *Proceedings CICC'97*, pp. 513–516.

[18] A. van der Ziel, *Noise in Solid Stase Devices and Circuits*, Wiley, New York, 1986.

[19] Y. Tsividis, *Operation and Modeling of The MOS Transistor*, McGraw-Hill, New York, 1987.

[20] A. Grebene, *Bipolar and MOS Analog Integrated Circuit Design*, Wiley, New York, 1984.

II RF Front-End Circuits

6 RF FRONT-END CIRCUITS

Qiuting Huang

Integrated Systems Laboratory
Swiss Federal Institute of Technology (ETH)
ETH-Zentrum, CH – 8092 Zurich, Switzerland

6.1 INTRODUCTION TO THE FOLLOWING PAPERS

To newcomers to the field of RF front-end design for wireless communications, the combination of rapid change in commercial standards and requirements, the choice of system architectures and frequency plans, and their relation to the performance of individual radio frequency circuits can be quite bewildering. The four contributions included in this chapter address these issues from different perspectives in varying degrees of detail.

The first two contributions come from academia and are biased towards basic theory. The second two contributions come from industry, providing more insight into the current state-of-the-art in wireless transceivers, and the overall system requirements in terms of cost, size (volume), and performance.

Qiuting Huang focuses on the fundamental issue of phase noise in RF oscillators, which is particularly important for wireless transceivers as it strongly affects the relative independence between different radio channels and services that transmit and receive signals of varying strengths simultaneously. In addition to avoiding unsound assumptions, the analyses of oscillation amplitude, oscillator response to interfering signals, and the identification of white noise sources have all been backed up with controlled experiments.

Behzad Razavi exposes the fundamental issues of frequency planning in transceiver design, as well as recent trends in highly integrated RF circuit design, in the context of a GSM/DCS dual-mode transceiver. The architecture is designed for a high level of integration, containing neither external image-reject filters nor IF filters. Quadrature mixers are therefore frequently used, and the LO signal is placed halfway between the GSM band and the DCS band to save hardware. In the transmitter part, the tutorial on GMSK modulation and influence of nonlinearity on the transmitted spectrum may be a very useful starting material to read for people new to the field. The emphasis in implementation is on CMOS, with extensive use of on-chip spiral inductors.

Petteri Alinikula shares with us the challenges of future wireless terminals from the viewpoint of one of the top three leading commercial companies in the market. Wideband CDMA transceivers appear to be the next target for most, especially in light of the recent agreement between Ericsson and Qualcomm over IPR issues. Linear power amplifiers, a sharp increase in DSP computing power, and multi-standard terminals are identified as major challenges facing designers of next-generation terminals. The author also shares with us his view of the evolution in state-of-the-art terminals, in terms of relative physical volume, component count and cost breakdown. Many readers will find the corresponding figures interesting. Also interesting is a concrete example of a direct-conversion architecture being used to save discretes and filters.

Stephan Heinen and Stephan Herzinger take us back to the GSM transceiver again, from the perspective of industry. They echo the view from Alinikula that overall system optimization is more important than the focus on a particular technology, the saving of a particular filter or the performance of a particular building block. In the discussion of GSM transmitters, the general requirements by type approval are presented in detail, as are the effects of nonlinearities on the transmitted spectrum. This again gives a newcomer to the field a nice introduction, from an angle differing slightly from Razavi's. Direct up-conversion, IF up-conversion and modulation-loop VCO architectures are discussed in terms of their relative merits and disadvantages, and the modulation loop is identified as the transmitter architecture of choice for constant-envelope modulation schemes such as GMSK and is presented in detail.

The contributions in this chapter provide a good overview of the current issues in RF front-end design, as the field of RF-IC design moves towards maturity, and the industry is poised to introduce the next generation of air-interface standards.

7 PHASE-NOISE-TO-CARRIER RATIO IN LC OSCILLATORS

Qiuting Huang

Integrated Systems Laboratory
Swiss Federal Institute of Technology (ETH)
ETH-Zentrum, CH–8092 Zürich
Switzerland

Abstract: The following questions are addressed in this chapter using a CMOS Colpitts oscillator as an example.

1. How does the carrier amplitude depend on the electrical parameters of the oscillator?

2. Does an additive interference, white noise for example, (always) become multiplicative with the carrier in such a way that it deserves the name *phase noise*? Is the response to such an interference always inversely proportional to its frequency offset from the carrier? Why?

3. How does the phase-noise-to-carrier ratio depend on the electrical parameters of the oscillator? Can such a dependence be described directly with the circuit parameters rather than intermediate quantities such as the signal power consumed by the tank, the mean-squared voltage amplitude, or the noise factor of the oscillator?

Answers to the above questions afford us a better understanding of LC oscillators and enable us to predict the exact oscillator performance without resorting to numerical simulation.

7.1 INTRODUCTION

The importance of phase noise in oscillators in RF and other communication circuits has made it one of the most extensively studied subjects in electronics. Journal papers on the subject [1–10] can be found from each of the last six decades, the earliest one [1] being as old as 1938, and the latest [10] as recent as 1998 The fact that papers on LC oscillators, too numerous in the last 60 years to cite exhaustively, keep appearing, serves to underline the fact that existing theories on phase noise have not been universally found satisfactory.

Experimentally, the qualitative behaviour of phase noise has been well known. An oscillator's output power spectrum consists of a peak at the carrier frequency (main oscillation, angular frequency ω_0), surrounded by a noise skirt symmetrical to the carrier frequency. Irrespective of the exact implementation, the noise skirt has the following characteristics:

- The noise spectral density is inversely proportional to the frequency offset from the carrier, except very close to the carrier frequency, where the influence of up-converted flicker noise dominates or the presence of the strong carrier begins to limit the measurement accuracy.

- The same noise manifests itself in the time domain as jitter around the oscillation's zero-crossing points, which can only be caused by noise in the phase of the oscillation rather than that super-imposed on its amplitude. Oscillator noise is therefore usually referred to as phase noise. The assumption of phase noise implies that the sideband spectrum above and below the carrier frequency must be equal in amplitude and opposite in sign. It also implies that output noise is multiplied with rather than added to the carrier.

- It is widely believed [11] that doubling the quality factor of the LC tank roughly improves the oscillator's carrier-to-noise spectral density (C/N) ratio by 6 dB, although systematic experimental verification of such a belief is hard to find in open literature.

Today, RF oscillators are implemented with high-quality inductors and capacitors outside the chip containing the active devices. This is the main reason for the widespread efforts in the integrated circuit community to improve the quality of fully integrated inductors on silicon, especially in the last few years [11–25]. Despite the avalanche of research papers, the quality factor Q of on-chip inductors at 1 GHz has hardly improved by a factor of two (from $Q \leq 3$ to $Q \leq 6$) in the last 10 years. Reported N/C ratios, however, vary by as much as 20 dB at 100 kHz offset. This underlines the fact that other factors than the Q of the tank also affect the N/C ratio strongly. Our understanding of these factors seems at times confused.

The confusion stems mainly from the inadequacy of existing phase noise models. In the next section, we will review typical existing models and analyze their weaknesses. This will be followed by our own analysis of a typical LC oscillator. The example used is the well-known Colpitts oscillator implemented with MOS transistors. The analysis starts in Section 7.3 with the general description of steady-state oscillation and the derivation of the exact oscillation amplitude as a function of circuit parameters. It will be followed by Section 7.4, describing the response of the oscillator to a small, deterministic interfering signal. Section 7.5 identifies the spectral density of all white noise sources, leading to an analytical expression of oscillator output noise-spectral-density-to-carrier ratio that is completely determined by circuit parameters. Experimental results will then be presented to validate the analytical expression, before the latter is used in the discussion of the exact design of RF oscillators.

7.2 THE WEAKNESSES OF EXISTING PHASE NOISE MODELS

Existing models in the last six decades can be grouped by the four typical methods used to derive them:

- Linear, frequency-domain analyses with noise sources additive to the carrier signal,
- linear, frequency-domain analyses with noise sources additive to the phase of the carrier,
- time-domain analyses with noise sources additive to the carrier signal,
- time-domain analyses with noise sources additive to the phase of the carrier.

Since all the models set out to fit the well-known experimental observations described in the previous section, the resulting expressions do qualitatively predict many aspects of phase noise behaviour. It is in the validity of the reasoning that led to those expressions that we find serious weaknesses.

7.2.1 Linear, frequency-domain analyses with noise sources additive to the carrier signal

Linear analyses simply assume that noise is superimposed on the carrier and the transfer function from the noise source to the oscillator output is that of the resonator tank, modified by the equivalent linear transconductance of the active device. Fig. 7.1 shows the equivalent circuit of the simplest case [9]. The V–I characteristic of a real active device, on the other hand, can generally be depicted as in Fig. 7.2, the ~ shape serving as a necessary feature to limit the oscillation amplitude [26]. The negative slope representing the negative (trans)conductance required for start-up is usually found in the middle. Since this V–I curve is usually highly nonlinear, the approaches to the equivalent linear transconductance (g_m) have been different.

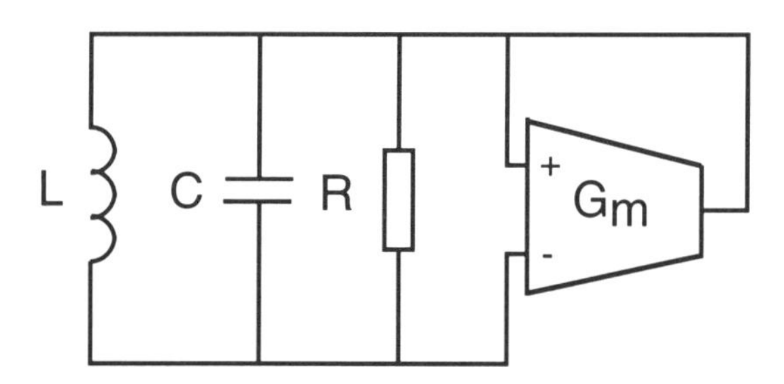

Figure 7.1 Simple model used in typical linear analyses.

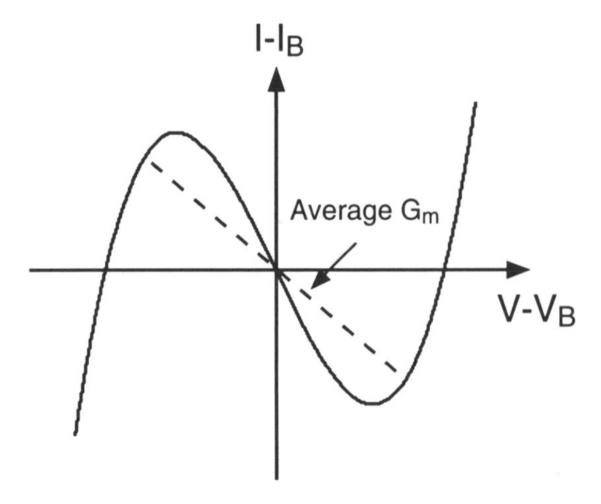

Figure 7.2 Typical description of V–I characteristic of an active device and its average gm for a large signal. Practical transistors can be even more nonlinear. I_B and V_B are the bias current and voltage, respectively.

Earlier analyses [2] allow the equivalent g_m to be quite different from the critical transconductance (g_{mc}) required to cancel the loss of the tank exactly. Such analyses result in models that generally predict a bandpass characteristic centered at the carrier frequency, flat at frequencies immediately adjacent to the carrier and rolling off at 6 dB per octave at higher offset frequencies. The corner between the two offset frequency ranges is typically of the order of the carrier frequency divided by the Q of the passive tank. One of the serious problems with such a model is that the predicted corner frequency, which can be 10 MHz for a 1 GHz oscillator and a Q of 100, is never observed in practice.

To fit them to experimental results, later analyses [8, 9] assume that the transconductance (or gain) of the active device, seen by the noise signal, is exactly the same as that for the carrier, which in turn must be the same as that required to cancel the tank loss in steady state. The resulting equivalent circuit is an ideal, zero-loss linear resonator. The inverse dependency on offset frequency is now present all the way to the carrier.

The fundamental objection to the use of the carrier's g_m is that the latter is only a crude large-signal concept indicating the ratio between the fundamental component of a distorted output current of a transistor and the voltage applied to its input. The noise component, being a smaller signal at a different frequency and phase, does not traverse the nonlinear V–I curve in Fig. 7.2 in the same way as the carrier, even though it is superimposed on the latter. The distortion it experiences, or the average g_m applicable, will therefore also be different from that of the carrier. A specific example of the CMOS Colpitts oscillator will be shown later to underline this point. If one still insists on linear superposition of the carrier and noise, there is no more justification left than the need to fit the theory to the observed $1/\Delta\omega$ characteristic. Even so, consistency then requires that the small-signal g_m at the center of Fig. 7.2 be used, because the noise signal is a very small signal. However, the small-signal g_m at the center is necessarily several times larger than the critical g_m (g_{mc}), otherwise oscillation would never have started! Further evidence that the type of linear analysis in [8, 9] is wrong lies in the fact that it predicts a response 6 dB too high at low offset frequencies. This last point will be shown later in this paper.

Fundamental objections can also be raised about the linear models for what they predict. If noise were merely added to the oscillation signal, it would not cause the zero-crossing-point jitter normally observed, even if the $1/\Delta\omega$ characteristic is given. This is because the odd symmetry in phase with respect to the carrier frequency, inherent to modulation in phase, is absent in the predicted oscillator noise spectrum.

7.2.2 Linear frequency domain analyses with noise sources additive to the phase of the carrier

Papers in this class also use a linear model, but consider the noise sources to be added directly to the phase of the carrier [6, 7]. The result is best known as Leeson's Model or Leeson's Formula. Although its original derivation is at best heuristic, it has become the most widely accepted model for phase noise in the last 30 years. In addition to being widely used in industry, it has been adopted in several reputable textbooks. The reasons for its success seem to be three-fold: 1. It is derived from phase so that the odd symmetry is assumed to have been taken care of. 2. The $1/\Delta\omega$ characteristic is there. 3. It predicts a perceived inverse dependence of the N/C ratio on the (loaded) Q of the tank. It is up to the users of the model to work out what the power level of the signal (P_s) and the effective noise figure (F) of the oscillator mean.

The fundamental problem with the model's derivation lies with the noise sources. For white noise to be additive to the phase directly, it must be located at places where it can influence the frequency or phase-setting elements of the oscillator, such as the junction diodes (intentional or parasitic) in series or parallel with the main linear, passive tank. Oscillators with negligible nonlinear LC elements would have no phase noise, or zero effective noise figure F, if we have to express the noise in terms of Leeson's model, but we know this would contradict reality. Even for oscillators with nonlinear capacitors, it is quite unclear why the phase or frequency shift caused by white noise on the reverse bias of a junction diode would have to do with anything other than the C–V characteristic of the capacitor and its importance relative to other linear capacitors in the tank in determining the oscillation frequency. The introduction of P_s into the "input phase noise" seems totally arbitrary, even though doing so results in a model that seems to reflect its role in the final expression of the N/C ratio qualitatively. Numerical verifications with practical oscillators also show that noise levels associated with junction diodes are too low compared to the noise observed at the oscillator output, not least because good oscillators are designed to minimize the influence of parasitic diodes, and the bias circuits of frequency-setting varactors usually have lowpass characteristics with cutoff frequencies well below the oscillation frequency.

7.2.3 Time-domain analyses with noise sources additive to the carrier signal

Typical analyses [4, 5] in this group model the nonlinearity of the active device with a third-order term and describe the dynamics of the oscillator with a Van-de-Pol differential equation driven by the noise source. Without pinpointing the precise flaws in those usually complicated derivations, it suffices to say that a constant term (γ in [4]) is usually defined which is of the order of magnitude of the relative difference between the small signal g_m of the active device and g_{mc}. This term divided by Q defines the corner of the relative offset frequency (i.e. relative to the carrier) of the final output noise spectrum which has a bandpass shape. Again, the flat portion of the spectrum below such a corner frequency is never observed in practical oscillator measurements.

7.2.4 Time domain analyses with noise sources additive to the phase of the carrier

In the most recent theoretical journal paper [10] on this subject, a general theory is proposed which claims to be applicable not only to LC oscillators, but also to all other types such as ring oscillators. We restrict the focus of our discussion on LC oscillators. The main weakness in the reasoning in this paper lies in the assumption that interference is added to the phase. The example given to support such an assumption is an ideal tank, excited by a current impulse. The sinusoidal signal after a charge has been deposited on the capacitor by the impulse is phase-shifted and changed in amplitude relative to the sinusoidal signal before the impulse, were the latter to continue. It is correct that both the phase shift and the amplitude change are periodic functions of the time of the impulse, although the statement that at maximum phase shift the amplitude change is zero is wrong, because it is not possible to add energy ($Q\Delta q/C + \Delta q^2/2C$, if $Q^2/2C$ is the total energy before the injection of charge) to an ideal tank without changing the amplitude of the oscillation, no matter when the small charge Δq is injected into the tank. The critical step of the paper's argument comes when it is stated that some form of AGC action always exists in an oscillator so that the amplitude finally settles back to the same value independent of the impulse. For the ideal LC tank this statement is clearly wrong. For the lossy tank in a practical oscillator this may be true, but we need to highlight the word *finally*, for it usually takes hundreds of oscillation cycles for the impulse to settle (precisely because AGCs, if needed, are designed to be insensitive to small-signal, high-frequency perturbations). This is in stark contrast to the case of ring oscillators constructed by CMOS inverters, where the AGC function may consist of clipping at a reference voltage such as the power supply, which may be achieved within the same oscillation cycle.

Since the impulse response of the amplitude does not decay to zero instantly, it contains as much information about the overall response of the LC tank to the injected current as the phase impulse response $h_\phi(t)$ constructed in [10]. The same comment holds for the convolution between the impulse responses and the injected continuous-time current signal, which results in the amplitude and phase responses defined in [10]. The two responses are also strongly correlated. Throwing away the amplitude response, as it has been done in [10], destroys a part of the information needed to re-construct the overall response. The nonlinearity required to convert additive noise to phase noise and generate the double-sideband spectrum is thus created more by the AGC assumption than any real transistors. To demonstrate further the fallacy of throwing away the amplitude response and using only the phase response to construct the output spectrum, we note that even for an ideal, passive LC tank with zero loss and resonance frequency ω_0, perturbed by an ideal sinusoidal current source at $\omega_0 + \Delta\omega$, a response will appear at $\omega_0 - \Delta\omega$ according to [10]! General theorems about linear systems and the specific solution to the linear differential equation with a sinusoidal driving function tell us this is not possible.

Even if an instantaneous AGC did exist, the model still has a serious problem. Since the derivation is independent of the offset frequency $\Delta\omega$, it should be valid even at offsets reasonably far outside the tank's passband, where the tank impedance becomes very low. For a high-Q tank, the voltage produced on the tank by an interfering current is so small at high offsets that the AGC should not have any influence. The attenuation is also going to be so high that the active device will not have sufficient g_m to produce enough current to counteract the injected current. In such a situation, one would expect the relationship between the interfering current and the voltage it produces on the tank to be the same as that of a passive tank without the active device. In this case, the model based on the linear analysis [9] must be valid, although the resulting interference can no longer be called phase noise due to the absence of the image sideband. The result of [10] in this case is wrong by a factor of two.

Most models described above attempt to be general. In doing so they fail to include information on parameters vital to the use of the models. A typical example is the effective noise figure F, which must be calculated taking into account all noise sources, their location in the circuit and the amount of distortion the noise experiences prior to reaching the tank. Another example is oscillation amplitude or power, which may depend on the Q of the tank. With the effective noise source and oscillation amplitude as intermediate variables, the exact relationships between N/C ratio and the circuit parameters such as tank loss, L and C values and transistor g_m are obscured. This sometimes makes the models open to interpretation, which is not entirely satisfactory to the designer.

In the following, we take the approach of analyzing a very specific oscillator, the Colpitts oscillator in Fig. 7.3a, in the time domain, so as to remove as much ambiguity about the assumptions as possible. Generalizations will be made only after the specific case is well established. Noise currents or voltages as they are introduced by the sources such as resistors and transistors are treated exactly as they are in the physical reality: additive to the carrier signal.

7.3 GENERAL DESCRIPTION OF LC OSCILLATOR OPERATION AND DETERMINATION OF OSCILLATION AMPLITUDE

Before one attempts the analysis of the influence of noise on oscillation, it is helpful to study first how a typical LC oscillator operates in steady-state. In the CMOS case, the study will also have the benefit of revealing an analytical expression for the oscillation amplitude [27]. The determination of the CMOS LC oscillator amplitude had always relied on some form of numerical solution. The most straight-forward method is to use a circuit simulator such as SPICE. An earlier analysis [28] shed some light on the way a MOS transistor operates in a Colpitts oscillator, but stopped short at parametric

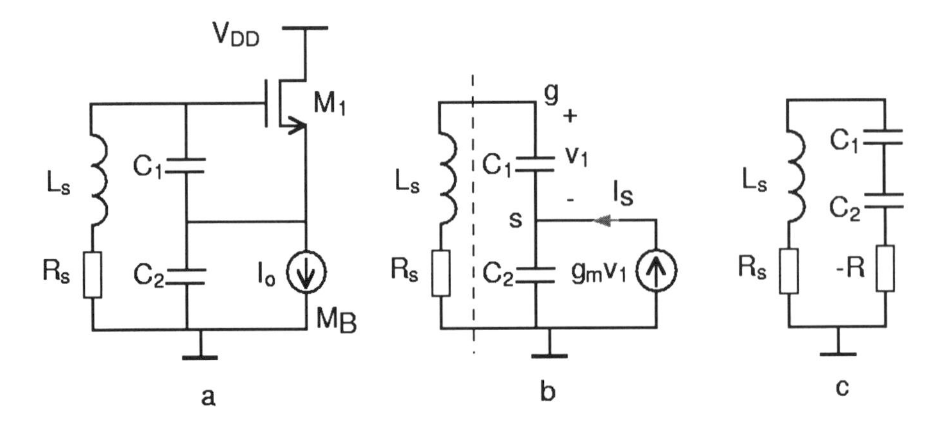

Figure 7.3 (a) Minimum representation of a Colpitts oscillator. (b) Equivalent circuit with transistor represented by a linear transconductance. (c) Equivalent negative resistance model of oscillator.

equations that require numerical solutions to obtain oscillation amplitude. The problem with numerical solutions, however, is that insight into the way circuit parameters affect the final amplitude is partially or completely lost.

Standard descriptions of LC oscillators are usually limited to the necessary condition for oscillation. For small signals, the MOS transistor can be modeled as a linear g_m, as shown in Fig. 7.3b. All the resistive losses of the tank have been lumped into a single resistor R_s. The transistor forms a positive feedback loop with capacitors C_1 and C_2. The equivalent impedance Z_R looking to the right of the dashed line in Fig. 7.3b contains a negative real part:

$$Z_R = \frac{V}{I} = \frac{1}{j\omega C} + (-R) = \frac{1}{j\omega C_1} + \frac{1}{j\omega C_2} - \frac{g_m}{\omega^2 C_1 C_2} \, . \tag{7.1}$$

The necessary condition for the oscillation to start is that the total resistance in the equivalent LC tank in Fig. 7.3c be negative,

$$g_m > g_{mc} = \omega_0^2 C_1 C_2 R_s \, , \tag{7.2}$$

where g_{mc} is defined as the critical transconductance for oscillation, and $\omega_0 = 1/\sqrt{LC}$ is the tank's angular resonance frequency. The negative middle term in the circuit's characteristic equation results in the latter's complex conjugate zeros having a positive real part. The zero-input response of the circuit is therefore an exponentially increasing sinusoidal oscillation. Equation (7.2) does not tell us whether the oscillation will stabilize to a particular amplitude.

During start-up, the average current in transistor M_1 may not be the same as the bias current. The difference between the two flows into the capacitive network formed by C_1 and C_2, causing the DC bias voltage between the gate and source of M_1, V_B, to shift. This will in turn change the average current in the direction of the bias current. The steady state is reached when the two are balanced. Although the non-linearity of M_1's V–I characteristic results in many harmonics in M_1's drain-source current, the latter is periodic in steady state. Fourier series expansion can therefore be used in the analysis. The steady-state oscillation amplitude between M_1's gate and source, V_m, and the DC bias V_B are bound by two constraints if the quality factor of the passive tank is high, so that the voltage on M_1's gate is practically a pure sine wave.

1. The average current through M_1 is equal to the bias current I_0. (I_0 corresponds to I_B in Fig. 7.2.)

2. The ratio between the amplitude of M_1's AC current and that of its gate-source voltage is exactly the transconductance of the passive tank between the V_{gs} branch and the I_s branch in Fig. 7.3b at resonance, which happens to be g_{mc}.

These two constraints allow us to set up two equations to solve for the two unknowns in steady-state oscillation, the bias voltage V_B and the oscillation amplitude V_m.

Since the values of the passive components in the tank and their resistive loss, which (we will show) have a direct impact on oscillation amplitude, typically have tolerances of 5–20 %, an analysis with much greater accuracy is unnecessary for practical purposes. This provides us with the justification to describe M_1's I–V relationship with the simple square law,

$$I_{DS}(t) = \frac{\beta}{2}\left[V_{gs}(t) + V_B - V_T\right]^2 , \tag{7.3}$$

where $V_{gs}(t)$ is the AC part of the gate-source voltage and V_T is the threshold voltage, below which the transistor cuts off and the current equals zero. The same justification also allows us to assume that because the tank's quality factor Q is usually high, harmonics in M_1's drain-source current I_{DS} have been sufficiently filtered by the tank, so that $V_{gs}(t)$ is a pure sinewave. We shall experimentally verify the errors caused by such simplifying assumptions later.

If the steady-state bias $(V_B - V_T)$ is greater than V_m, as depicted in Fig. 7.4a, so that $V_{gs}(t) = V_m \cos(\omega_0 t)$ never causes the transistor to cut off, then the average current through M_1 is given by:

$$\begin{aligned} I_{ave} &= I_0 = \frac{\omega_0}{\pi}\int_0^{\frac{T}{2}} \frac{\beta}{2}\left[V_m \cos\omega_0 t + V_B - V_T\right]^2 dt \\ &= \frac{\omega_0 \beta V_m^2}{2\pi}\int_0^{\frac{T}{2}} \left[\cos\omega_0 t - x\right]^2 dt = \frac{\beta V_m^2}{4}(1+2x^2)\,, \end{aligned} \tag{7.4}$$

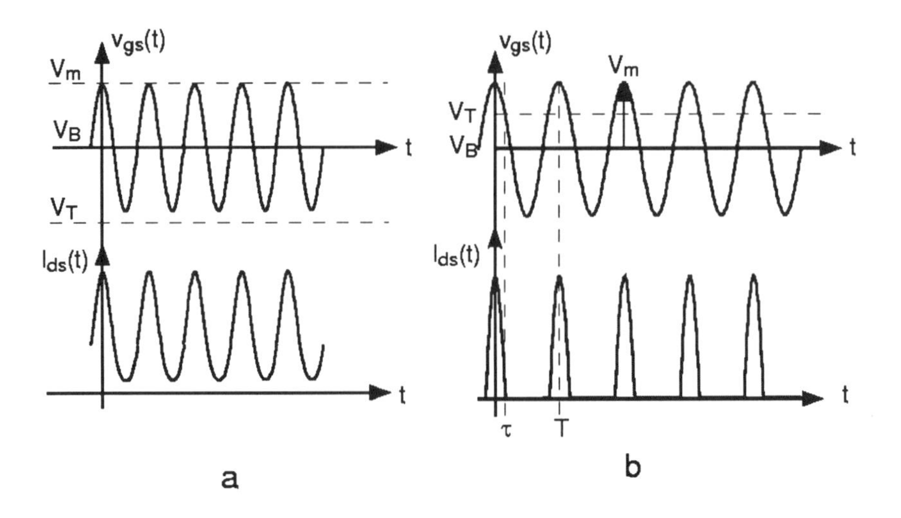

Figure 7.4 (a) Voltage and current waveforms when M_1 is in amplifier mode. (b) Voltage of $f(x)$ and its approximation.

where x is defined as $\frac{V_T - V_B}{V_m}$ and $x \leq -1$. The fundamental component of the AC current at ω_0, on the other hand, is given by:

$$I_{\omega_0} = \frac{2\omega_0}{\pi} \frac{\beta V_m^2}{2} \int_0^{\frac{T}{2}} [\cos\omega_0 t - x]^2 \cos\omega_0 t \, dt = -x\beta V_m^2 \,. \tag{7.5}$$

Multiplying I_{ω_0} by the transimpedance between the I_s branch and V_{gs} branch in Fig. 7.3b, we obtain the steady-state oscillation amplitude:

$$V_m = I_{\omega_0} H(\omega_0) = \frac{-x\beta V_m^2}{g_{mc}} \,. \tag{7.6}$$

Solving Eqs. (7.3) and (7.6) jointly, we obtain:

$$V_m = \frac{\sqrt{2}}{\beta}\sqrt{2\beta I_0 - g_{mc}^2} \text{ and } x = -\frac{g_{mc}}{\beta V_m} \leq -1 \,. \tag{7.7}$$

Re-organizing (7.7), the same relations can be expressed as:

$$\beta V_m = \sqrt{2}\sqrt{2\beta I_0 - g_{mc}^2} \leq g_{mc}, \text{ or } g_{mc}^2 \leq g_{m0}^2 = 2\beta I_0 \leq \frac{3}{2} g_{mc}^2 \,. \tag{7.8}$$

From (7.7), we see that if one wanted to keep M_1 from cutting off, a combination of a small transconductance coefficient and a large bias current would be the best way to maintain a sufficiently large amplitude. The maximum V_m is reached when the second half of (7.7) is satisfied with equality, in which case $V_{\max} = g_{mc}/\beta$. Condition (7.8) shows that the range for the nominal transconductance g_{m0} is very small, so that a 20% increase in g_{mc} can easily kill the oscillation.

To ensure start-up, as well as adequate amplitude, practical oscillators are designed to have a g_{m0} several times larger than g_{mc}. In this case the conditions in Eqs. (7.7) and (7.8) are no longer satisfied, and transistor M_1 does get cut off during part of the oscillation, as depicted by Fig. 7.4b. Eqs. (7.3) and (7.5) must therefore be modified to account for the time in which M_1 is cut off. This is done by reducing the limit of integration to the time M_1 is on:

$$\tau = \frac{1}{\omega_0}\cos^{-1}\left(\frac{V_T - V_B}{V_m}\right) = \frac{1}{\omega_0}\cos^{-1}(x) \qquad -1 \le x \le 1\,. \tag{7.9}$$

The average current through M_1 is now given by:

$$I_0 = \frac{\omega_0 \beta V_m^2}{2\pi}\int_0^{\frac{\cos^{-1}(x)}{\omega_0}} [\cos(\omega_0 t) - x]^2\,dt = \frac{\beta V_m^2}{2\pi}\left[\frac{1+2x^2}{2}\cos^{-1}(x) - \frac{3}{2}x\sqrt{1-x^2}\right]. \tag{7.10}$$

The oscillation amplitude, on the other hand, is given by:

$$\begin{aligned} V_m = I_{\omega_0} H(\omega_0) &= \frac{1}{g_{mc}}\frac{\omega_0 \beta V_m^2}{\pi}\int_0^{\frac{\cos^{-1}(x)}{\omega_0}} [\cos(\omega_0 t) - x]^2 \cos(\omega_0 t)dt \\ &= \frac{\beta V_m^2}{3\pi R_s \omega_0^2 C_1 C_2}\left[(2+x^2)\sqrt{1-x^2} - 3x\cos^{-1}(x)\right]. \end{aligned} \tag{7.11}$$

Re-arranging Eqs. (7.10) and (7.11), we obtain:

$$\begin{aligned} V_m = \frac{I_0}{g_{mc}} f(x) &= \frac{4I_0}{3g_{mc}}\frac{\left[(2+x^2)\sqrt{1-x^2} - 3x\cos^{-1}(x)\right]}{\left[(1+2x^2)\cos^{-1}(x) - 3x\sqrt{1-x^2}\right]} \\ &\approx \frac{I_0(5+x)}{3R_s\omega_0^2 C_1 C_2} \end{aligned} \tag{7.12}$$

$$\frac{2\beta I_0}{g_{mc}^2} = \left(\frac{g_{m0}}{g_{mc}}\right) = g^2(x) = \frac{9\pi}{2}\frac{\left[(1+2x^2)\cos^{-1}(x) - 3x\sqrt{1-x^2}\right]}{\left[(2+x^2)\sqrt{1-x^2} - 3x\cos^{-1}(x)\right]^2}. \tag{7.13}$$

Even without solving Eq. (7.13) explicitly for x, it can be seen from Eq. (7.12) that the oscillation amplitude is only a weak function of x, and thus also of the duty cycle of M_1. As x varies within its limits ± 1, the amplitude V_m varies by no more than a third. One can thus say that V_m is roughly proportional to the bias current I_0, and inversely proportional to the critical g_m for oscillation, g_{mc}. Fig. 7.5 compares the exact function $f(x)$ and its approximation $(5+x)/3$ in Eq. (7.12). The maximum difference is less than 2 %, occurring at $x = 0$, which is a negligible error in terms of amplitude prediction. The approximate version of Eq. (7.12) has the advantage of much greater simplicity.

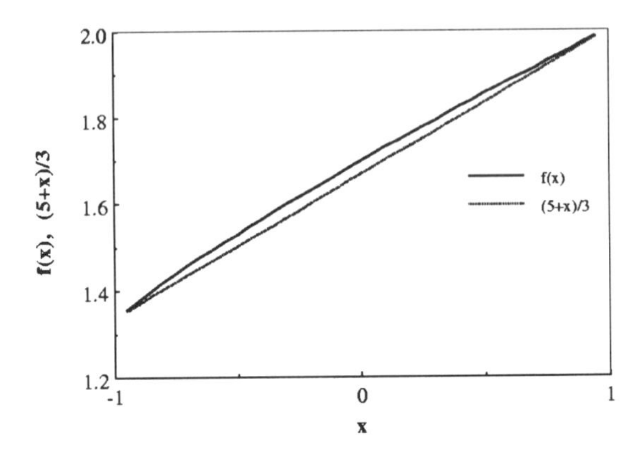

Figure 7.5 Comparison of $f(x)$ with its approximation.

Although a solution for x in explicit form is more difficult to obtain from Eq. (7.13), it is only a function of M_1's nominal transconductance g_{m0} normalized to the g_{mc} of the passive tank. Thus x needs to be solved only once numerically versus a useful range of normalized g_{m0} and can be looked up once the ratio between g_{m0} and g_{mc} is known. Fig. 7.6 shows the calculated x, as well as the corresponding duty cycle $\alpha = 2\tau/T = \cos^{-1}(x)/\pi$ versus (g_{m0}/g_{mc}). Note that the horizontal axis is on logarithmic scale to allow the low g_{m0}/g_{mc} region to be better displayed. It can be seen that both x and the duty cycle are relatively steep functions when the normalized g_{m0} of M_1 is low. For a g_{m0}/g_{mc} ratio of 3, which can easily be expected for practical designs, $x = 0.2$ and $\alpha = 44\,\%$. The improvements of x and α are much slower beyond $g_{m0}/g_{mc} > 5$, resulting only in $0.5 < x < 1$ and $\alpha < 33\,\%$ so that setting M_1's transconductance much higher than the g_{mc} of the tank is inefficient as far as achieving better amplitude is concerned. The limit in V_m is achieved when g_{m0}/g_{mc} tends to infinity, in which case x tends to unity and

$$V_m = \frac{2I_0}{g_{mc}}\,. \tag{7.14}$$

It is interesting to note that this limit is independent of the detailed parameters of M_1. In fact, the limit is exactly the same if M_1 in Fig. 7.3a is replaced by a bipolar transistor [26]. It is also worth noting that the duty cycle of M_1 is 0 % in the limit and only 33 % for $g_{m0}/g_{mc} = 5$. This indicates that the active device, be it MOS or BJT, operates deeply in class-C mode in a typical LC oscillator. In other words, the active device behaves more like a switch than an amplifier. To demonstrate this, we replace the active transistor in the Colpitts oscillator by a switch in series with a current source I_{pulse}, as shown in Fig. 7.7a. The threshold of the switch is set relative to the s-node of the circuit. If I_{pulse} is larger than I_0, the DC voltage across capacitor

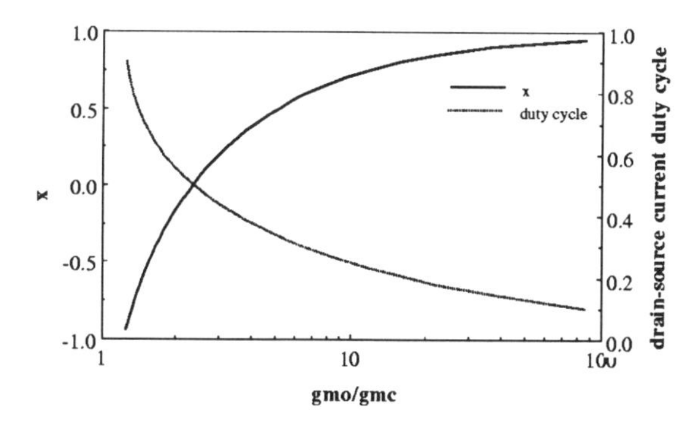

Figure 7.6 Parameter x and duty cycle as a function of normalized nominal g_m of M_1.

C_1 will change relative to the threshold of the switch so as to restrict the period the switch is turned on. The constraint is that the average current through the switch equals the bias current, $I_{\text{pulse}} \cdot 2\tau/T = I_0$. If we increase the available current I_{pulse} further and further, the duty cycle of the switch becomes smaller and smaller, as depicted by Fig. 7.7b, until the current through the switch becomes a Dirac function of area I_0T. Thus, irrespective of how the switch is constructed, as long as a large current is available during its on-period, the limit of the oscillation amplitude is:

$$V_m = \left(\frac{1}{g_{mc}}\right)\frac{2}{T}\int_0^T I_0 T\delta(t)e^{-j\omega_0 t}\,dt = \frac{2I_0}{g_{mc}}. \tag{7.15}$$

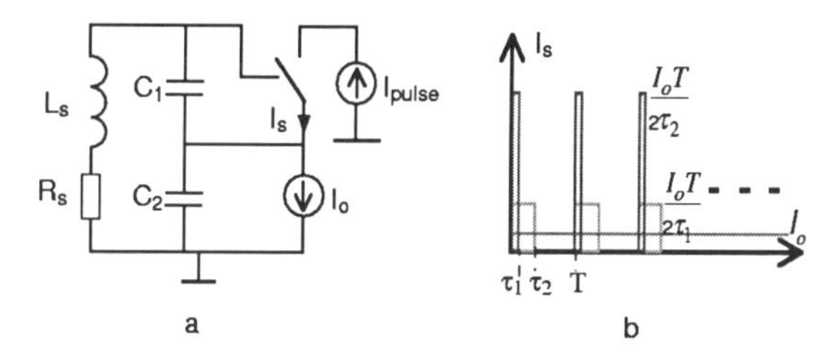

Figure 7.7 Thought experiment: Colpitts oscillator with ideal switch (a). Current vs. duty cycle (b).

The examples of BJT and MOS implementations show that the details of the switch construction determine how fast the limit in Eq. (7.15) is reached. As long as the active device, be it BJT, MOS or any other switch-like transistors, has sufficient transconductance, the details are unimportant, as the amplitude will be close to the limit set by Eq. (7.15). In practical designs, insight into the approximate amplitude dependence on the circuit parameters can be obtained from Eq. (7.15), whereas a more precise estimate can be obtained from Eq. (7.12).

To verify Eq. (7.12), SPICE simulations as well as experiments have been performed. The simulated amplitude agrees with Eq. (7.12) to within 1 % for $Q > 50$. For a moderate Q of around 10, the simulated amplitude is about 1–2 dB below that predicted by Eq. (7.12). To perform controlled experiments, a 78 MHz Colpitts oscillator is constructed as shown in Fig. 7.8, where the values of the main oscillator, as well as those of biasing and coupling components, can be found. The choice of the oscillation frequency was made to be sufficiently low to allow values of the resonator inductor and capacitors to be much higher than board-level parasitics, and yet sufficiently high so that the conclusions from the measurements can be extrapolated to high-frequency oscillators in the low gigahertz range. Other reasons against choosing a much lower frequency include suitability to high-frequency instruments that prefer circuits with low impedance levels in general, and oscillation frequency drift during the required measurement time for the same relative accuracy. Avoiding interferences caused by strong broadcast stations was also a consideration in the choice of the oscillator frequency.

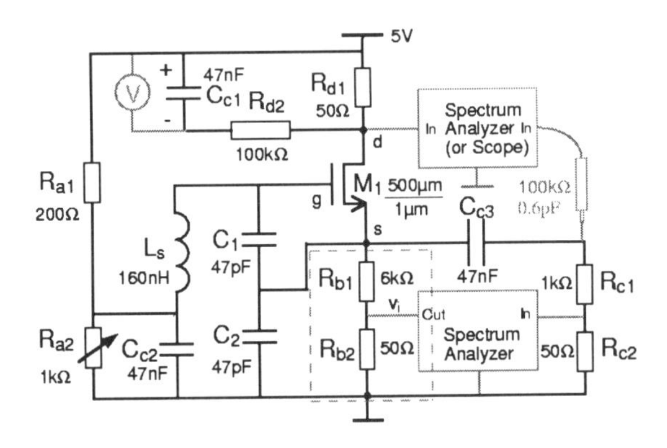

Figure 7.8 Colpitts oscillator constructed for measuring amplitude and duty cycle, as well as the responses to injected sinusoidal current.

The transistor M_1 has a transconductance coefficient of 90 μA/V^2 and was produced in a 1 μm CMOS process. Before mounting the transistor, the equivalent parallel resistance R_p of the tank at resonance has been measured, and the value of the parallel resistance, R_{c1}, is selected accordingly to make the resulting R_p exactly 500 Ω (or $g_{mc} = 2$ mS) at resonance. The gate bias voltage is adjusted by means of R_{a2} to ensure that the measured DC current through M_1 is 0.5 mA during steady-state oscillation. The corresponding $g_{m0} = 6.7$ mS. According to Fig. 7.6, $x = 0.275$ and the duty cycle $\alpha = 0.41$, whereas the oscillation amplitude is 0.44 V according to Eq. (7.12). Fig. 7.9 shows the measured steady-state source and drain voltages of M_1. The oscillation amplitude at the source is 450 mV, whereas the duty cycle, measured as the time between the two cursor positions of the drain voltage waveform divided by the oscillation period, is 0.414. Further measurements of a similar oscillator for four different

g_{mc}'s versus bias current from 200 μA to 5 mA [27] show that the amplitude prediction by Eq. (7.12) is always within 1 dB of measured values. Since normal tolerances of both the resonance and the loss of LC tanks, whether made of SMD components or integrated on silicon, can easily be 5–20 %, the tolerance in g_{mc} can easily be a couple of dBs. The accuracy of Eq. (7.12) is therefore as good as necessary for practical design purposes.

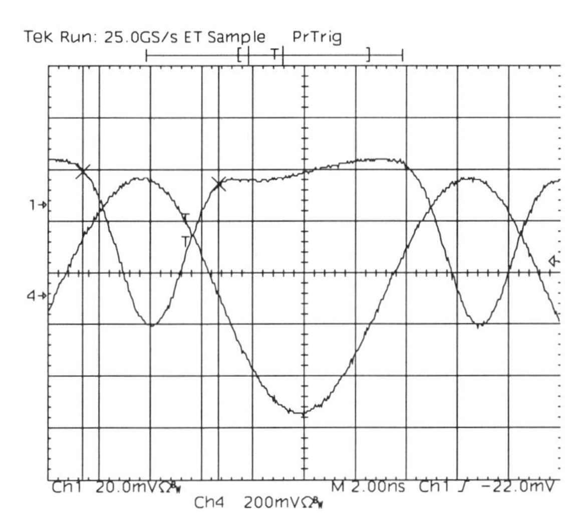

Figure 7.9 Measured oscillation waveforms. Channel 1: M_1's drain voltage representing its drain current. Channel 3: M_1's source voltage.

In the above oscillation amplitude derivation, the fact that signals are all periodic in the oscillator is exploited. An alternative way to derive the oscillation amplitude, which will also prove useful for analyzing small interfering signals at frequencies different from the main oscillation, is to represent the Colpitts oscillator in steady state by a conceptually equivalent circuit, as shown in Fig. 7.10. To avoid the difficulty associated with the piece-wise representation of M_1's V–I characteristic, we represent the cut-off period of M_1 with the switch shown in Fig. 7.10, driven by a gating function having the same frequency as the main oscillation, but with an unknown duty cycle $\alpha = 2\tau/T = \cos^{-1}(x)/\pi$. The current fed into the passive tank, $I_s(t)$, can now be represented as:

$$I_s(t) = \frac{\beta}{2}[V_m \cos\omega_0 t + V_B - V_T]^2 \cdot \left[\frac{2\tau}{T} + \sum_{i=1}^{\infty} \frac{2\sin(i\omega_0\tau)}{i\pi} \cos(i\omega_0 t)\right]. \quad (7.16)$$

The first term is the usual square-law V–I relationship for MOSFETs, except that we assume it is valid even if the sum inside the first brackets (the effective gate overdrive) is negative. Since the second term in Eq. (7.16), which represents the gating function, becomes zero whenever the effective gate overdrive is negative, Eq. (7.16) is the same as the usual piece-wise representation of the MOS transistor's V–I relationship. This is illustrated in Fig. 7.11.

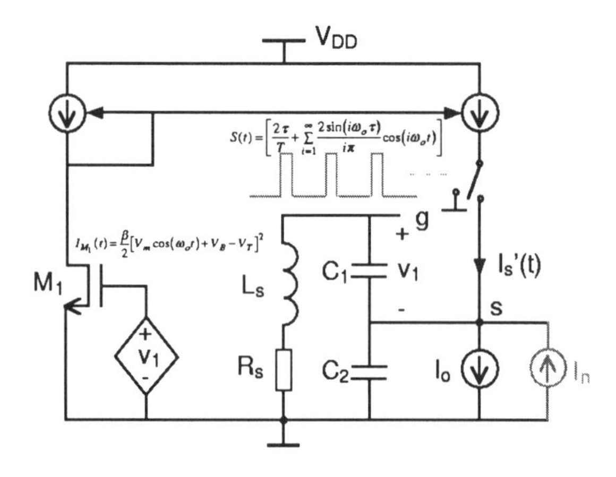

Figure 7.10 Equivalent representation of the Colpitts oscillator in Fig. 7.3a.

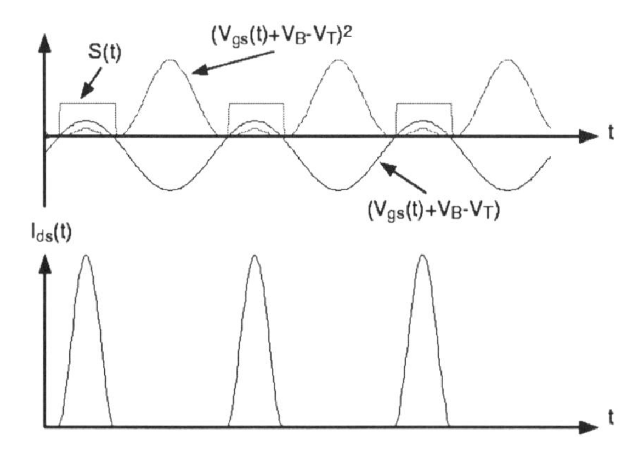

Figure 7.11 Voltage and current waveforms showing that the piece-wise nature of M_1's current can be represented by Eq. (7.2).

With Eq. (7.16), the DC and fundamental frequency components of the feedback current can be worked out by pure trigonometrical manipulations. It can be shown that the amplitude and duty cycle worked out this way are identical to those obtained by the earlier Fourier series expansion method. The advantage of Eq. (7.16), however, is that signals that are additive to the main oscillation but are not necessarily periodic in the latter's period can be analyzed, as long as their amplitudes are sufficiently small compared to the main oscillation, so that the frequency and duty cycle of the gating function remain practically the same.

7.4 OSCILLATOR RESPONSE TO AN INTERFERING CURRENT

Having established the oscillation amplitude as a function of the tank and transistor parameters as well as the bias current, and re-emphasized the fact that the active device in most high-Q LC oscillators operates in class-C mode, we are now better placed to tackle the phase noise in LC oscillators. Earlier on we objected to the linear analysis of phase noise, which assumes that an interfering signal (sinusoidal current for example) will generate a response (voltage) on the tank, and that the transistor acts like a linear transconductance equal to the critical g_m of the tank so that the equivalent tank loss is zero. To demonstrate that this is wrong with a concrete example, we return to the Colpitts oscillator and assume that an interfering voltage, $V_u \cos[(\omega_0 + \delta\omega)t + \phi_u]$, is present at the gate:

$$\begin{aligned} I_s(t) &= \frac{\beta}{2}\{V_m \cos(\omega_0 t) + V_u \cos[(\omega_0 + \Delta\omega)t + \phi_u] + V_B - V_T\}^2 \\ &\cdot \left[\frac{2\tau}{T} + \sum_{i=1}^{\infty} \frac{2\sin(i\omega_0\tau)}{i\pi} \cos(i\omega_0 t)\right]. \end{aligned} \tag{7.17}$$

Expanding Eq. (7.16) and collecting current terms only at $\omega_0 + \Delta\omega$, we find:

$$\begin{aligned} I_s[(\omega_0 + \Delta\omega)t] &= \left\{\frac{2}{3\pi}\left[(2+x^2)\sqrt{1-x^2} - 3x\cos^{-1}(x)\right] + \frac{2}{3\pi}(1-x^2)^{\frac{3}{2}}\right\} \\ &\cdot \frac{\beta V_m}{2} V_u \cos[(\omega_0 + \Delta\omega)t + \phi_u]. \end{aligned} \tag{7.18}$$

Combining (7.18) with (7.11), we obtain:

$$I_s[(\omega_0 + \Delta\omega)t] = \left[g_{mc} + \frac{\beta V_m}{3\pi}(1-x^2)^{\frac{3}{2}}\right] V_u \cos[(\omega_0 + \Delta\omega)t + \phi_u]. \tag{7.19}$$

Equation (7.19) clearly shows that, except in the very special case of $|x| = 1$, the equivalent transconductance of M_1 to the interfering voltage V_u is not exactly the critical g_m required to cancel the tank loss. In fact, for small values of x, the second term in the square brackets in (7.19) is usually much larger than the first term.

Equation Eq. (7.16) also shows the more fundamental flaw of the linear analysis. If an interfering voltage at $\omega_0 + \Delta\omega$ indeed appears, the resulting current will not only have a component at the same frequency, but also one at its image frequency, $\omega_0 - \Delta\omega$, due to the gating function. This second component will be fed back to the tank and generate a voltage component at the image frequency, so that a double-sideband response will result even if the interference is only single sideband. The voltage at the image frequency will in turn be modulated by the gating function, to create its own image, that is at the original interfering frequency $\omega_0 + \Delta\omega$. It is the interaction between the component at $\omega_0 + \Delta\omega$, which we refer to as the *response*, and the component at $\omega - \Delta\omega$, which we refer to as the *image*, that holds the key to all the phase noise behaviour which we have been trying to understand.

To analyze this double-sideband response to a single-sideband interference, let us inject a small current interference (shown in grey in Fig. 7.10), $I_n \cos[(\omega_0 + \Delta\omega)t + \phi_n]$, into the s-node in Fig. 7.10. The main oscillation (carrier) serves as the reference, so its phase is assumed to be zero. Although the phase of the injected current is not important, we assume it is ϕ_n for the sake of generality. The process in which the response and image are modulated back and forth is aided by the catalyst of the gating function, so that it is reasonable to assume that the steady state is reached within a small number of oscillation cycles. The imprecise word *small* is to be understood as much smaller than the number of clock cycles it takes for the response to an impulsive interference such as individual bursts caused by white noise to decay significantly on a high-Q (low loss) tank, so that steady-state analysis is justified for bursty noise sources. At this stage we only know that there will be a voltage response at both $\omega_0+\Delta\omega$ and $\omega_0-\Delta\omega$, but their amplitude and phase in steady state, V_u, ϕ_u, and V_L, ϕ_L for the upper and lower sidebands, respectively, are unknown.

$$\begin{aligned} V_{gs}(t) &= V_1(t) - V_B \\ &= V_m\cos(\omega_0 t) + V_u\cos[(\omega_0+\Delta\omega)t+\phi_u] + V_L\cos[(\omega_0-\Delta\omega)t+\phi_L]\,. \end{aligned} \tag{7.20}$$

Assuming that V_u and V_L are much smaller than V_m and referring to Fig. 7.10, the feedback current I_s in steady state is given by

$$\begin{aligned} I_s(t) = &\frac{\beta}{2}\{V_m\cos(\omega_0 t) + V_u\cos[(\omega_0+\Delta\omega)t+\phi_u] \\ &+V_L\cos[(\omega_0-\Delta\omega)t+\phi_L] + V_B - V_T\}^2\cdot\left[\frac{2\tau}{T}+\sum_{i=1}^{\infty}\frac{2\sin(i\omega_0\tau)}{i\pi}\cos(i\omega_0 t)\right] \\ &\left(0<\alpha=\frac{2\tau}{T}<100\%, \rightarrow |x|<1\right). \end{aligned} \tag{7.21}$$

Expanding Eq. (7.21), we find that only the terms at the carrier frequency ω_0, interference frequency $\omega_0+\Delta\omega$ and the image frequency $\omega_0-\Delta\omega$ are worth collecting, because they will be fed back to the gate of the tank to sustain the oscillation and interfering voltages. Those terms containing second-order combinations of V_u and V_L can be neglected because they are much smaller than the rest. The terms around DC and higher harmonics of ω_0 can also be neglected, because they will be heavily attenuated by the high-Q resonator and will not produce much voltage at the gate. Thus the significant part of I_s, I_s' is given by:

$$\begin{aligned} I_s(t) = &\frac{\beta V_m}{3\pi}\left[(2+x^2)\sqrt{1-x^2} - 3x\cos^{-1}(x)\right] \\ &\cdot\{V_m\cos(\omega_0 t) + V_u\cos[(\omega_0+\Delta\omega)t+\phi_u] + V_L\cos[(\omega_0-\Delta\omega)t+\phi_L]\} \\ &+\frac{\beta V_m}{3\pi}(1-x^2)^{\frac{3}{2}}\{V_u\cos[(\omega_0+\Delta\omega)t+\phi_u] + V_L\cos[(\omega_0-\Delta\omega)t+\phi_L] \\ &+V_L\cos[(\omega_0+\Delta\omega)t-\phi_L] + V_u\cos[(\omega_0-\Delta\omega)t-\phi_u]\}\,. \end{aligned} \tag{7.22}$$

Applying Kirchhoff's current law to the s-node in Fig. 7.10, we have:

$$I_s(t) + I_n(t) + C_1\frac{dV_1}{dt} - C_2\frac{dV_2}{dt} = 0 \tag{7.23}$$

It can be shown that for $\Delta\omega \ll \omega_0$,

$$C_1\frac{dV_1}{dt} - C_2\frac{dV_2}{dt} \approx$$
$$2(C_1+C_2)\Delta\omega\{V_u \sin[(\omega_0+\Delta\omega)t+\phi_u] - V_L \sin[(\omega_0-\Delta\omega)t+\phi_L]\}$$
$$- R_s C_1 C_2 \omega_0^2\{V_m \cos(\omega_0 t) + V_u \cos[(\omega_0+\Delta\omega)t+\phi_u] + V_L \cos[(\omega_0-\Delta\omega)t+\phi_L]\}\ . \tag{7.24}$$

Combining Eqs. (7.11), (7.22) and (7.24) into Eq. (7.23), we obtain:

$$\begin{aligned}
&\frac{\beta V_m}{3\pi}(1-x^2)^{\frac{3}{2}}\{V_u \cos[(\omega_0+\Delta\omega)t+\phi_u] + V_L \cos[(\omega_0-\Delta\omega)t+\phi_L] \\
&+V_L \cos[(\omega_0+\Delta\omega)t-\phi_L] + V_u \cos[(\omega_0-\Delta\omega)t-\phi_u]\} \\
&+2(C_1+C_2)\Delta\omega\{V_u \sin[(\omega_0+\Delta\omega)t+\phi_u] - V_L \sin[(\omega_0-\Delta\omega)t+\phi_L]\} \\
&+I_n \cos[(\omega_0+\Delta\omega)t+\phi_n] = A\cdot\cos[(\omega_0+\Delta\omega)t] + B\cdot\sin[(\omega_0+\Delta\omega)t] \\
&+C\cdot\cos[(\omega_0-\Delta\omega)t] + D\cdot\sin[(\omega_0-\Delta\omega)t] = 0\ .
\end{aligned} \tag{7.25}$$

For Eq. (7.25) to be valid at all times, the coefficients for the sine and cosine terms at both frequencies must be zero. This provides us with four simultaneous equations to solve for the four unknowns in steady state, V_u, ϕ_u, V_L, and ϕ_L.

$$\begin{aligned}
A =& \frac{\beta V_m}{3\pi}(1-x^2)^{\frac{3}{2}}[V_L\cos(\phi_L) + V_u\cos(\phi_u)] + 2(C_1+C_2)\Delta\omega V_u \sin(\phi_u) \\
&+ I_n \cos\phi_n = 0\ ,
\end{aligned} \tag{7.26}$$

$$\begin{aligned}
B =& \frac{\beta V_m}{3\pi}(1-x^2)^{\frac{3}{2}}[V_L\sin(\phi_L) - V_u\sin(\phi_u)] + 2(C_1+C_2)\Delta\omega V_u \cos(\phi_u) \\
&- I_n \sin\phi_n = 0\ ,
\end{aligned} \tag{7.27}$$

$$C = \frac{\beta V_m}{3\pi}(1-x^2)^{\frac{3}{2}}[V_L\cos(\phi_L) + V_u\cos(\phi_u)] - 2(C_1+C_2)\Delta\omega V_L \sin(\phi_L) = 0\ , \tag{7.28}$$

$$D = \frac{\beta V_m}{3\pi}(1-x^2)^{\frac{3}{2}}[V_u\sin(\phi_u) - V_L\sin(\phi_L)] - 2(C_1+C_2)\Delta\omega V_L \cos(\phi_L) = 0\ . \tag{7.29}$$

Solving Eqs. (7.26)–(7.29) jointly, we find:

$$\phi_L = \tan^{-1}\left[\frac{\beta V_m(1-x^2)^{\frac{3}{2}}}{3\pi(C_1+C_2)\Delta\omega}\right] - \phi_n + k\pi \qquad k = 0,1,2\ldots \tag{7.30}$$

$$\phi_u = \pi - \phi_L - \phi_c \text{ where } \phi_c = \tan^{-1}\left[\frac{6\pi(C_1+C_2)\Delta\omega}{\beta V_m(1-x^2)^{\frac{3}{2}}}\right] \tag{7.31}$$

$$V_L = V_u \cos\phi_c \tag{7.32}$$

$$\begin{aligned}
V_u^2 &= \frac{I_n^2}{[2(C_1+C_2)\Delta\omega]^2[1+3\cos^2(\phi_c)]} \\
&= \frac{I_n^2}{[2(C_1+C_2)\Delta\omega]^2}\,\frac{[\beta V_m]^2(1-x^2)^3 + [6\pi(C_1+C_2)\Delta\omega]^2}{\left\{4\cdot[\beta V_m]^2(1-x^2)^3 + [6\pi(C_1+C_2)\Delta\omega]^2\right\}}\ .
\end{aligned} \tag{7.33}$$

Usually the offset frequency is relatively small, $6\pi(C1+C2)\Delta\omega \ll \beta V_m(1-x^2)^{\frac{3}{2}}$, so that the upper- and lower-sideband responses are equal in amplitude and 180° out of phase, as shown by Eqs. (7.31) and (7.33). This is just the same as what would be created by a modulation in phase, so that additive interference does create modulation in phase, due to the nonlinear and switching behaviour of M_1. For the same low offset frequencies,

$$V_u^2 \approx \frac{I_n^2}{4[2(C_1+C_2)\Delta\omega]^2} \quad \text{or} \quad V_u \approx \frac{I_n}{4\cdot(C_1+C_2)\Delta\omega} \tag{7.34}$$

Note that V_u in Eq. (7.34) is half of what would be predicted by a linear analysis. The upper-sideband response is indeed inversely proportional to the offset frequency, as has been well known experimentally. This inverse dependency does, however, not continue uninterruptedly, according to Eq. (7.33). At sufficiently high offset frequencies outside the passive tank's passband, $6\pi(C1+C2)\Delta\omega \gg 2\beta V_m(1-x^2)^{\frac{3}{2}}$, and the transistor no longer has sufficient g_m to provide any significant feedback to counter the interfering current, so that the relationship between the latter and the voltage response is purely that of a passive tank, and the image sideband disappears, according to Eqs. (7.32) and (7.33). The result is now identical to that predicted by a linear analysis:

$$V_u \approx \frac{I_n}{2\cdot(C_1+C_2)\Delta\omega}\,. \tag{7.35}$$

The transition point where the image sideband response V_L becomes 3 dB smaller than the upper sideband response V_u can also be derived from (7.32):

$$\Delta\omega^{-3\,\mathrm{dB}} = \frac{\beta V_m(1-x^2)^{\frac{3}{2}}}{6\pi(C_1+C_2)}\,. \tag{7.36}$$

For oscillators with 100 % current duty cycle, $x \leq -1$, the corner frequency is zero and the noise remains additive, because there is no more gating function and the squaring nonlinearity of the MOSFET only translates the upper-sideband interfering signal to frequencies around DC and the second harmonic of the carrier, which will be suppressed by the carrier. When x tends to 1, as in the case of a lossless resonator, the corner frequency also tends to zero and the output noise remains additive. For most practical oscillators $|x| < 1$ and noise becomes multiplicative, affecting the phase of oscillation. The -3 dB offset frequency given in Eq.(7.36) is usually too high to be observed in normal phase noise measurements, because the sideband spectral density at such a frequency is already well below the noise floor of the measuring instruments.

It is important to note from Eqs. (7.34)–(7.36) that the detailed transistor parameters are only important in determining the boundary between the interferences that will merely be super-imposed on the main oscillation and those that will cause the phase of the oscillation to jitter. In the former case, the $1/\Delta\omega$ characteristic is merely due to the impedance of the passive tank, decreasing outside its passband. In the latter case, it is the creation of the image response and the effective negative feedback loop between the signals at the two frequencies that are responsible for canceling any excess g_m of M_1 that is not needed for compensating the loss of the passive tank. Although in both cases the response is inversely proportional to the offset frequency, the two cases differ by 6 dB and cause very different disturbances to the oscillator. In the phase jitter case, both the image response and its interaction with the response at the frequency of interference are due to the active device's nonlinearity, especially the switching action in most oscillators. Analyses attempting to bypass the explicit description of the transistor's nonlinearity (a slowly varying AGC is a linear device in this context) therefore bypass the most fundamental mechanism of amplitude-to-phase-noise conversion. This tends to result in shaky foundations even though the resulting models of such analyses are good guesses for what a real oscillator's noise spectrum would look like. Our analysis shows that the modulation process is essential, yet the exact amount of M_1's g_m, and indeed the detailed shape of its large-signal V–I characteristic, are unimportant. Equations similar to (7.26)–(7.29) can be set up for transistors with quite different characteristics. While the corner frequency will now be different from that given by Eq. (7.36), Eq. (7.34) will still result at low offset frequencies. In this sense our analysis is general and is applicable to all high-Q LC oscillators.

To verify the results in this section, the oscillator in Fig. 7.8 is used to test its voltage response to an injected deterministic current. Since the equivalent input impedance at the s-node can be very high close to the resonance frequency of the tank, injecting a well-known current is difficult. Instead of applying a current directly, we apply a sinusoidal voltage as shown, and measure the voltages V_s and V_i. The Norton equivalent of the circuit within the dashed box in Fig. 7.8 is shown in Fig. 7.12. Since the effective injected current is given by V_i/R_{b1}, the impedance Z looking into the s-node of the oscillator from the equivalent current source is the measured node voltage V_s divided by $V_i/R_{b1}, Z = (V_s/V_i)R_{b1}$. Since for low-loss tanks the current flowing through C_1 and C_2 close to resonance is much higher than other current components flowing into the s-node, the latter can be neglected and $V_{gs}(t) = (C_1/C_2)V_s(t)$. The transimpedance $Z_T = V_{gs}/I_n$ can therefore be measured reliably, at least in the close vicinity of the carrier, via the voltage ratio (V_s/V_i). Close to the carrier, the applied signal levels are kept very small to avoid pulling the main oscillation. As the offset frequency is gradually increased, the voltage response gradually becomes so small that it sinks under the noise floor of the spectrum analyzer. The interfering signal level is then increased whenever necessary just to keep the spectral peaks representing the voltage response above the noise floor by a few dBs. Also measured is the ratio between the voltage response at the image frequency and the injected current, which we refer to as the image transimpedance.

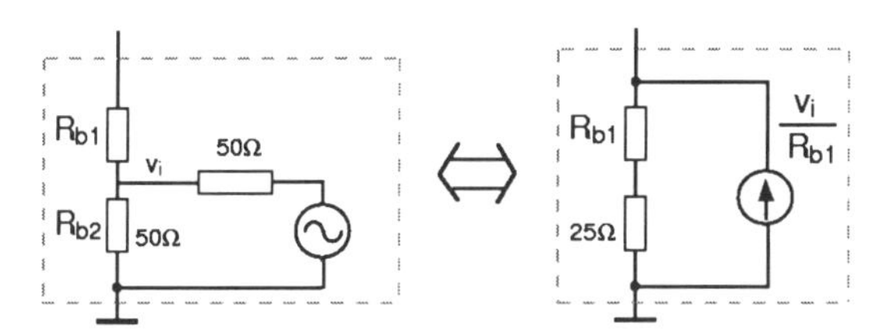

Figure 7.12 The Norton equivalent representation of the subcircuit devised to inject controlled current into the oscillator.

The measured transimpedances are shown in Fig. 7.13, against those calculated using Eqs. (7.32) and (7.33). Also shown is the straight line to which the upper-sideband transimpedance is asymptotic at higher offset frequencies, which would result from a linear analysis. The measurements and calculations agree to within 1 dB at low offset frequencies and 2 dB at high offset frequencies, which is remarkable over nearly 6 decades of offset frequency and 7 decades of impedance levels. Even more remarkable is the accurate prediction of the cross-over point at 1.3 MHz. This shows that as long as the parasitics are small and the oscillator can be truly represented by Fig. 7.3a, — the formulas derived above are very accurate.

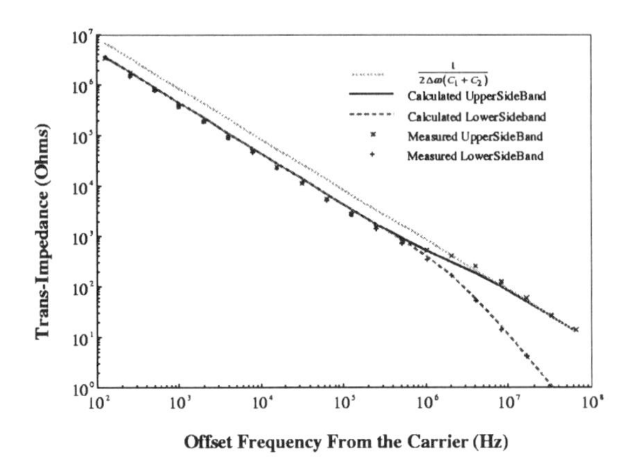

Figure 7.13 Calculated and measured voltage responses at both the upper sideband and lower sideband, to injected current at the upper sideband (the circuit parameters are slightly different from those in Fig. 7.9: V_m=0.37V and x=0.2.).

The significance of the theoretical prediction of the cross-over point and its accurate experimental verification lies not so much in the 6 dB difference in noise prediction at higher frequencies as in being a litmus test of the rigorousness of the formulation of our noise analysis. The cross-over point is usually so far away in relative offset frequency from the carrier that the noise there and beyond is so low that it does not affect any practical design. The fact that nobody ever noticed its existence before is ample proof of this. However, the two observations embodied in the conclusions of our analysis and experiments are so fundamental and common sense that analyses leading to opposite conclusions to any of the two must be based on false assumptions. The two observations are that interfering noise close to the carrier causes phase modulation that results in a double-sideband response, and that interference far away from the oscillation frequency is no more than an additive noise. In the latter case, the response must be very similar to that of a passive tank alone.

A further point to note from Fig. 7.13 is that the $1/\Delta\omega$ characteristic continues all the way down to 100 Hz offset, where further measurement was prevented by the drift of the free-running oscillator. An early measurement of a Clapp oscillator at a similar carrier frequency, locked to a crystal oscillator by a very-narrow-band PLL to reduce drift, showed that such a $1/\Delta\omega$ characteristic continues nearly to 10 Hz. The experiments therefore demonstrate that analyses predicting a much higher corner frequency are wrong. Below such offsets, pulling must be dominant so that the exact shape of the phase noise should be immaterial. At the high offset end, our measurements show that the $1/\Delta\omega$ characteristic continues until 64 MHz, further than 2/3 of the carrier frequency itself! No flattening is observed, so that part of Leeson's description of the phase noise spectrum is incorrect. Our experimental results are to be expected, though, for it would be very hard to imagine what noise mechanism would create a response of constant amplitude on a passive tank, while the latter's impedance gradually declines to zero! The flat noise spectrum we sometimes do see in practical oscillators must have come from buffers rather than the LC oscillator proper.

7.5 NOISE-TO-CARRIER RATIO IN A CMOS COLPITTS OSCILLATOR

The analysis of the oscillator response to a single sinusoidal current can easily be extended to multiple signals at different frequencies. If the sum of all interfering signals is small compared to the carrier, then one can assume that the interfering current at each frequency will generate a pair of response voltages at the gate, and the expression of M_1's feedback current is similar to Eq. (7.21), except now there are many pairs of response voltages at the gate that are unknown to us. As long as we can neglect the second-order combinations of interfering voltage components after expanding the square-law function, the subsequent mathematical manipulations can be considered linear, because the multiplication by the gating function only results in frequency translations of each component. Thus each pair of voltage responses can be solved individually, and will be related to the individual current component in the same way as described by Eqs. (7.32) and (7.33). In this sense, the oscillator response to composite current of different frequencies is the superposition of the responses to current components at each frequency. This enables our noise analysis to follow all the reasoning of noise analyses of common linear networks.

If the offset frequency is much higher than $\Delta\omega^{-3\,\mathrm{dB}}$ (Eq. (7.36)), the noise spectral density of the oscillator's gate-source voltage at each frequency is simply given by Eq. (7.35) times the spectral density of equivalent noise current at that frequency. In the much more important case that the offset frequency is below $\Delta\omega^{-3\,\mathrm{dB}}$, the response is given by Eq. (7.34) times the spectral density of equivalent noise current at the same frequency plus the contribution due to the image of noise current component at the other side of the carrier.

$$\begin{aligned} V_n^2(\omega_0+\Delta\omega) &= \frac{I_n^2(\omega_0+\Delta\omega)}{16\cdot(C_1+C_2)^2\Delta\omega^2} + \frac{I_n^2(\omega_0-\Delta\omega)}{16\cdot(C_1+C_2)^2\Delta\omega^2} \\ &\quad + C'\frac{I_n(\omega_0+\Delta\omega)I_n^*(\omega-\Delta\omega)}{8(C_1+C_2)^2\Delta\omega^2}\,. \end{aligned} \tag{7.37}$$

In (7.36), C' represents the cross correlation factor between the two noise components. For white noise it is reasonable to assume that noise components at different frequencies are uncorrelated to each other, i.e. $C'=0$, so that the noise voltage spectral density is given by:

$$V_n^2(\omega_0+\Delta\omega) = \frac{I_n^2}{8\cdot(C_1+C_2)^2\Delta\omega^2}\,. \tag{7.38}$$

Experiments will be compared to Eq. (7.38) later to verify the assumption $C' = O$. Now that we have established the relationship between the phase noise spectrum and the input white noise current, what remains to be done is to identify all major noise sources in the Colpitts oscillator. In the order of increasing importance, they are the noise current associated with the bias transistor M_B, the noise current due to the loss resistance R_s of the tank, and the noise current due to the main switching transistor M_1. If we can convert all three sources into equivalent current sources as shown in Fig. 7.14, then the total output noise can be easily obtained by applying Eq. (7.38).

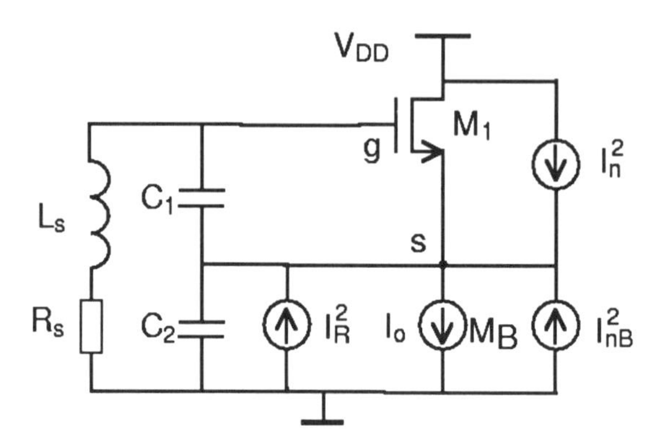

Figure 7.14 Colpitts oscillator with all noise sources represented as current into the s-node.

Since the drain of M_B is directly connected to the s-node in Fig. 7.3a and its current is constant, the noise current due to M_B is the most straightforward, $I_{nB}^2 = \frac{8kT}{3} g_{M_B}$. The noise contribution of R_s can be worked out via the Norton equivalent current at the s-node. Representing the noise of R_s as a voltage source and short-circuiting the s-node, we can obtain the mean-squared current noise power spectral density (PSD) of the Norton equivalent circuit near the resonant frequency, $I_R^2 \approx (\omega C_2)^2 \cdot 4kTR_s = (C_2/C_1) \cdot 4kT \cdot g_{mc}$.

The time-varying nature of its drain-source current makes M_1 a more complex noise source. Referring to Fig. 7.7b, the average drain-source current through M_1 during the time it is on is given by Eq. (7.39).

$$I'_{m1} = \frac{I_0}{\alpha} = I_0\left(\frac{T}{2\tau}\right) = I_0\frac{\pi}{\cos^{-1}(x)} \tag{7.39}$$

$$I_{n'}^2 = \frac{8kT}{3}g'_{m1} = \frac{8kT}{3}\sqrt{2\beta I_0\frac{\pi}{\cos^{-1}(x)}} = \frac{8kT}{3}g_{m0}\sqrt{\frac{\pi}{\cos^{-1}(x)}}. \tag{7.40}$$

The average white noise power spectral density during M_1's on-time is therefore given by Eq. (7.40). If a time-domain noise current $n'(t)$ has the same power spectral density as that given by Eq. (7.40), then the time-domain noise current for M_1 is equivalent to

$n'(t)$ gated by the switching function $s(t)$ with duty cycle $\alpha = \cos^{-1}(x)/\pi$.

$$n(t) = n'(t)\left[\frac{2\tau}{T} + \sum_{i=1}^{\infty}\frac{2\sin(i\omega_0\tau)}{i\pi}(\cos i\omega_0 t)\right]. \tag{7.41}$$

Since $n'(t)$ itself has a bandwidth much wider than the frequency of the gating function, the power spectral density of $n(t)$ around the oscillation frequency is approximately that of $n'(t)$, given by Eq. (7.40), scaled by the duty cycle of the gating function [29].

$$I_n^2 = \frac{\cos^{-1}(x)}{\pi}I_{n'}^2 = \frac{8kT}{3}g_{m0}\sqrt{\frac{\cos^{-1}(x)}{\pi}}, \tag{7.42}$$

A more accurate estimate of I_n^2 is possible by splitting M_1's on-time into infinitely small intervals, each $d\tau$ wide, to allow the exact current to be used for each interval instead of the average in Eq. (7.39). Since the noise contribution of each $d\tau$ interval can be calculated with a gating duty cycle of $d\tau/T$, the overall noise PSD is the sum of the individual contributions, which amounts to an integration that leads to

$$I_n^2 = \frac{8kT}{3}\beta V_m\frac{\sqrt{1-x^2} - x\cos^{-1}(x)}{\pi}. \tag{7.43}$$

Since most practical oscillators already have fairly low duty cycles, the difference between Eqs. (7.42) and (7.43) is fairly small. For the circuit in Fig. 7.8, the overall noise estimate using Eq. (7.43) is about 0.5 dB below that using Eq. (7.42). Equation (7.42) is therefore more preferable because it gives more insight into the relationship between noise and the nominal transconductance, as well as the current duty cycle. Both are likely to be design parameters.

Replacing the noise current term in Eq. (7.38) with the sum of contributions from the bias transistor, the resistive loss of the tank, and that of the switching transistor M_1, we obtain:

$$V_n^2(\omega_0 + \Delta\omega) = V_n^2(\omega - \Delta\omega)$$
$$= \frac{kT}{2\cdot(C_1+C_2)^2\Delta\omega^2}\left[\frac{2}{3}g_{mB} + \left(\frac{C_2}{C_1}\right)g_{mc} + \frac{2}{3}g_{m0}\sqrt{\frac{\cos^{-1}(x)}{\pi}}\right]. \tag{7.44}$$

The ratio between the noise spectral density and the signal power can now be derived by combining Eqs. (7.12) and (7.44):

$$\frac{\mathrm{N}}{\mathrm{C}}(\Delta\omega) = \frac{2\cdot V_n^2(\omega_0+\Delta\omega)}{V_m^2}$$
$$= \left(\frac{3}{5+x}\right)^2\frac{kT}{I_0^2Q^2}\left(\frac{\omega_0}{\Delta\omega}\right)^2\left[\frac{2}{3}g_{mB} + \left(\frac{C_2}{C_1}\right)g_{mc} + \frac{2}{3}g_{m0}\sqrt{\frac{\cos^{-1}(x)}{\pi}}\right]. \tag{7.45}$$

For small duty cycles, x is close to unity, and Eq. (7.45) is then approximately given by:

$$\frac{\mathrm{N}}{\mathrm{C}}(\Delta\omega) = \frac{kT}{I_0^2 Q^2}\left(\frac{\omega_0}{2\Delta\omega}\right)^2 \left[\frac{2}{3} g_{mB} + \left(\frac{C_2}{C_1}\right) g_{mc} + \frac{2}{3} \cdot g_{m0} \sqrt{\frac{\cos^{-1}(x)}{\pi}}\right] \tag{7.46}$$

To verify Eq. (7.45), the parameters in Fig. 7.8 were used to calculate the oscillator output spectral density due to thermal noise sources. The noise contributions in the square brackets in Eq. (7.45) are 0.17 mS for the 6 kΩ bias resistor, 2 mS for the rest of the resistive loss of the passive tank that had been measured before M_1 was mounted, and 2.86 mS due to M_1. The quality factor Q of the tank is slightly under 25. This gives a calculated phase noise of -115 dBc/Hz at 10 kHz offset. The calculated phase noise is compared with measured results in Fig. 7.15, covering 4 octaves of offset frequency range.

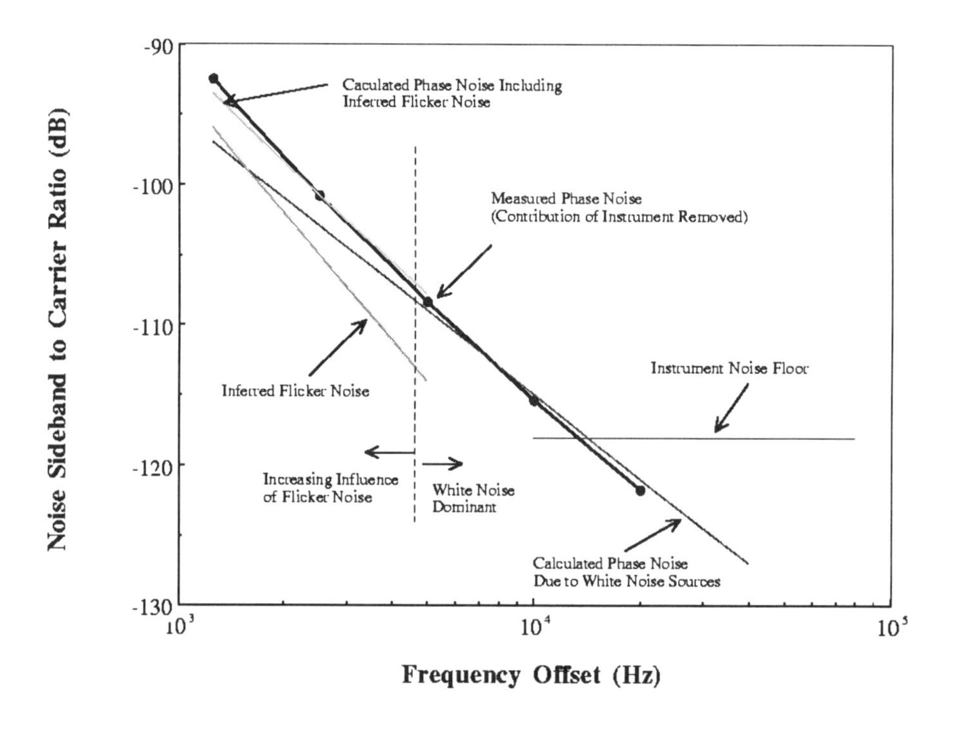

Figure 7.15 Calculated and measured phase noise for the oscillator in Fig. 7.8.

Because the oscillator was designed to be relatively high-Q (to be close to the practical situation that has been assumed in the derivations of this paper), the overall phase noise is very low. The noise floor of the instrument and the onset of up-converted flicker noise therefore limit the measurable thermal noise region to the last two octaves. This gives us three measurement points that are clearly on the 6 dB/oct line. To

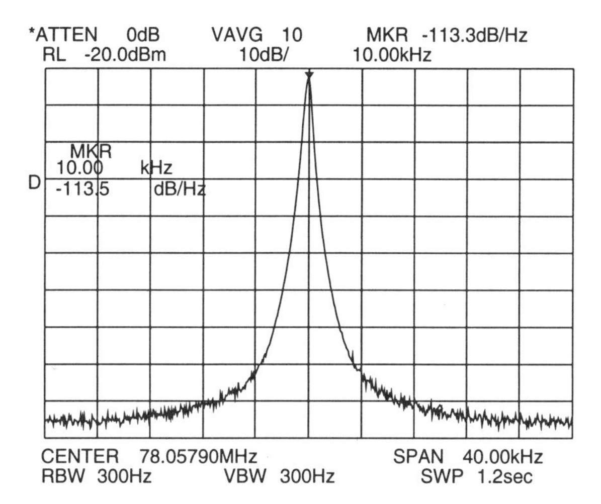

Figure 7.16 Measured output power spectrum for the oscillator in Fig. 7.8.

remove the influence of the noise from the instrument (HP 8563 E Spectrum Analyzer), the latter is calibrated by measuring an 80 MHz crystal oscillator at the corresponding offset frequencies. Since the phase noise of an 80 MHz crystal oscillator is better than −110 dB at 300 Hz offset [30], the measured noise at offsets above 1 kHz is effectively that of the instrument. This uncorrelated noise contribution of the instrument is then subtracted from the measured noise power at the 10 kHz and 20 kHz offset points. As is typical of most noise measurements, the repeatability of the measured noise power spectral density (with averaging) is about within 0.5–1 dB. The measurement accuracy of the points shown in Fig. 7.15 can therefore be considered to be 1 dB. This is the same accuracy with which the calculated thermal noise line in Fig. 7.15 predicts the measurement. The measurement at 1.25 kHz is 16 dB higher than that at 5 kHz, showing this region to be strongly influenced by flicker noise. From the slope of the measured noise PSD we can infer that the cross-over point between thermal noise and flicker noise is around 1.5 kHz. Adding the inferred flicker noise to the calculated white noise contribution, we have a calculated phase noise that also agrees with measurement within 1 dB for the first two octaves of measured offset frequency. This can be seen as conclusive evidence that Eq. (7.45) accurately predicts the sideband noise-to-carrier ratio, not least because there are no parameters used in the calculation of Eq. (7.45) that have not been independently verified by measurement to the desired accuracy first. Since the relationship between an injected current and measured voltage response has also been accurately verified in the previous section, the significance of the measurement in Fig. 7.15 lies in verifying Eqs. (7.36) and (7.38), as well as the power spectral densities of all the white noise sources in the oscillator, especially that of M_1. Equation (7.38) is 3 dB below what would be predicted by a linear analysis similar to Eq. 17 in [9]. Fig. 7.16 shows a measured spectrum of the oscillator.

7.6 EXACT DESIGN OF RF OSCILLATORS

In addition to having been derived on a rigorous basis, Eq. (7.45) is completely based on circuit parameters known to the designers. This makes the discussion of the exact design of RF oscillators most straightforward. Before this discussion, however, it is interesting to compare Eq. (7.45) or Eq. (7.46) to the well-known Leeson's formula, which designers have been using for the last 30 years, to examine the dependence on the resonator's quality factor Q in particular. In Leeson's original paper, the signal level P_s at the input of the active device is not very clearly defined, so that it can either be interpreted as the square of the oscillation voltage, or the power dissipated by the resonator. In the former case we can see that for a constant oscillation voltage, the noise-to-carrier ratio will only scale with the noise power given in Eq. (7.44), which clearly does not scale with Q^2. In the latter case, $P_s = (C1/C2)g_{mc}(V_m)^2/2$. Equation (7.46) can thus be rewritten as

$$\begin{aligned}\frac{\mathrm{N}}{\mathrm{C}}(\Delta\omega) &= \frac{2\cdot V_n^2(\omega_0+\Delta\omega)}{(C_1/C_2)g_{mc}V_m^2}\left(\frac{C_1}{C_2}\right)g_{mc} \\ &= \frac{2kT}{(Q)^2P_s}\left(\frac{\omega_0}{2\Delta\omega}\right)^2\left(\frac{C_1}{C_2}\right)\left[\frac{2}{3}\frac{g_{mB}}{g_{mc}}+\left(\frac{C_2}{C_1}\right)^2+\frac{2}{3}\frac{g_{m0}}{g_{mc}}\sqrt{\frac{\cos^{-1}(x)}{\pi}}\right] \\ &= \frac{2kTF}{(Q)^2P_s}\left(\frac{\omega_0}{2\Delta\omega}\right)^2, \end{aligned} \tag{7.47}$$

which apparently is Leeson's formula with an inverse dependence on Q^2! Closer examination of Eq. (7.47), however, shows that both the signal power P_s and the noise figure F so defined, far from being independent parameters, are usually also strong functions of Q. No conclusions could therefore really be drawn about the Q dependence of the oscillator noise performance on the basis of Leeson's formulation.

In practical situations, one does sometimes find that doubling the Q of the tank improves the N/C by exactly 6 dB, while the transistor and its bias current are unchanged. This can arise when the change of Q is purely due to improvement in the loss resistance R_s, and the middle term in the square brackets of Eq. (7.46) is much smaller than the other two terms. This may have been the case especially for designers using discrete MOSFETs or BJT transistors, where the transconductance for a given current is not within the designer's control. For an integrated circuit designer this need not be so, because the size of the transistor should be scaled down with g_{mc} to maintain just the necessary g_{m0}/g_{mc} range of 3–5 as explained earlier. In this case, the noise factor F in Leeson's formulation is unchanged, while the signal power P_s has doubled. The improvement of N/C will be 9 dB each time Q is doubled by halving R_s.

On the other hand, N/C can be significantly changed by different combinations of tank parameters with the same quality factor Q. Assuming, as a typical example, that the loss resistance R_s of the inductor dominates the loss of the tank and scales linearly with the inductance L_s. Doubling the inductance and halving the capacitances then leaves both the Q of the tank and its resonance frequency unchanged. The combined change in resistance and capacitance, however, causes g_{mc} in Eq. (7.46) to be halved, as well as the required g_{m0}, so that 3 dB can be gained either in achievable N/C or in the required current consumption without improving the quality of the inductor! Indeed, we note that most recently published oscillators with integrated inductors have very small capacitors, which gave them respectable phase noise performance at a reasonable current consumption even though the integrated inductors continue to have a poor quality factor between 3 and 6 at 1 GHz. One's ability to improve phase noise by reducing capacitance, however, is limited by the parasitic capacitors associated with the circuit, which have higher losses and which tend to increase the required tuning range for the desired frequency. A higher tuning range in capacitance may also result in higher variability of oscillation amplitude, which in turn affects N/C.

An additional advantage of the phase noise model described by Eq. (7.45) is that the role of current consumption (as opposed to power consumption P_{DC}, or the signal power P_s) is explicitly stated. Indeed, there is an almost one-to-one relationship between N/C and the bias current I_0. This point is important in RF oscillator designs for mobile communications, where power consumption is extremely important.

For a local oscillator (LO) used in radio frequency circuits, especially those used in cellular telephones, the three most important parameters are oscillation amplitude, phase-noise-to-carrier ratio and power consumption. For a given voltage supply, the latter reduces to current consumption. A well-defined LO amplitude is required by the mixers, which tend to exhibit a higher noise figure and less conversion gain both when the LO amplitude is too low and too high. Limiting amplifiers (buffers) are frequently used to remove variations of the LO amplitude due to the imprecise nature of the tank loss as well as the tuning of capacitors, but a well-designed worst-case amplitude range still results in the best power consumption. The phase noise of the oscillator relative to the carrier amplitude is responsible for reciprocal mixing of out-of-band, strong interfering signals to the same IF frequency as the desired RF signal. It also reduces the signal-to-noise ratio in the desired radio channel. Power consumption, on the other hand, directly affects the lifetime of the battery. The latter is usually the most expensive item in a cellular telephone [31, 32]. Bearing the above performance parameters and Eqs. (7.12), (7.45) and Fig. 7.6 in mind, we can formulate the following systematic design procedure for RF oscillators:

1. Determine the lowest capacitance that can be used based on the knowledge of circuit parasitics, variability of the capacitors themselves, required RF tuning range (for channel selection, for example) and the available variable capacitor tuning range. (Note that this may be substantially larger for a production circuit than that for a laboratory experiment.)

2. Select or implement (in the case of fully integrated circuits) the inductor in such a way that ensures the best worst-case Q for the resonator. The value of the inductance is determined by capacitances chosen in the previous step and the desired nominal frequency.

3. Set the value of the bias current to achieve the desired amplitude according to Eq. (7.12) and using the worst-case g_{mc}. (If the difference between best- and worst-case amplitudes exceeds what is acceptable by the mixer, then either a limiting amplifier must be used, or step 1 must be repeated to reduce the difference.)

4. Use Eq. (7.12) and the worst-case g_{mc} to calculate the ratio N/C at critical offset frequencies, assuming that g_{m0} will be 3–5 times g_{mc}. If it exceeds the prescribed

limit by k dB, and a limiting amplifier is used to limit the LO amplitude, then bias current should be increased by $0.5k$ dB, to reduce N/C by k dB.

5. If no limiting amplifier is used, the bias current should then be increased by k dB in step 4, which reduces N/C by $2k$ dB. The design procedure must then be repeated from step 1 to ensure that the best-case amplitude is not too high. This is done by increasing the capacitances by $0.5k$ dB, so that g_{mc} is increased by k dB to bring amplitude down to the desired level. This increases the N/C by another k dB so that the acceptable limit is just met.

6. Size M_1 according to the bias current, so that its nominal transconductance g_{m0} is 3–5 times g_{mc}.

7. The transconductance g_{mB} of the bias transistor M_8 should be made as small as possible to reduce noise. Since the bias current is fixed, this means reducing its W/L ratio. The latter is limited by the available voltage headroom for a given supply to keep the transistor in the saturation region of operation.

With the accurate formulas derived in this paper, the Colpitts oscillator design becomes a routine for the designer as of step 3 above. The designer can then concentrate on the less well-defined steps 1 and 2, estimate of parasitics and variability as well as implementation of better inductors.

7.7 CONCLUSIONS

Despite numerous journal and conference papers published in the past 60 years, our understanding of why the phase-noise-to-carrier ratio is inversely proportional to the offset frequency from the carrier has been limited. This paper seeks to argue why the reasoning behind most of the phase-noise models has been based on shaky foundations. As a result, the models' predictions of noise are wrong in one portion of the spectrum or the other for LC oscillators.

In this contribution, the oscillator is described in the time domain without any *a priori* linearity assumption. Noise sources are introduced where they are first added to the signal, to avoid assuming a conclusion (such as noise in phase) before the start of any analysis. The only assumption based more on intuition than rigour was the assumption that for bursty type of sinusoidal interferences, modulation by switching helps the responses reach steady state in much less than the natural decay time of the passive tank. Experiments seem to validate this assumption. All basic equations set up in this paper are based on the Kirchhoff laws. The approximations that have been made include assuming high-Q resonators that will attenuate harmonics of the oscillation voltage on the resonator to negligible levels, a piece-wise square-law V–I characteristic for the MOS transistor, and that the interfering signals are sufficiently small compared to thc main oscillation such that thc duty cycle in which the transistor turns on remains virtually unchanged.

The active device in a typical LC oscillator operates heavily in class-C mode, turning on for only a short fraction of each oscillation period to inject a current pulse into the LC tank to maintain the steady state. In the limit, the oscillation amplitude of any branch voltage within the tank is two times the average current through the branch of the switching active device divided by the transconductance between the two branches at resonance. This limit is independent of the exact V–I characteristic of the active device. Practical high-Q LC oscillators, such as those implemented with BJT or CMOS transistors, have oscillation amplitudes very close to this limit.

The highly nonlinear V–I characteristic of the active device that causes the latter to be switched by the main oscillation modulates any interfering signal to generate a response not only at the frequency of the interferer, but also at its mirror frequency with respect to the carrier. This image response is in turn modulated by the oscillation through the transistor, which creates its own image at the frequency of the original interfering signal. This forms an additional negative feedback loop that cancels any excess positive feedback of the transistor. The response at the interfering frequencies close to the carrier thus becomes inversely proportional to its frequency offset from the latter, irrespective of the exact V–I characteristic and the duty cycle of the switching device, nor the passband width of the passive tank. Such a response and its image are practically equal in amplitude and opposite in polarity. Their combined effect is therefore equivalent to a modulation of the carrier in phase. At higher offset frequencies, the tank impedance becomes very small so that very little voltage is produced at the transistor input. The transistor's feedback current, which contains the image response due to modulation, is insignificant compared to the interfering current. The voltage response to the interference is therefore now virtually the same as that of a passive tank, which still decreases with the inverse of the offset frequency, with negligible image component. The effect of noise — far away from the carrier is therefore merely additive. The cross-over point between dominance of phase noise and dominance of additive noise is a function of the transistor and tank parameters.

Because of the existence of image response, phase noise due to white noise is 3 dB higher than that due to noise components at frequencies above (or below) the carrier alone. The switching nature of the transistor also makes the noise generated by the latter depend on the duty cycle, and hence on the parameters of the rest of the oscillator. All noise sources, including that of the switching transistor, have been calculated in this chapter. The resulting expression for the phase-noise-to-carrier ratio comprises therefore only original circuit parameters instead of intermediate variables that require determination by simulation or experiments.

Controlled experiments have been carefully constructed to verify the theory derived in the paper. The predicted oscillation amplitude is consistently within 1 dB of measurements. The predicted duty cycle also matches measured values by the same accuracy. The response of the oscillator to an injected interfering current as well as the image response agree with what the theory predicts within the accuracy with which parameters of the passive tank can be measured. In the 80 MHz range, this accuracy is better than 1 dB. The cross-over frequency between phase and additive noise, and indeed its very existence, have been confirmed by experiments. in practical designs Measurements of the phase-noise-to-carrier ratio also show that the agreement with the formula derived in this paper is within 1 dB, the repeatability of any phase noise measurement and the accuracy within which circuit parameters have been independently measured. The phase noise model in this paper can therefore be used with confidence in practical designs.

Although the specific Colpitts oscillator has been used in deriving the formulas in this paper, the basic reasoning behind the setting up of basic equations is not specific to Colpitts oscillators. Although for other LC oscillators, the determination of the amplitude and duty cycle as well as the critical g_m will be different, the basic ideas about why the oscillator's response to interference is phase noise for low and additive noise for high offset frequencies and the role of the image in the inverse dependence of phase noise on offset frequency will remain the same. In this sense, the analysis in this paper is general for all LC oscillators.

Acknowledgments

The assistance of P. Basedau and R. Rheiner in carrying out the experiments presented in this paper is gratefully acknowledged. The author also wishes to thank F. Piazza for useful discussions on oscillator phase noise during the past two years.

References

[1] I. L. Berstein, "On fluctuations in the neighborhood of periodic motion of an auto-oscillating system", *Doklady Akad. Nauk.*, vol. 20, p. 11, 1938.

[2] A. Spälti, "Der Einfluss des thermischen Widerstandsrauschens und des Schrot-teffectes auf die Störmodulation von Oscillatoren", *Bulletin des Schweizerischen Elektrotechnischen Vereins*, vol. 39, pp. 419–27, June 1948.

[3] A. Blaquiére, "Spectre de Puissance d'un Oscillateur Non-Linéaire Perturbé per le Bruit", *Ann. Radio Elect.*, vol. 8, pp. 153–179, August 1953.

[4] Hafner, "The Effect of Noise in Oscillators", *Proc. IEEE*, vol. 54, pp. 179–198, Feb.1966.

[5] A. van der Ziel, *Noise in Solid-State Devices and Circuits*, W. 240–242, Wiley-lnterscience, New York, 1986.

[6] D. B. Leeson, "A Simple Model of Feedback Oscillator Noise Spectrum", *Proc. IEEE*, pp. 329–330, Feb. 1966.

[7] G. Sauvage, "Phase Noise in Oscillators: A Mathematical Analysis of Leeson's Model", *IEEE Trans. on Instrum. and Meas.*, vol. IM-26, Dec. 1977.

[8] J. Everard, "Low-noise power-efficient oscillators: theory and design", *IEE Proceedings*, vol.133, Pt. G, No. 4, pp. 172–180, August 1986.

[9] J. Craninckx and M. Steyaert, "Low Noise Voltage-Controlled Oscillators Using Enhanced LC-Tanks", *IEEE Trans. Circ. & Syst.*, vol. 42, pp. 794–804, Dec.1995.

[10] A. Hajimiri and T. Lee, "A General Theory of Phase Noise in Electrical Oscillators", *IEEE J. Solid-State Circ.*, vol. 33, No. 2, pp. 179–194, Feb. 1998.

[11] N. Nguyen and R. Meyer, "A 1.8-GHz Monolithic LC Voltage-Controlled Oscillator", *IEEE J. Solid-State Circ.*, vol. 27, No. 3, pp. 444–450, March 1992.

[12] N. Nguyen and R. Meyer, "Si IC-Compatible Inductors and LC Passive Filters", *IEEE J. Solid-State Circuits*, vol. 25, No. 4, pp. 1028–1031, Aug. 1990.

[13] P. Basedau and Q. Huang, "A 1GHz, 1.5V Monolithic LC Oscillator in 1-μm CMOS", *Proc. 1994 European Solid-State Circuits Conf.*, pp. 172–175, Ulm, Germany, Sept. 1994.

[14] M. Soyuer et al, "A 2.4-GHz silicon bipolar oscillator with integrated resonator", *IEEE J. Solid-State Circuits*, vol. 31, No. 2, pp. 268–270, Feb. 1996.

[15] A. Ali and J. Tham, "A 900MHz Frequency Synthesizer with Integrated LC Voltage-Controlled Oscillator", *ISSCC Digest of Tech. Papers*, pp. 390–391, San Francisco, USA, Feb. 1996.

[16] A. Rofougaran et al, "A 900MHz CMOS LC-Oscillator with Quadrature Outputs", *ISSCC Digest of Tech. Papers*, pp. 392–393, San Francisco, 1996.

[17] M. Soyuer et al, "A 3V 4GHz nMOS Voltage-Controlled Oscillator with Integrated Resonator", *ISSCC Digest of Tech. Papers*, pp. 394–395, San Francisco, USA, Feb. 1996.

[18] B. Razavi, "A 1.8GHz CMOS Voltage-Controlled Oscillator", *ISSCC Digest of Tech. Papers*, pp. 388–389, San Francisco, USA, Feb. 1997.

[19] L. Dauphinee, M. Copeland and P. Schvan, "A Balanced 1.5GHz Voltage Controlled Oscillator with an Integrated LC Resonator", *ISSCC Digest of Tech. Papers*, pp. 390–391, San Francisco, USA, Feb. 1997.

[20] J. Jansen et al, "Silicon Bipolar VCO Family for 1.1 to 2.2 GHz with Fully-Integrated Tank and Tuning Circuits", *ISSCC Digest of Tech. Papers*, pp. 392–393, San Francisco, USA, Feb. 1997.

[21] J. Craninckx and M. Steyaert, "A 1.8 GHz Low-Phase-Noise CMOS VCO Using Optimized Hollow Spiral Inductors", *IEEE J. Solid-State Circuits*, vol. 32, No. 5 pp. 736–744, May 1997.

[22] J. Craninckx and M. Steyaert, "A Fully Integrated Spiral-LC CMOS VCO Set with Prescaler for GSM and DCS-1800 Systems", *Proc. Custom Integrated Circuit Conference*, pp. 403–406, Santa Clara, USA, May 1997.

[23] J. Parker and D. Ray, "A Low-Noise 1.6 GHz PLL with On-Chip Loop Filter", *Proc. CICC*, pp. 407–410, Santa Clara, USA, May 1997.

[24] M. Zannoth et al, "A Fully Integrated VCO at 2 GHz", *ISSCC Digest of Tech. Papers*, pp. 224–225, San Francisco, USA, Feb. 1998.

[25] P. Kinget, "A Fully Integrated 2.7 V 0.35μm CMOS VCO for 5 GHz Wireless Applications", *ISSCC Digest of Tech. Papers*, pp. 226–227, San Francisco, USA, Feb. 1998.

[26] D. Pederson and K. Mayaram, *Analog Integrated Circuits for Communication*, Kluwer Academic Publishers, Boston, 1991.

[27] Qiuting Huang, "Power Consumption vs LO Amplitude for CMOS Colpitts Oscillators", *Proc. CICC*, pp. 255-258, Santa Clara, USA, May 1997.

[28] K. Mayaram and D. Pederson, "Analysis of MOS transformer-coupled oscillators", *IEEE J. Solid-State Circ.*, vol. 22, No. 6 pp. 1155–1162, Dec. 1987.

[29] J. Fischer, "Noise Sources and Calculation Techniques for Switched Capacitor Filters", *IEEE J. Solid-State Circ.*, vol. SC-17, pp. 742–752, Aug. 1982.

[30] Q. Huang and P. Basedau, "Design Considerations for High-Frequency Crystal Oscillators Digitally Trimmable to Sub-ppm Accuracy", *IEEE Trans. On VLSI*, vol. 5, No. 4, pp. 408–416, Dec. 1997.

[31] S. Rogerson, "Where less is more", *Mobile Europe*, vol. 7, pp. 27–30, Feb. 1997.

[32] Q. Huang et al, "The Impact of Scaling Down to Deep Submicron on CMOS RF Circuits", *IEEE J. Solid-State Circuits*, vol. 33, No. 7, pp. 1023–1036, July 1998.

8 DESIGN STUDY OF A 900 MHz/1.8 GHz CMOS TRANSCEIVER FOR DUAL-BAND APPLICATIONS

Behzad Razavi

Electrical Engineering Department
University of California
Los Angeles, USA

Abstract: This chapter describes the design of a CMOS transceiver targeting dual-band applications in the 900 MHz and 1.8 GHz bands. The receiver is based on a modified Weaver architecture, and the transmitter incorporates new methods of quadrature up-conversion. The receiver and the transmitter have been fabricated in a 0.6μm CMOS technology.

8.1 INTRODUCTION

The emergence of various wireless standards within the US and around the world has created a demand for RF transceivers that can operate in more than one mode. The standards deployed in the 900 MHz band are the TDMA standard IS-54 and the CDMA standard IS-95 (US) and the GSM standard (Europe). Additionally, with the availability of the 1.8 GHz band for cellular networks, new standards such as PCS1900 and DCS1800 have been introduced. Furthermore, the Global Positioning System (GPS) (1.5 GHz) and the European cord-less standard DECT (1.9 GHz) provide attractive services that can be incorporated in mobile telephones. In addition to providing "roaming" capability, the use of two or more standards that operate in different bands boosts the capacity of the network, thus lowering the probability of unsuccessful calls during peak traffic hours. Thus, multi-standard RF transceivers are predicted to play a critical role in wireless communications in the 900–2000 MHz range.

This paper describes circuit and architecture design techniques for dual-standard RF transceivers with emphasis on CMOS implementations. Particular attention is paid to issues such as cost, level of integration, form factor, power dissipation, and time to market. These techniques are presented in the context of a dual-band system specifically designed to operate with GSM and DCS1800 standards. Both the receiver and the transmitter have been fabricated in a 0.6μm CMOS technology. The experimental results obtained for each circuit are described in [1–3].

8.2 RECEIVER DESIGN CONSIDERATIONS

Accommodating two or more standards in one receiver generally requires substantial added complexity in both the RF section and the baseband section, leading to a high cost. Thus, the system must be designed so as to share as much hardware as possible. For this reason, both the architecture design and the frequency planning of a multi-standard transceiver demand careful studies and numerous iterations.

The GSM and DCS1800 standards incorporate the same modulation format, channel spacing, and antenna duplexing. Summarized in Table 8.1 are the characteristics of each standard, indicating that a dual-band transceiver can exploit the properties common to both so as to reduce the off-chip hardware.

Table 8.1 System characteristics of GSM and DCS1800.

	GSM	*DCS 1800*
Modulation	Gaussian Minimum Shift Keying	
Multiple Access	Time-Division Multiple Access	
Duplexing	Frequency-Division Duplexing	
Receive Band	935–960 MHz	1805–1880 MHz
Transmit Band	890–915 MHz	1710–1785 MHz
Channel Spacing	200 kHz	200 kHz
Number of Channels	124	350

In order to minimize the number of oscillators and synthesizers, the receivers and transmitters in a dual-band system must be designed concurrently, with the frequency planning carried out such that the receive and transmit paths are driven by the same synthesizers. Although GSM and DCS1800 use frequency-division duplexing (FDD) at the front end, their actual operation is somewhat similar to time-division duplexing (TDD) because their receive and transmit time slots are set off by 1.73 ms (three time slots) [4]. Thus the receiver and the transmitter can time-share the frequency synthesizers.

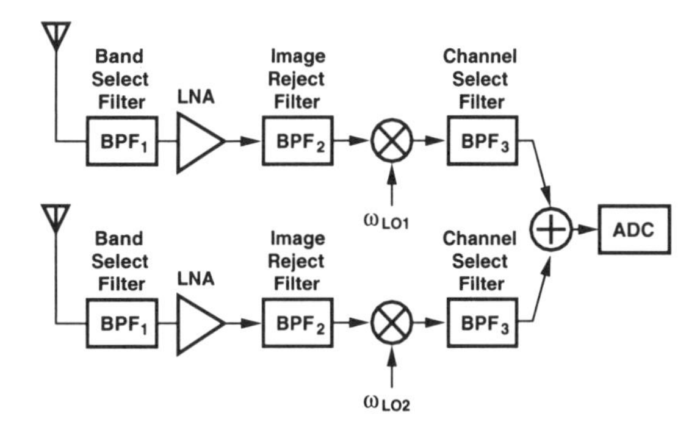

Figure 8.1 Simple dual-standard receiver.

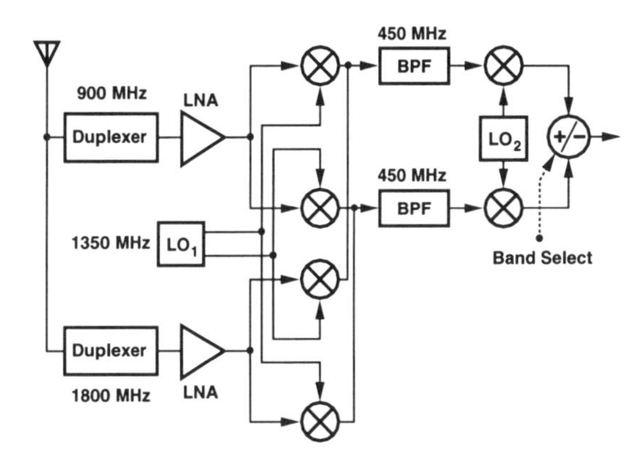

Figure 8.2 Dual-band receiver.

Fig. 8.1 illustrates an example of a dual-standard receiver with two different input frequency bands. After down-conversion to a single intermediate frequency (IF), the two signals are added and digitized. The remaining IF and baseband processing is subsequently performed in the digital domain. Note that the dynamic range and noise of the A/D converter are critical issues in such an environment.

The receiver of Fig. 8.1 requires many off-chip filters. Alternatively, the receiver can be configured as shown in Fig. 8.2 [1], where the 900 MHz and 1.8 GHz bands are processed by two paths containing a low-noise amplifier (LNA) and a quadrature down-conversion mixing operation. The results are subsequently added and, after simple on-chip bandpass filtering, are applied to a second set of quadrature mixers. Note that each quadrature pair of the first set of mixers constitutes a Weaver image-reject receiver together with the second set [5]. Also note that the first local oscillator (LO) frequency is set midway between 900 MHz and 1.8 GHz, i.e., the first mixing operation is with high-side injection for 900 MHz and low-side injection for 1.8 GHz. This choice of frequencies makes the two bands images of each other, allowing the Weaver architecture to select one and reject the other by addition or subtraction of the outputs of the second set of mixers.

Owing to gain and phase mismatches, the image rejection ratio of the Weaver architecture is limited to 30 to 40 dB, inadequate for cellular applications such as GSM and DCS1800. Nevertheless, since for each band the image is 900 MHz away, the front-end duplexer filters provide more than 40 dB of rejection.

The distribution of gain, noise, and nonlinearity in the receiver chain plays a key role in the overall performance, necessitating iterations between architecture design and circuit design. Since channel-selection filtering is postponed to the stages following the IF mixers, the third-order intercept point (IP_3) of each stage must scale according to the total gain preceding that stage. With the initial estimate of the IP_3, the corresponding circuit is designed so as to minimize its noise contribution.

The problem of secondary images [4] constrains the choice of the second IF in the Weaver topology. We consider two scenarios here. The present design provides quadrature down-conversions to allow translation to the baseband. This approach, however, suffers from some of the difficulties encountered in direct-conversion receivers. For example, DC offsets due to the self-mixing of the second LO, flicker noise in the analog baseband circuits, and I and Q mismatch corrupt the down-converted signals [6,7].

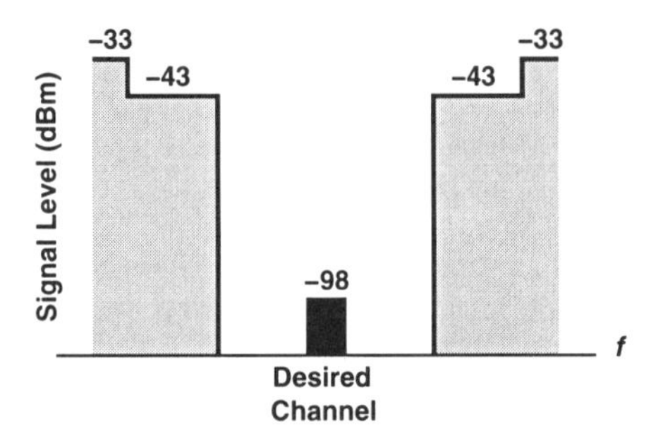

Figure 8.3 GSM receiver interference mask.

Another possibility for the second down-conversion exists by virtue of the specific channelization in GSM and DCS1800. As shown in the receiver interference mask of Fig. 8.3, the adjacent channels in a cell are unoccupied. This suggests that, if the second LO frequency is placed at the edge of the desired channel, then the secondary image coincides with one of the adjacent channels. Illustrated in Fig. 8.4, this technique translates the center of the desired channel to 100 kHz instead of zero. Thus, flicker noise corrupts primarily the edge of the channel, and elimination of DC offsets by AC coupling becomes feasible. Furthermore, the down-conversion in the analog domain need not separate the signal into I and Q phases, thereby avoiding related mismatches.

The above approach does introduce a difficulty: the channel-select filter now requires much sharper selectivity because 43 dBm interferers appear closer to the desired channel. This issue may be resolved by partitioning the filtering between analog and digital domains, albeit at the cost of tightening the linearity requirements of the A/D converter.

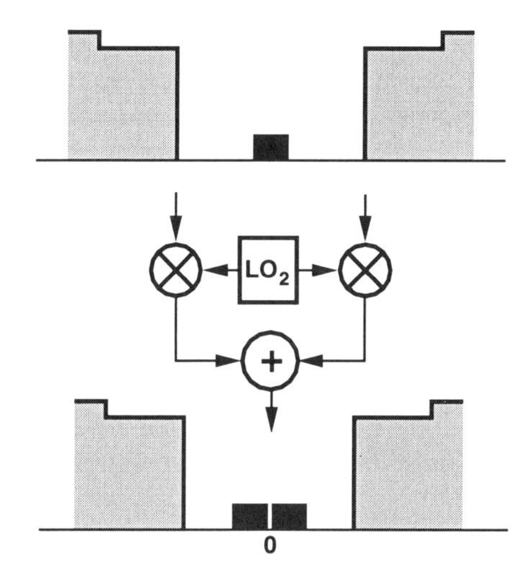

Figure 8.4 Near-baseband down-conversion.

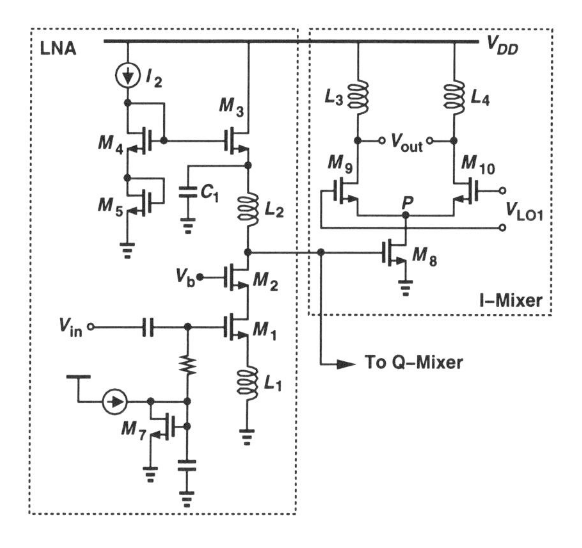

Figure 8.5 LNA/mixer combination.

8.3 RECEIVER BUILDING BLOCKS

8.3.1 LNA/mixer

In order to achieve a relatively low noise figure and a reasonable input match, the LNA employs a common-source cascode stage with inductive degeneration (Fig. 8.5). To avoid uncertainties due to bond-wire inductance, both the source inductor and the drain inductor are integrated on the chip. The LNA draws approximately 5 mA from the supply and exhibits a noise figure of less than 2.5 dB and an IP_3 of greater than -2 dBm in each band. The parasitic capacitance of L_2, the drain junction and overlap capacitance of M_2, and the input capacitance of the mixers resonate with L_2 at the frequency of interest. With a Q of about 3, this resonance lowers the image signal by approximately 10 dB.

The LNA directly drives the quadrature RF mixers, which are configured as single-balanced circuits. Employing inductive loads to minimize thermal noise, each mixer drains 2 mA to achieve a reasonable trade-off between noise and nonlinearity.

With 22 dB of voltage gain in the LNA, it is desirable to realize an IP_3 of greater than +15 dBm (1.58 V_{rms}) in the mixer, while maintaining its input-referred noise voltage below roughly $5\,\mathrm{nV}/\sqrt{\mathrm{V}}$. The dimensions of M_8–M_{10} in Fig. 8.5 has a strong impact on the performance of the RF mixer. Transistor M_8 is sized such that its over-drive voltage is sufficiently large to guarantee the required IP_3. This is in contrast to bipolar implementations, where enormous emitter degeneration would be necessary to achieve an IP_3 greater than +15 dBm. The key point here is that for a given bias current and IP_3, a properly sized MOS transistor exhibits a much higher transconductance than a degenerated bipolar structure. Transistors M_9 and M_{10} in Fig. 8.5 also influence the noise and conversion gain of the mixer. The choice of the width of these devices is governed by a trade-off between their switching time and the parasitic capacitances they introduce at node P. For a given (sinusoidal) LO swing, M_9 and M_{10} are simultaneously on for a shorter period of time as their width increases. A compromise is thus reached by choosing $(W/L)_{9,10} = 400\,\mu\mathrm{m} \times 0.6\mu\mathrm{m}$, allowing the pair to turn off with a differential swing of 100 mV while degrading the conversion gain by less than 1 dB.

The interface between the LNA and the mixer merits particular attention. As shown in Fig. 8.5, to achieve a well-defined bias current in the mixer, the LNA incorporates the DC load M_3 with diode-connected devices M_4 and M_5. Neglecting the DC drop across L_2, we note that $V_{GS4} + V_{GS5} = V_{GS3} + V_{GS8}$. Thus, proper ratioing of M_3 and M_8 with respect to M_4 and M_5 defines I_{D8} as a multiple of I_2. Capacitor C_1 provides AC ground at the source of M_3 so that the output resistance of $M_3[= 1/(g_{m3} + g_{mb3})]$ does not degrade the Q of L_2. Realized as an NMOS transistor, C_1 consists of a large number of gate fingers to reduce the channel resistance, achieving a Q of greater than 30 at the frequency of interest.

In contrast to AC coupling techniques, the above approach incurs no signal loss, but it consumes some voltage headroom. Interestingly, M_3 can serve as the current source for another circuit, e.g. an oscillator, thus re-using the bias current of the LNA.

8.3.2 IF mixer

The differential output of the RF mixers in Fig. 8.5 is capacitively coupled to the input port of the IF mixers, allowing independent biasing. With an overall voltage gain of about 26 dB in the LNA and the RF mixers, the nonlinearity of the IF mixers tends to limit the performance of the receiver. To this end, we note that a differential pair with a constant tail current exhibits higher third-order nonlinearity than a grounded-source pair using the same current and device dimensions. Shown in Fig. 8.6, the IF mixer is configured as a double-balanced circuit consisting of an input pair, a current multiplexer, and a switching quad. The input devices draw a drain current of 1 mA and are sized to sustain an over-drive voltage of 500 mV, thereby achieving an IP_3 of approximately +18 dBm (3.2 V_{rms}). The low transconductance of these transistors together with voltage headroom limitations ultimately results in a slight voltage conversion loss (about −2 dB) in the IF mixer. The switches in the current multiplexer negate the signal current according to the logical state of band while sustaining a drain-source voltage of approximately 35 mV.

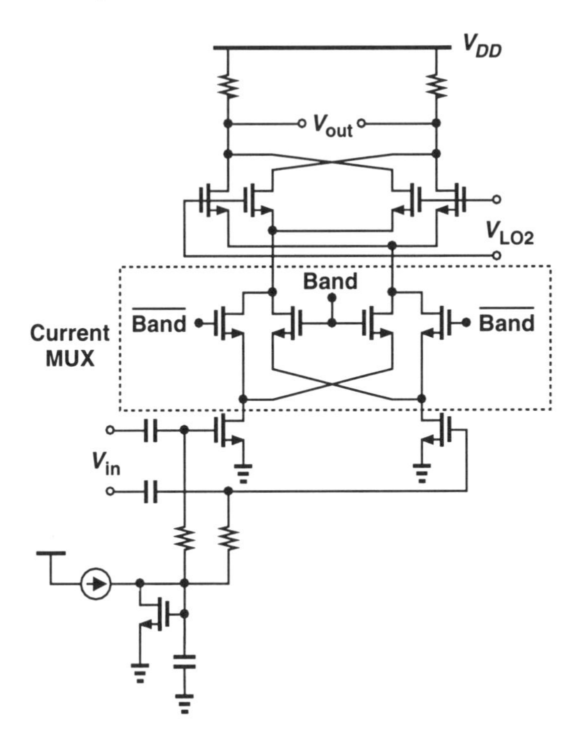

Figure 8.6 IF mixer.

An important concern in heterodyne and image-reject receivers is the translation of various interferers to the desired channel frequency after down-conversion. Owing to nonlinearities and switching operations in each mixer, an interferer at ω_{int} results in components at $k\omega_{int} \pm m\omega_{LO}$. With two down-conversions using ω_{LO1} and ω_{LO2}, the down-converted spurs appear at $k\omega_{int} \pm m\omega_{LO1} \pm n\omega_{LO2}$ (Fig. 8.7), many of which may fall into the desired baseband channel. Since in-band interferers are not filtered before channel selection, and since they are located on the same side of ω_{LO1} as the desired signal, they are not suppressed by the image-rejection technique used in the receiver. It is also important to note that the spurious response is not exercised in a simple noise figure measurement, but it reveals the performance of the receiver in a realistic environment. The spurious response of the dual-band receiver has been examined with the aid of a spreadsheet program. Five interference frequencies in each band were found to be the most significant sources of down-converted spurs.

Another interesting phenomenon that results from the choice of the two LO frequencies is the in-band leakage to the antenna. The LO_1–IF_1 feed-through and the

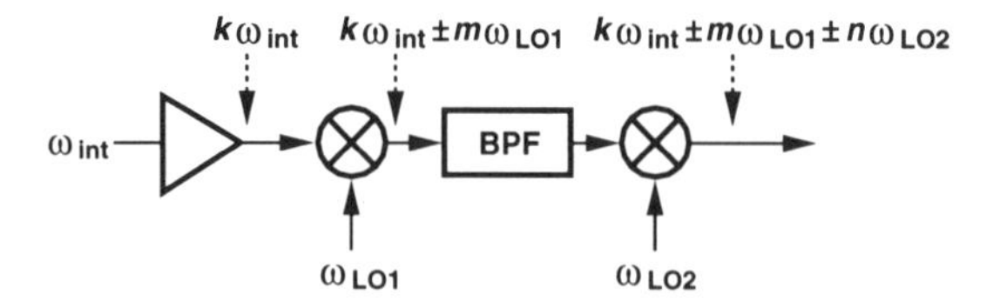

Figure 8.7 Generation of spurs.

low Q of the BPF give rise to a significant 1350 MHz component at the input of the IF mixer. Upon second mixing, this component is translated to both 900 MHz and 1800 MHz, potentially appearing as in-band leakage to the antenna(s). However, by virtue of differential signaling from the first IF onward, and by proper low-pass filtering at the output, this type of leakage can be suppressed to acceptably low values.

8.4 TRANSMITTER DESIGN CONSIDERATIONS

In order to minimize the number of oscillators and synthesizers, the receivers and transmitters in a dual-band system must be designed concurrently, with the frequency planning made such that the receive and transmit paths are driven by the same synthesizers. Although GSM and DCS1800 use frequency-division duplexing (FDD) at the front end, their actual operation is somewhat similar to time-division duplexing (TDD) because their receive and transmit time slots are set off by 1.73 ms (three time slots) (Fig. 8.8). Thus, frequency synthesizers can be time-shared between the receiver and the transmitter. Since the dual-band receiver incorporates LO frequencies of 1350 MHz and 450 MHz, it is desirable to utilize the same frequencies for the transmit path as well.

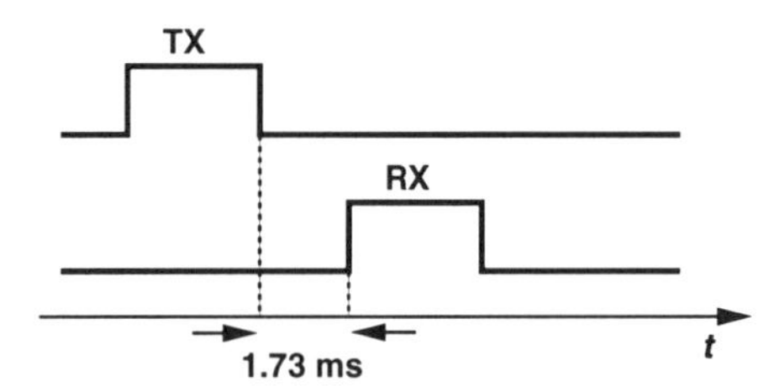

Figure 8.8 Time offset between receive and transmit time slots.

Before considering suitable transmitter architectures, we review Gaussian minimum shift keying (GMSK) modulation to arrive at some of the design implications. Fig. 8.9 conceptually illustrates frequency shift keying (FSK) and GMSK. In FSK, rectangular baseband pulses are directly applied to a frequency modulator, e.g. a volt-

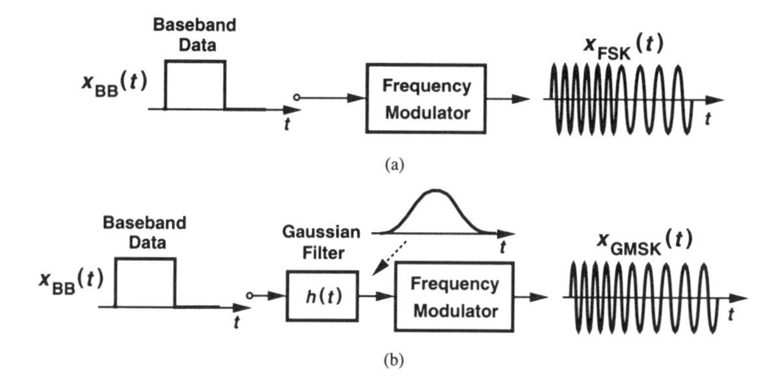

Figure 8.9 Generation of FSK and GMSK signals.

age-controlled oscillator (VCO), thereby creating an output waveform given by

$$x_{FSK}(t) = A\cos[\omega_c t + K_0 \int x_{BB}(t)dt], \tag{8.1}$$

where K_0 is a constant denoting the "depth" of modulation and $x_{BB}(t)$ represents the baseband signal.

An important drawback of FSK is the large bandwidth occupied by the modulated signal, partly because of the abrupt transitions in the frequency introduced by the sharp edges of the baseband pulses. We expect that if the frequency changes more smoothly from one bit to the next, then the required bandwidth decreases. In fact, the spectrum of the signal expressed by Eq. 8.1 decays in proportion to $f^{2(n+3)}$, where n denotes the highest continuous derivative of $x_{BB}(t)$ [9]. Based on this observation, GMSK modulation alters the shape of the baseband pulses so as to vary the frequency gradually. As shown in Fig. 8.9(b), the rectangular pulses are first applied to a Gaussian filter, thereby generating smooth edges at the input of the frequency modulator. The resulting output is expressed as

$$x_{GMSK}(t) = A\cos[\omega_c t + K_0 \int x_{BB}(t) * h(t)dt], \tag{8.2}$$

where $h(t) = \exp(-t^2/\tau^2)$ is the impulse response of the Gaussian filter.

The conceptual method described by Fig. 8.9(b) and Eq. 8.2 is indeed employed in some transmitters, e.g. for the Digital European Cordless Telephone (DECT) standard. However, if the amplitude of the baseband signal applied to the VCO or the gain of the VCO are poorly controlled, so is the bandwidth of the modulated signal. For this reason, in high-precision systems such as GSM, the waveform in Eq. 8.2 is rewritten as

$$x_{GMSK}(t) = A\cos\omega_c t\cos\theta - A\sin\omega_c t\sin\theta, \tag{8.3}$$

where $\theta = K_0 \int x_{BB}(t) * h(t)dt$, and $\cos\theta$ and $\sin\theta$ are generated using accurate mixed-signal techniques [4, 10]. Equation 8.3 forms the basis for our transmitter design.

In order to employ 450 MHz and 1350 MHz LO frequencies, we postulate that the transmitter must incorporate two up-conversion steps: from baseband to an intermediate frequency (IF) of 450 MHz and from 450 MHz to 900 MHz or 1.8 GHz. We also recognize that a simple mixer driven by the 450 MHz IF and the 1350 MHz LO generates the 900 MHz and 1.8 GHz signals with *equal* amplitudes, necessitating substantial filtering to suppress the unwanted component. It is therefore desirable to perform the second up-conversion by single-sideband (SSB) mixing.

With the foregoing observations, we consider the topology shown in Fig. 8.10 as a possible solution. The baseband I and Q signals are converted up to 450 MHz and subsequently separated into quadrature phases by means of an RC–CR network, resulting in $V_M = A\cos(\omega_1 t - \theta - \pi/4)$ and $V_N = A\sin(\omega_1 t - \theta - \pi/4)$. Single-sideband mixing of the IF and the second LO signals is then carried out by two mixers, with their outputs added or subtracted so as to produce the 900 MHz or 1800 MHz output according to the band select command.

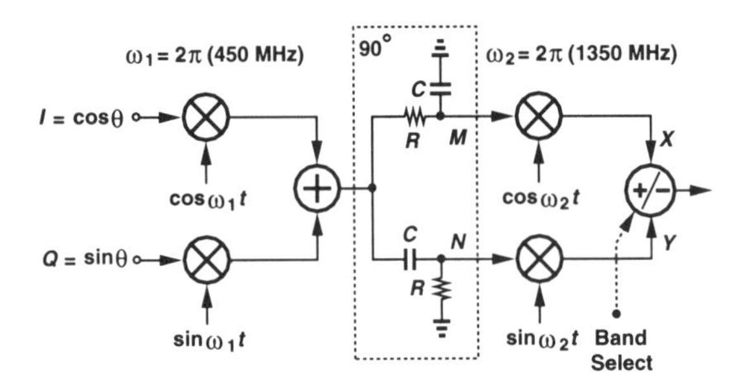

Figure 8.10 Simple dual-band transmitter.

The architecture of Fig. 8.10 provides a compact solution for dual-band operation, but it suffers from several drawbacks. First, the RC–CR network introduces a loss of 3 dB in the signal path and, more importantly, loads the first up-converter. Second, both of the outputs appear at the same port, making it difficult to utilize tuned narrowband amplification at this port. Third, even with perfect matching between the quadrature paths in the SSB mixer, the variation of the absolute value of RC with process and temperature leads to a considerable amplitude mismatch between V_M and V_N, thereby creating a significant unwanted sideband at the output. For example, a 20 % error in RC results in an unwanted sideband only 20 dB below the wanted component.

The existence of an unwanted sideband 900 MHz away from the desired signal may seem unimportant, because various filtering operations in the following power amplifier (PA) and matching network provide further suppression. However, second-order distortion in the PA — a significant effect, because PAs are typically single-ended — may lead to a troublesome phenomenon in the generation of DCS1800 signals. Illustrated in Fig. 8.11, the issue arises because the second harmonic of the 900 MHz sideband falls into the transmitted DCS1800 channel. Since, from Carson's rule [8], the second harmonic of a frequency-modulated signal occupies roughly twice as much bandwidth as the first harmonic, the 1800 MHz output may exhibit substantial

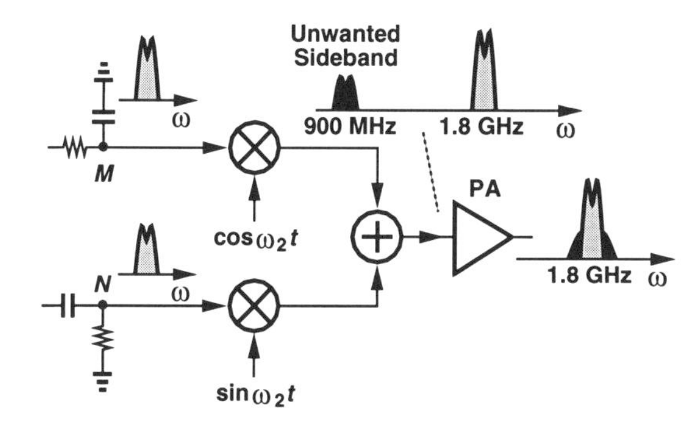

Figure 8.11 Effect of unwanted sideband on DCS1800 output.

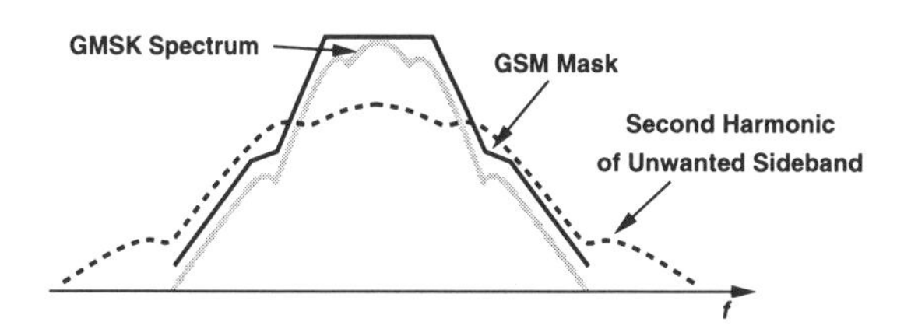

Figure 8.12 Increase in adjacent-channel power due to second harmonic of unwanted sideband.

adjacent-channel power (ACP), violating the transmission mask (Fig. 8.12). Thus, the unwanted sideband produced by the SSB mixer must be sufficiently small.

In summary, the architecture of Fig. 8.10 requires two modifications: (1) the IF quadrature generation must avoid the use of RC–CR networks, and (2) the GSM and DCS1800 paths must be separated at some point such that each can incorporate narrowband tuning.

In order to produce the quadrature phases of the 450 MHz IF signal, we recognize that the *baseband* signal is available in quadrature phases, namely, $\cos\theta$ and $\sin\theta$ in Eq. 8.3. The IF signal can thus be generated in quadrature form as depicted in Fig. 8.13, where proper choice of the phases together with addition or subtraction at the output yields both $\cos(\omega_1 t + \theta)$ and $\sin(\omega_1 t + \theta)$. Compared to the circuit of Fig. 8.10, this configuration both provides higher gain balance between the two paths and avoids the loss and loading of the RC–CR network while using two more mixers. The additional mixers consume more power, but since the conversion gain is higher in this case, the following SSB mixers require less power, leading to an overall power dissipation comparable to that of Fig. 8.10.

The 450 MHz outputs of the first up-converter can now be multiplied by the quadrature phases of the second LO and added or subtracted to generate the GSM and DCS1800 signals. As mentioned above, it is preferable to design the second up-conversion modulators such that the 900 MHz and 1.8 GHz waveforms appear at the outputs of two different circuits, thus allowing efficient narrowband amplification. This is

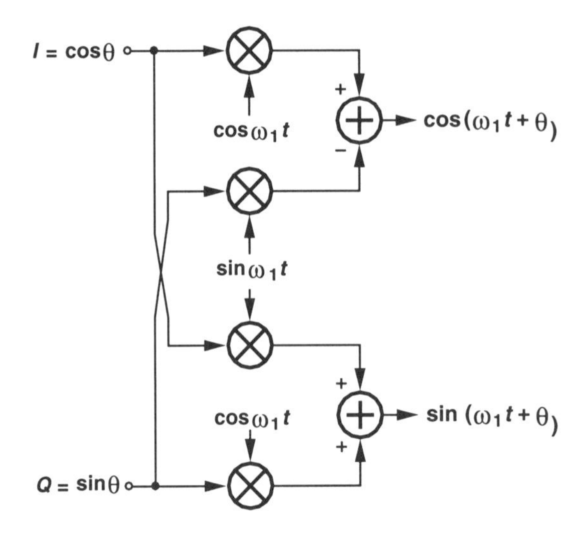

Figure 8.13 Up-conversion with quadrature outputs.

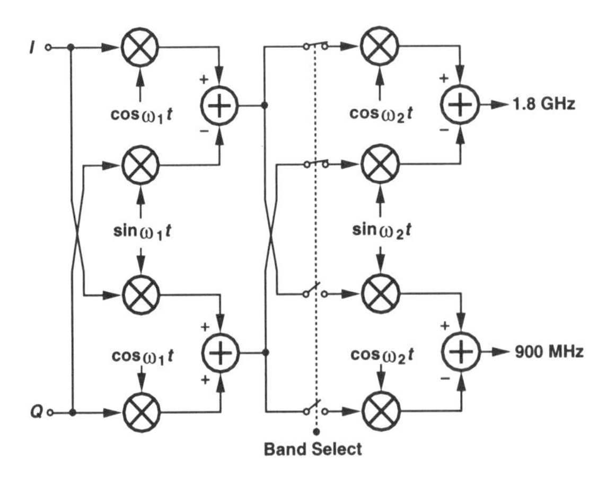

Figure 8.14 Two-step up-conversion generating 900 MHz and 1.8 GHz outputs.

accomplished as shown in Fig. 8.14, where two independent SSB modulators produce the two bands according to the band select command. Note that narrowband tuning also suppresses the unwanted sideband resulting from mismatches in the SSB mixers. To save power consumption, only one of the modulators is active in either mode.

The overall architecture of the dual-band transmitter is shown in Fig. 8.15. Since all of the signals up to ports A and B are differential, each band incorporates a differential to single-ended (D/SE) converter, applying the result to an output buffer. The transmitter requires external power amplifiers to deliver the high power levels specified by GSM and DCS1800.

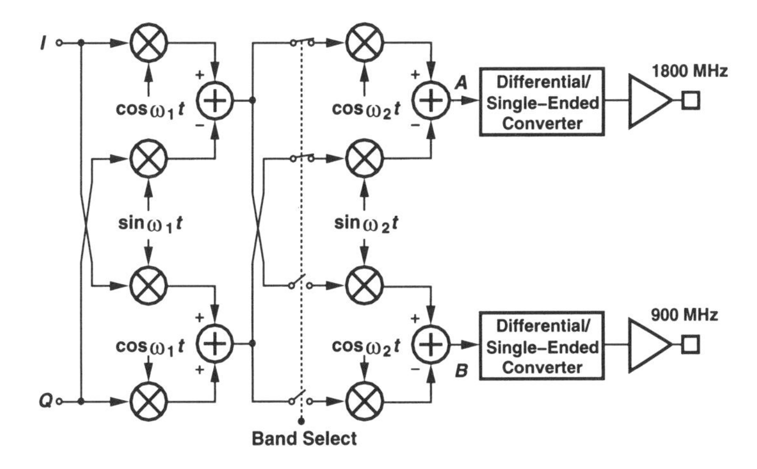

Figure 8.15 Dual-band transmitter architecture.

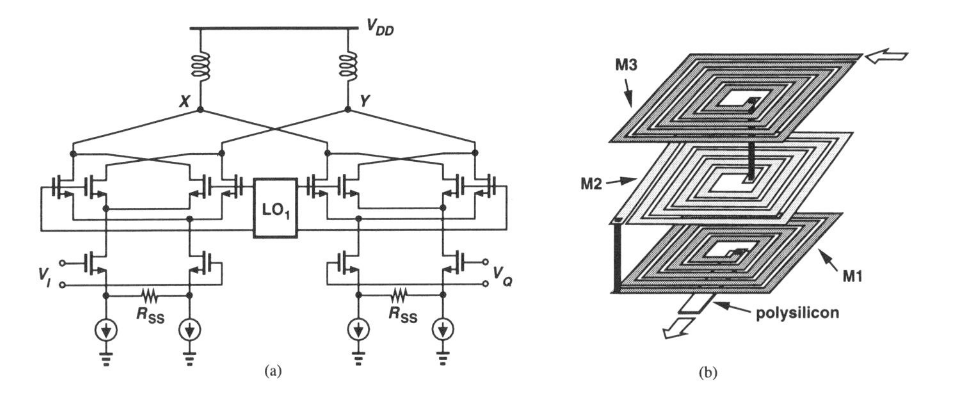

Figure 8.16 (a) First up-conversion modulator, (b) implementation of load inductors.

8.5 TRANSMITTER BUILDING BLOCKS

In this section, we describe the transistor-level implementation of each building block, emphasizing the design constraints imposed by the architecture. The circuit topologies are identical for both bands, but device dimensions and bias currents are chosen to optimize the performance of each.

8.5.1 First up-conversion

The 450 MHz up-conversion modulator consists of two Gilbert cell mixers whose outputs are added in the current domain. Shown in Fig. 8.16, the circuit utilizes resistive source degeneration, thereby improving the linearity in the baseband port of each

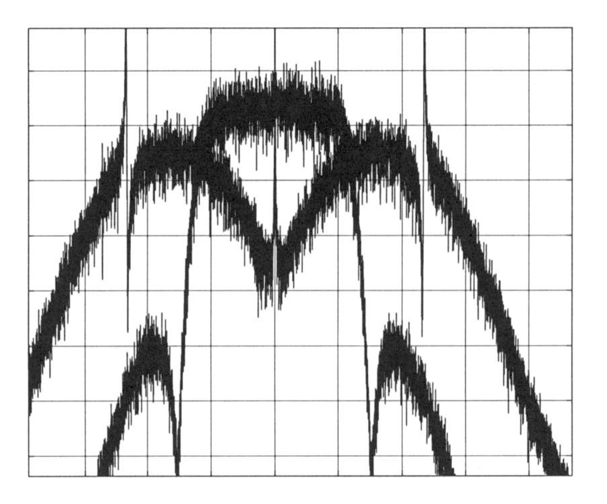

Figure 8.17 Simulated spectra of the two terms in Eq. 8.5 (Horiz. scale normalized to bit rate, vert. scale: 5 dB/div).

mixer. Two 100 nH inductors convert the output current to voltage. To minimize the area occupied by each inductor, a stack of three spiral structures made of three metal layers (Fig. 8.16(b)) is used [1], reducing the area by approximately a factor of 8 [11]. Since the polysilicon connection and the bottom spiral suffer from substantial parasitic capacitance to the substrate, this node is connected to the supply voltage, increasing the self-resonance frequency of the inductor. The quality factor of the inductor is estimated to be about 4 and the self-resonance frequency is about 600 MHz.

Why must the baseband ports be linearized? Let us return to Eq. 8.3 with the assumption that $\cos\theta$ and $\sin\theta$ experience third-order distortion. The resulting IF signal can then be expressed as

$$x_{\mathrm{GMSK}}(t) = A\cos\omega_c t[\cos\theta + \alpha\cos(3\theta)] - Asin\omega_c t[\sin\theta + \alpha\sin(3\theta)], \tag{8.4}$$

where α represents the amount of third-order nonlinearity. Grouping the terms in Eq. 8.4, we obtain

$$\begin{aligned} x_{\mathrm{GMSK}}(t) &= A\cos[\omega_c t + K_0 \int x_{\mathrm{BB}}(t) * h(t)dt] \\ &+ \alpha A\cos[\omega_c t + 3K_0 \int x_{\mathrm{BB}}(t) * h(t)dt]. \end{aligned} \tag{8.5}$$

Equation 8.5 reveals that third-order distortion gives rise to a component centered around ω_c, but with a modulation index three times that of the ideal GMSK signal. Invoking Carson's rule, we postulate that the second term occupies roughly three times the bandwidth, raising the power transmitted in adjacent channels. Fig. 8.17 shows the simulated spectra of the two components in Eq. 8.5 with $\alpha = 1$, indicating that the unwanted signal indeed consumes a wider band. For this reason, as depicted in Fig. 8.18, α must be small enough such that the transmission mask is not violated.

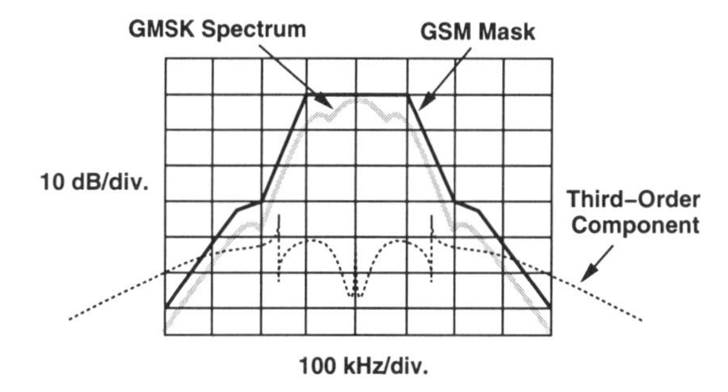

Figure 8.18 Effect of harmonic distortion at baseband ports.

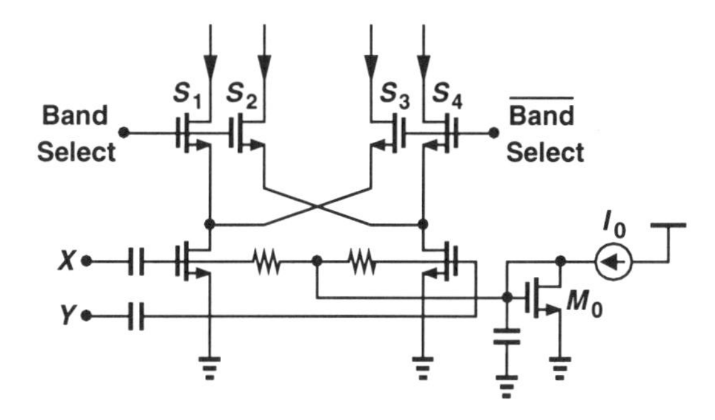

Figure 8.19 Voltage-to-current converter with output switching.

In this design, the resistive degeneration and tail currents are chosen so as to ensure $\alpha \approx 0.01$ with a 0.5 V_{pp} baseband input, yielding a third-order component 40 dB below the desired signal.

8.5.2 SSB modulator

The signals generated at nodes X and Y in Fig. 8.16 must be "routed" to one of the SSB modulators according to the band select command. As illustrated in Fig. 8.19, the routing is performed in the current domain to minimize signal loss due to the addition of the switches. Capacitively coupled to the output of the 450 MHz up-converter, the voltage–to–current converter employs grounded-source input devices to save the voltage headroom otherwise consumed by a tail current source. The bias current of the circuit is defined by M_0 and I_0. Note that S_1–S_4 operate in deep triode region, sustaining a small voltage drop. Also, the linearity of this and subsequent stages is not critical, because GMSK signals display a constant envelope and are quite insensitive to spectral regrowth [4].

With the quadrature phases of the IF signal available in the current domain, SSB mixing assumes a simple topology. Shown in Fig. 8.20, the circuit senses the differential current signals routed from each 450 MHz up-converter, performs mixing with the

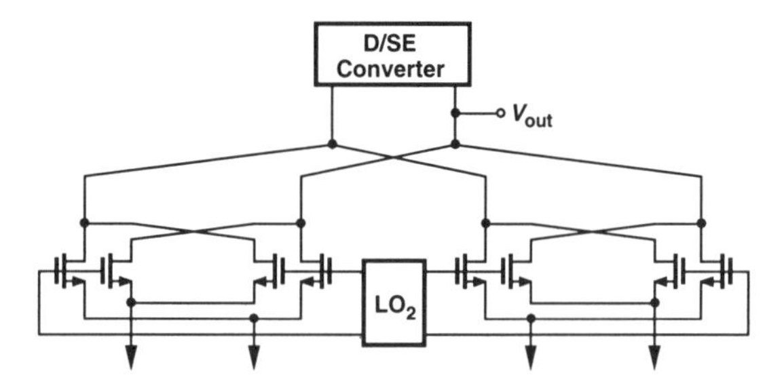

Figure 8.20 Single sideband modulator circuit.

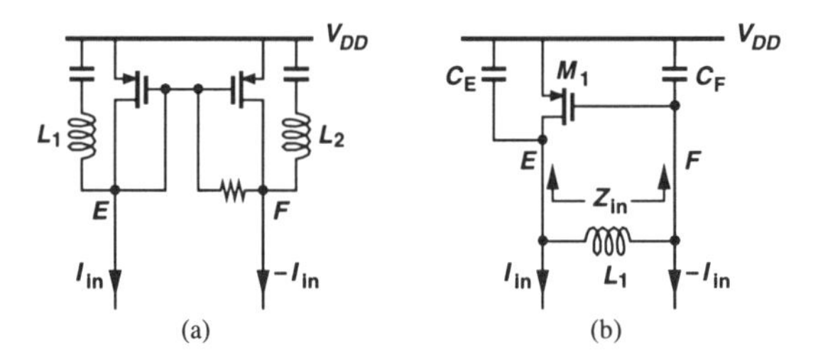

Figure 8.21 Differential to single-ended conversion using (a) tuned current mirror, (b) negative resistance generator.

second LO, adds the resulting currents with proper polarity, and converts the output to single-ended form.

8.5.3 Differential to Single-Ended Converter

In order to achieve a reasonable gain, it is desirable to employ tuning in the D/SE converter. Fig. 8.21 depicts two realizations of such a circuit. In Fig. 8.21(a), a current mirror together with two inductors creates resonance at nodes E and F, reducing the effect of device capacitances. The difficulty here is the large gate–source capacitance of the two PMOS devices, mandating a small value for L_1 and hence a low conversion gain. Fig. 8.21(b) presents an alternative topology where a PMOS device introduces a negative resistance in parallel with a floating inductor. It can be shown that

$$Z_{in} = \frac{g_{m1}}{C_E C_F s^2} + \frac{1}{C_E s} + \frac{1}{C_F s}, \tag{8.6}$$

whcrc C_E and C_F denote the total capacitance at nodes E and F, respectively [4]. As a compromise between margin to oscillation and boost in gain, the negative resistance, $-g_{m1}/(C_E C_F \omega^2)$, is chosen so as to increase the Q of the inductor by approximately a factor of two. Note that L_1 can assume a relatively large value because it sees C_E and C_F in *series*. In this design, the signal is sensed at F because this port exhibits a lower output impedance. Simulations indicate that the topology of Fig. 8.21(b) provides about three times the voltage gain of the circuit in Fig. 8.21(a).

8.5.4 Output buffer

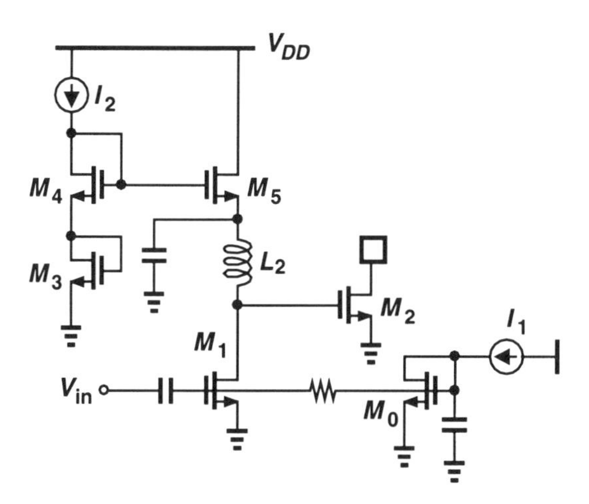

Figure 8.22 Output buffer.

The output buffer is shown in Fig. 8.22. Two common-source stages, M_1 and M_2, boost the signal level, driving the 50 Ω impedance of the external instrumentation. The bias current of M_1 is defined by I_1 and M_0 and that of M_2 by I_2 and M_3. Neglecting the AC drop across the inductor, we have $V_{GS3} + V_{GS4} = V_{GS5} + V_{GS2}$; that is, I_{D2} can be ratioed with respect to I_2.

8.6 CONCLUSION

The design of dual-band transceivers poses many challenges in terms of frequency planning and circuit building blocks. An image-reject receiver and a two-step transmitter architecture have been introduced that provide dual-band operation with 450 MHz and 1350 MHz LO frequencies. Also, circuit techniques for low-noise amplification, down-conversion mixing, generation of the quadrature phases of the transmit IF signal, and differential to single-ended conversion have been presented.

References

[1] S. Wu and B. Razavi, "A 900-MHz/1.8-GHz CMOS Receiver for Dual-Band Applications", *IEEE J. Solid-State Circuits*, vol. 33, pp. 2178–2185, Dec. 1998;

[2] B. Razavi, "A 900-MHz/1.8-GHz CMOS Transmitter for Dual-Band Applications", *Symposium on VLSI Circuits Dig. of Tech. Papers*, pp. 128–131, June 1998;

[3] B. Razavi, "A 900-MHz/1.8-GHz CMOS Transmitter for Dual-Band Applications", *IEEE Journal of Solid-State Circuits*, vol. 34, pp. 573–579, May 1999;

[4] B. Razavi, *RF Microelectronics*, Upper Saddle River, NJ: Prentice-Hall, 1998;

[5] D. K. Weaver, "A Third Method of Generation and Detection of Single-Sideband Signals", *Proc. IRE*, vol. 44, pp. 1703–1705, Dec. 1956;

[6] B. Razavi, "Design Considerations for Direct-Conversion Receivers", *IEEE Trans. Circuits and Systems, Part II*, vol. 44, pp. 428–435, June 1997;

[7] A. A. Abidi, "Direct-Conversion Radio Transceivers for Digital Communications", *IEEE Journal of Solid-State Circuits*, vol. 30, pp. 1399–1410, Dec. 1995;

[8] L. W. Couch, *Digital and Analog Communication Systems*, Fourth Edition, New York: Macmillan Co., 1993;

[9] J. B. Anderson, T. Aulin, and C.-E. Sundberg, *Digital Phase Modulation*, New York: Plenum Press, 1986;

[10] K. Feher, *Wireless Digital Communications*, New Jersey: Prentice-Hall, 1995;

[11] R. B. Merril, *et al.*, "Optimization of High Q Inductors for Multi-Level Metal CMOS", *Proc. IEDM*, pp. 38.7.1–38.7.4, Dec. 1995;

9 INTEGRATED WIRELESS TRANSCEIVER DESIGN WITH EMPHASIS ON IF SAMPLING

Mihai Banu[1], Carlo Samori[3], Jack Glas[1], and John Khoury[2]

[1]Bell Laboratories, Lucent Technologies
600 Mountain Ave., Murray Hill, NJ 07974, USA

[2]The Microelectronics Group, Lucent Technologies
1247 S Cedar Crest Blvd., Allentown, PA 18103, USA

[3]Politecnico di Milano, Dipartimento di Elettronica e Informazione
Piazza L. da Vinci 32, 20133 Milano, Italy

9.1 INTRODUCTION

Digital wireless systems such as GSM are in wide use worldwide and will continue to expand and replace the remaining analog systems at an increasing rate. The technical factors that have contributed to this success are fundamental capabilities such as easy use of SSB (single-side-band) signals, efficient channel multiplexing (time or code division), increased capacity through compression, resilience against interference and noise, and secure transmission. In addition, the development of relatively inexpensive and low-power wireless transceivers has triggered a great proliferation of consumer portable units. Several key technologies are enabling the successful design and manufacture of these devices. Examples include GHz-band Si RF ICs, mixed-signal CMOS

VLSI ICs, high-quality RF and IF passives, inexpensive plastic packaging, and DSP-based processing.

The block diagram of a typical modern transceiver is shown in Fig. 9.1. It consists

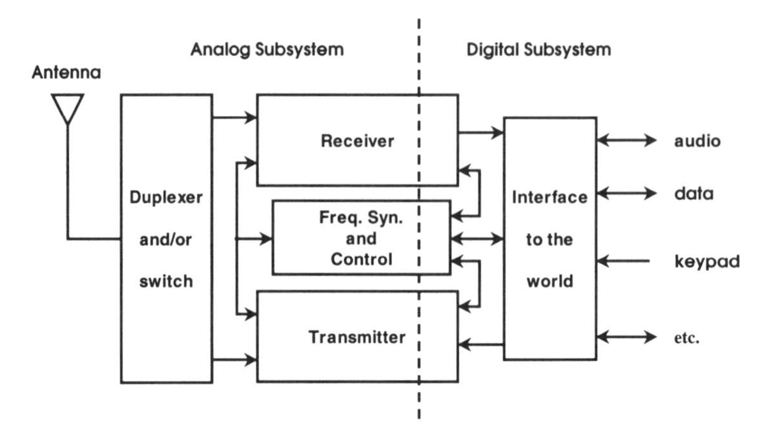

Figure 9.1 A typical wireless transceiver.

of analog RF, IF, and baseband sections and a digital section. The analog/digital interface (i.e. A/D and D/A converters) is placed at baseband, a choice dictated primarily by cost and power dissipation considerations. Despite the usual preference for "digital whenever possible" designs, it is still to be shown that moving the analog/digital interface away from DC and closer to the antenna results in superior transceivers for hand held applications, for which cost and power dissipation are essential design specifications. The hope is that, in the near future, advances in circuit fabrication technology or new radio design techniques will lead to an increased digitization of the transceiver implementation. In turn, this will bring more robust products with increased reliability and additional programmable features.

Assuming that IF sampling will eventually prevail in portable applications as a step forward in the "more digital" design direction, this paper presents some basic issues and tradeoffs of this approach. Similar aspects have been discussed in other places such as in [1–3]. The emphasis will be on the receive path. It is shown that a major limitation in moving the A/D conversion closer to the antenna comes from the filtering requirements of the front end, a severe condition for inexpensive IC designs. However, contrary to a common belief, IF-sampled receivers do not require more stringent oscillator jitter specifications than conventional receivers. Band-pass sigma-delta A/D converters with noise notch shaping at one quarter of the sampling frequency are well suited for IF sampling applications. In order to explain these facts, it is necessary to review the key aspects of transceiver architectures and design methods.

9.2 INTEGRATED TRANSCEIVER CONVENTIONAL WISDOM

A successful GSM-quality transceiver front-end is shown in Fig. 9.2. The receiver path consists of a single IF superheterodyne tuner and an I/Q IF-to-baseband direct-conversion stage. The transmitter includes a direct modulator with offset. This architecture uses only two oscillators and allows frequency plans with relatively few spurious signals in the RF and IF bands. A high level of integration is possible, but only in the presence of several high-quality off-chip filters. The dividing line between analog and digital is at baseband, where extensive digital processing is applied.

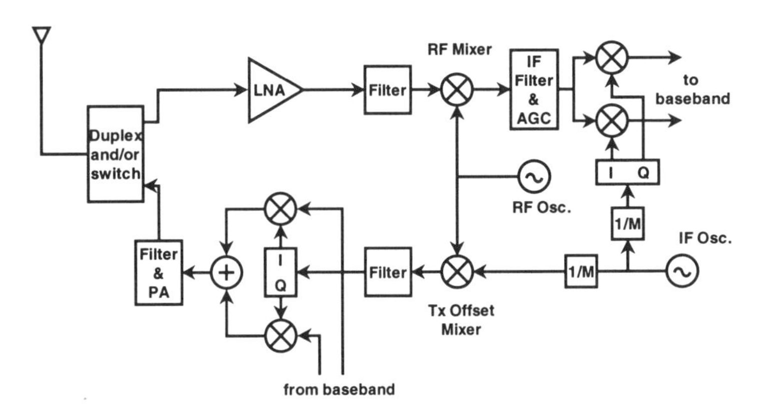

Figure 9.2 A GSM-quality integrated transceiver architecture.

Focusing on the receiver, the typical processing of the worst-case antenna signal is illustrated in Figures 9.3a) and 9.3b). Using analog filtering, AGC, and frequency translation, a narrow frequency band of several channel widths centered on the desired channel is separated from the RF band and shifted to DC. Here, after more analog filtering and amplification, the signal is digitized. The final channel selection, equalization, and demodulation are done in the digital domain. The exact amount of processing at each stage of the receiver is carefully selected to optimize the overall transceiver performance, power dissipation, and cost. For example, it is not economical to separate the desired channel totally at IF since the proper IF filters would be too expensive.

Next, the most important analog design considerations of the receiver will be discussed briefly. The backbone of the receiver's analog part is a cascade of filter-and-amplify sections. In Fig. 9.4 we see a common design tradeoff between filter attenuation and amplifier gain in these sections. The input of each filter or gain block is allowed to contain a certain maximum level of out-of-channel interference according to the signal-swing capabilities of the actual circuits. The amount of post filter amplification is designed such that the maximum interference at the filter output and the maximum interference at its input are equal. In this way the circuit linearity requirements are uniform from section to section. Of course, since the various blocks operate at different frequencies and bandwidths, this is only a first-order rule.

The frequency-conversion mixers connecting the filter-and-amplify sections are directly responsible for several possible sources of receiver errors. The image and IF side-band generation problems are shown in Fig. 9.5 for the RF section. While the IF side bands are very far from the actual IF and can be easily removed, the rejection of the image frequency is a major design consideration. Active image rejection is possible but limited in effectiveness (30–40 dB maximum) and power consuming. The most common practice is to use passive filters, as shown in Fig. 9.6. The nonlinear

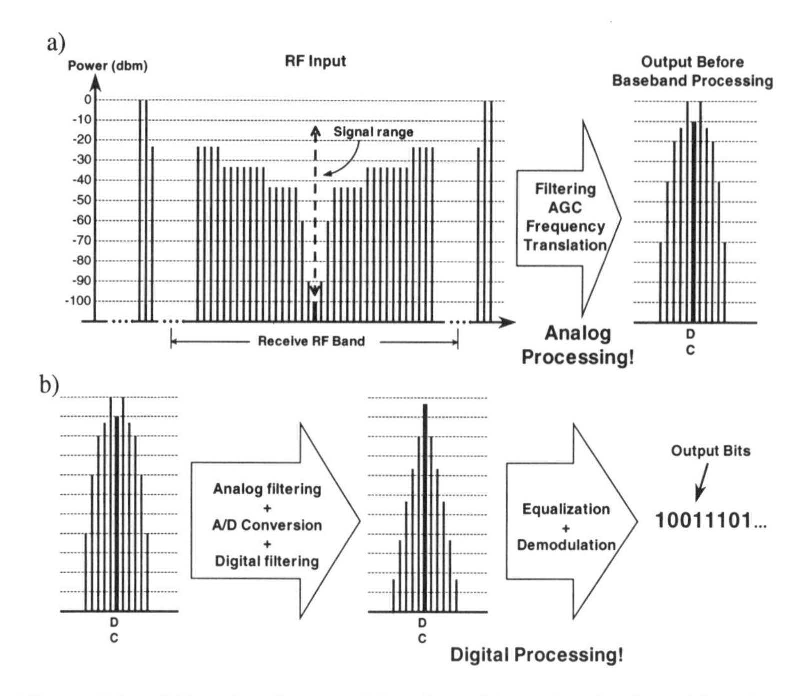

Figure 9.3 a) Receiver front-end functions, b) receiver back-end functions.

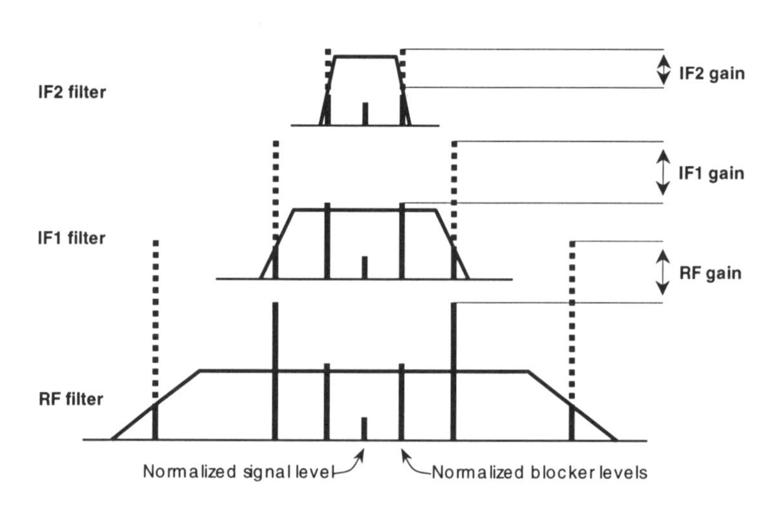

Figure 9.4 Filtering-gain tradeoff (qualitative schematic).

character of the actual mixer creates a further "image" problem due to mixing of LO harmonics with input signal harmonics. Fig. 9.7 demonstrates the generation of such spurious responses at the first IF. Therefore, the choice of IF frequencies is of utmost importance and involves compromises between filtering complexity and mixer spuri-

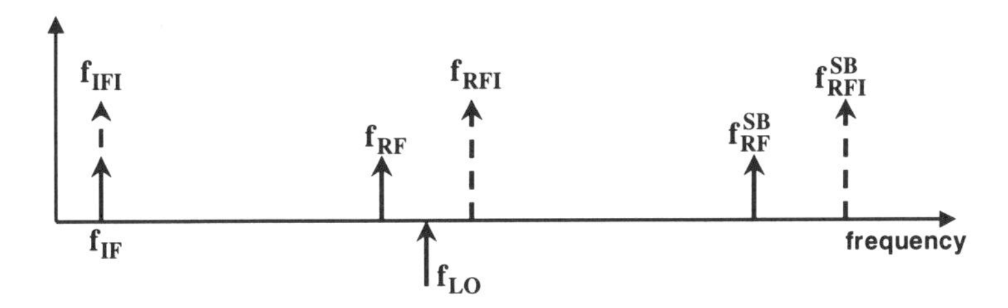

Figure 9.5 Mixer image and side bands problems.

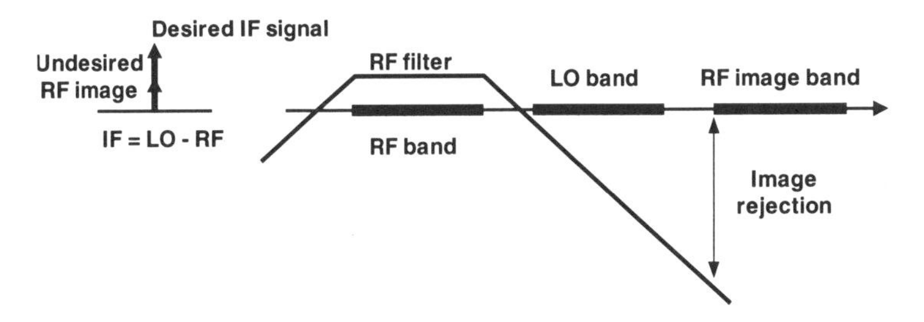

Figure 9.6 Mixer passive image rejection.

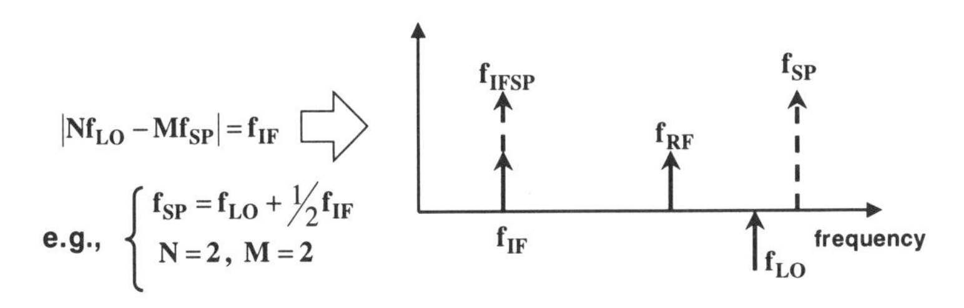

Figure 9.7 Mixer spurious IF response.

ous characteristics. Furthermore, the presence of DC offsets and practical mismatches limit the performance of the final I/Q conversion step.

Another important aspect in receiver design is the reciprocal mixing effect. Fig. 9.8 shows how the LO noise at offset frequencies is mixed to IF by large RF interfering signals (blockers). This is an important factor in the calculations of oscillator noise requirements and receiver sensitivity in the presence of many channels.

9.3 IF SAMPLING: CONDITIONS AND LIMITATIONS

The natural evolution of the receiver from Fig. 9.2 into an IF-sampled architecture is shown in Fig. 9.9. A first superficial examination would show that it is reasonable to expect this receiver to maintain all advantages of the original design with the additional features of perfect I/Q matching, no baseband DC offsets, and requiring a single

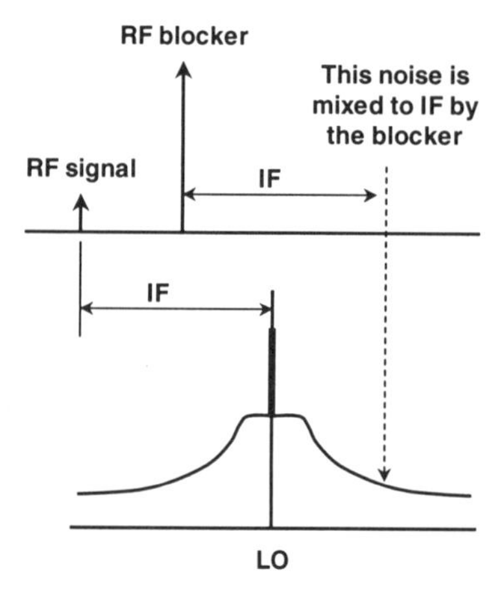

Figure 9.8 Reciprocal mixing.

A/D converter. These attractive characteristics are traded for higher speed in the A/D conversion operation and the addition of digital processing. It appears that, ultimately, the power dissipation required in actual implementations will determine the success of this approach. The power dissipation depends on circuits used and will improve with the scaling of IC technologies. However, even without assuming any specific IC capabilities, we can make several important observations about IF sampling including the realization that the previous examination of Fig. 9.9 is overly simplified.

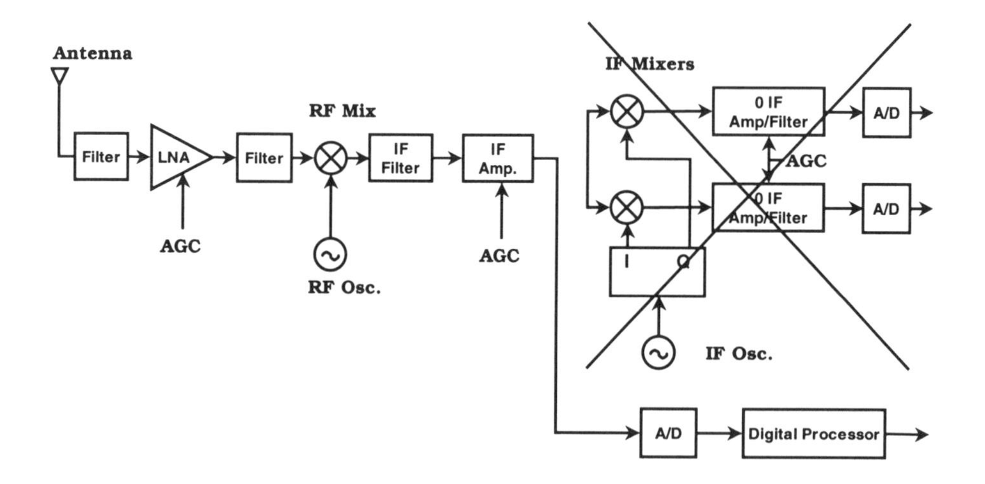

Figure 9.9 IF sampling receiver architecture.

The mixing of IF to DC, as shown in Fig. 9.2, has the important advantage of not requiring image rejection of any interfering signals. In addition, IF mixers and baseband anti-aliasing filters are relatively easy to design. When a sampler replaces these mixers, the IF filter is required to provide substantially more stop-band attenuation to avoid aliasing of blockers. This affects the design and the cost of the transceiver. The sampler implementation, which is more difficult than that of the mixer, has similar consequences. We have fundamentally conflicting requirements: a high IF makes anti-aliasing and RF image rejection filtering easy and the sampler design difficult, and vice versa.

For a given sampling frequency f_S, the maximum IF bandwidth is obtained for IF at $f_S/4$. This is shown in Fig. 9.10. Since the first IF is usually 70 MHz or more, 4×-sampling circuits are difficult to implement. A better possibility is to apply under-sampling. If the receiver's first IF is placed at an appropriate integer multiple of $f_S/4$, where f_S is a reasonably low sampling frequency, we obtain the same signal spectrum as in Fig. 9.10.

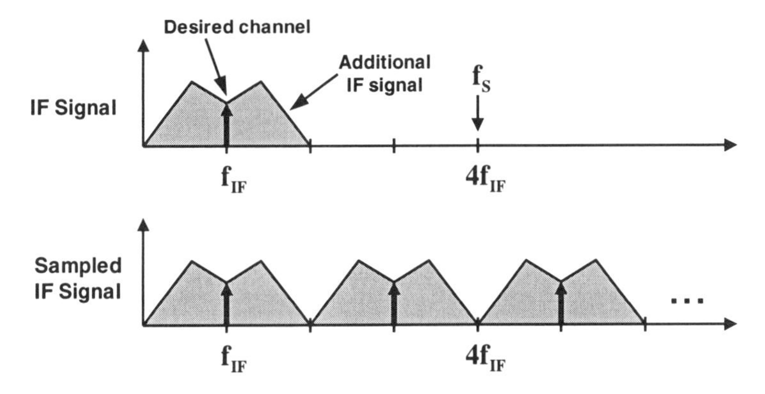

Figure 9.10 4X sampling of IF signals.

This is equivalent to down-converting the signal to $f_S/4$ and then sampling. Naturally, the anti-aliasing requirements of the IF filter become increasingly difficult to meet at higher under-sampling ratios, as illustrated in Fig. 9.11.

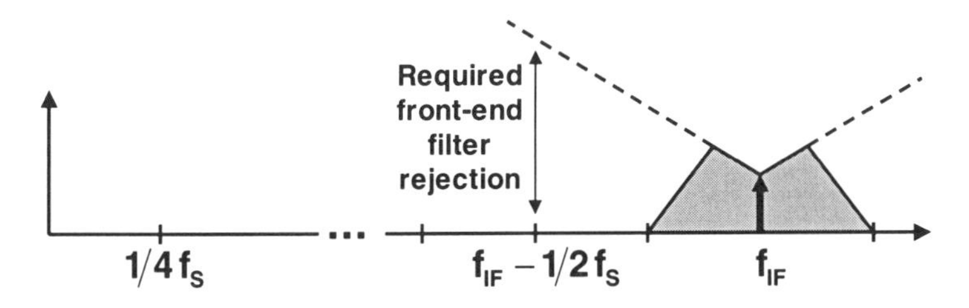

Figure 9.11 Sub-sampling tradeoffs.

An additional important constraint is related to the relationship between the IF and the symbol rate. In classical architectures (sampling at DC), the two quantities are independent, but in the IF sampling case, they are related by the $f_S/4$ condition. This assumes an integer number of samples per symbol for easy digital demodulation. For example, a GSM IF-sampled receiver should use an appropriate multiple of 13/4 MHz for the IF and a multiple of 13 MHz for f_S. A 52 MHz sampling rate and a 65, 91, or 117 MHz IF would fit this rule.

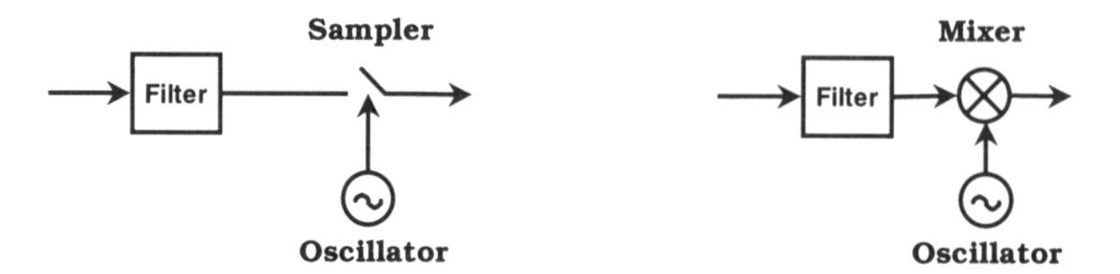

Figure 9.12 Sampler versus mixer.

It is instructive to analyze further the potential limitations of under-sampling. In particular, the oscillator jitter is commonly alleged to be a trouble spot for this technique. As illustrated in Fig. 9.12, we will do a comparison with the conventional mixer. One can calculate the effect of oscillator jitter with the help of Figures 9.13 and 9.14.

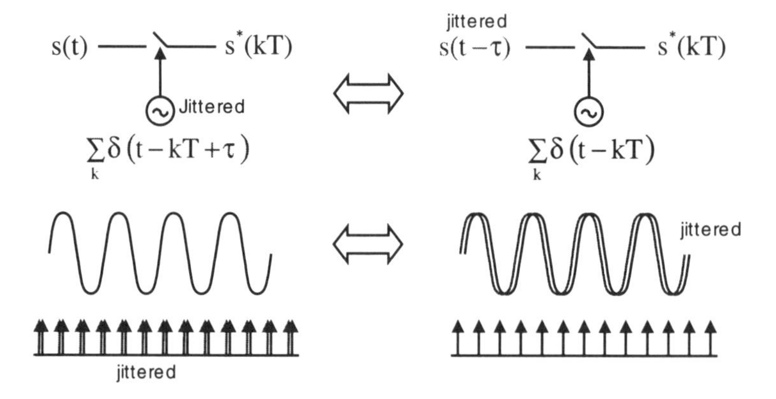

Figure 9.13 Modeling sampling-oscillator noise.

The output of the sampler $S^*(kT)$, produced in reality by the noisy sampling of a pure input signal $S(t)$, can be regarded as the noiseless sampling of an effectively jittery input $S(t-\tau)$. Fig. 9.13 shows this equivalence, with τ being a low-pass random process describing the oscillator jitter. If this process has the spectral density function $S_{\delta t}(\Delta f)$, the oscillator and effective input SSCRs (signal-sideband-to-carrier ratios) are related as in Fig. 9.14. This is easily derived by observing that the variation $\delta O(t)$ of the oscillator output $O(t) = A_o \sin(\omega_o t + \phi_o)$ is proportional to its time derivative $A_o \omega_o \cos(\omega_o t + \phi_o)$ and $\tau = \delta t$ while the variation $\delta S(t)$ of the effective input signal $S(t) = A_s \sin(\omega_s t + \phi_s)$ is proportional to $A_s \omega_s \cos(\omega_s t + \phi_s)$ and the same δt. Eliminating this variable between the $\delta O(t)$ and $\delta S(t)$ expressions, the relationship between

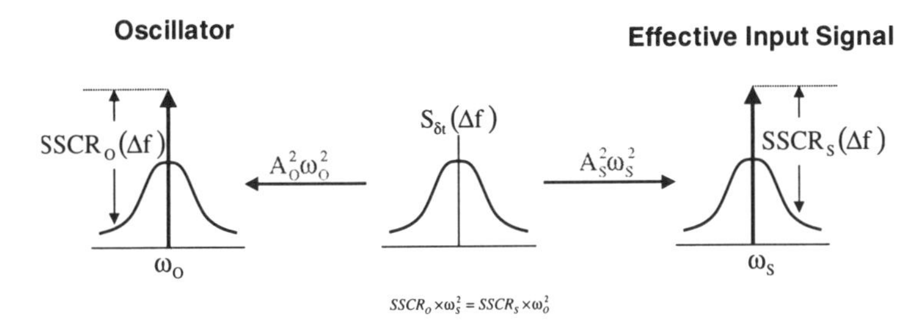

Figure 9.14 Effective input noise signals.

the actual oscillator noise spectrum and the effective input signal spectrum is obtained. Graphically this dependence is represented in Fig. 9.15.

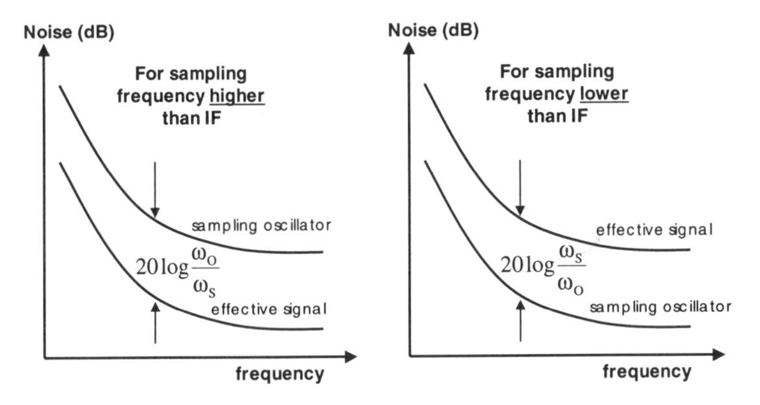

Figure 9.15 Effective input noise spectra of sampler.

Applying the previous development to the RF input case in Fig. 9.16 shows clearly that sampling large blockers superimposes substantial oscillator noise over the desired RF signal.

This is similar to reciprocal mixing in conventional mixers. Furthermore, when we compare the two effects in the equivalent frequency plans of Fig. 9.17 for the conventional and the IF-sampled cases, the oscillator jitter requirements become identical, as seen in Fig. 9.18.

An interesting observation is that under-sampling effectively acts as down-conversion to a second IF at $f_S/4$. In this view, the anti-aliasing conditions of the first IF filter become "image rejection" conditions for the down-conversion to the second IF. This interpretation shows intuitively why under-sampling and mixing require identical oscillator specifications.

Comparing the intrinsic noise performance of the two circuits in Fig. 9.12, we observe that input noise folding is present in both cases, albeit through different (but related) mechanisms. The IF filters reduce these effects. A more severe sampler problem absent from the mixer is the aliasing of self thermal noise. The in-channel portion

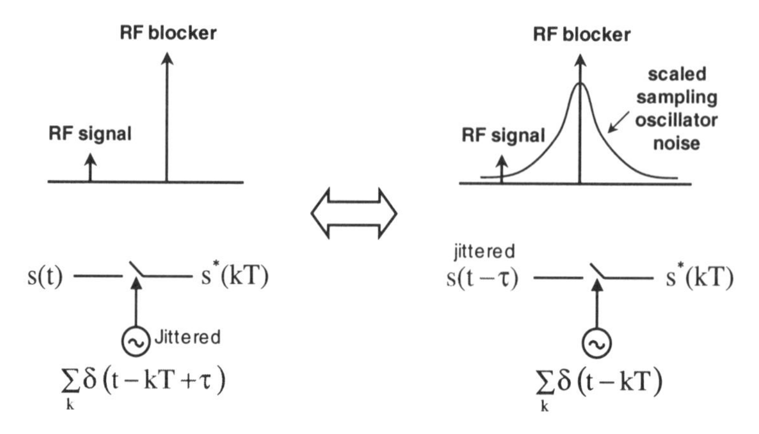

Figure 9.16 Sampling blockers.

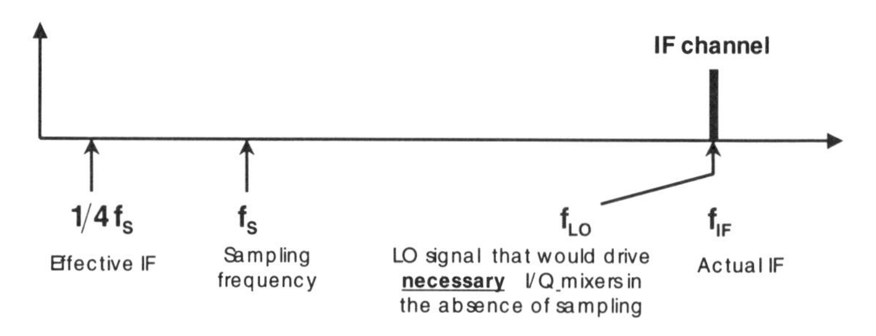

Figure 9.17 Equivalent IF sampling and conventional frequency plans.

of this noise for a given sampling frequency is KT/C divided by the over-sampling ratio with respect to the channel width. A potentially serious practical limitation for large under-sampling ratios is the use of small sampling capacitors dictated by circuit bandwidth requirements. Fortunately, in a well designed receiver, 70–80 dB of gain is present between the antenna and the sampler, reducing the KT/C noise effect to insignificant levels. This is similar to conventional sampling at the baseband.

9.4 BAND-PASS A/D CONVERSION

The discussions regarding Fig. 9.10 and 9.17 point out that band-pass sigma-delta A/D conversion is naturally suited for IF sampling. The circuit resources for obtaining high dynamic range are used only within the desired channel. The interfering input signals and the A/D quantization noise are filtered out in the decimator, whose digital signal processing capabilities mimic the conventional analog approach. The structure of Fig. 9.19 is identical to that of the discarded block in Fig. 9.9.

The block diagram of a conventional band-pass sigma-delta modulator is shown in Fig. 9.20. It is important to verify the performance of this circuit in the presence of blockers. Computer simulations, which have been checked with extensive measure-

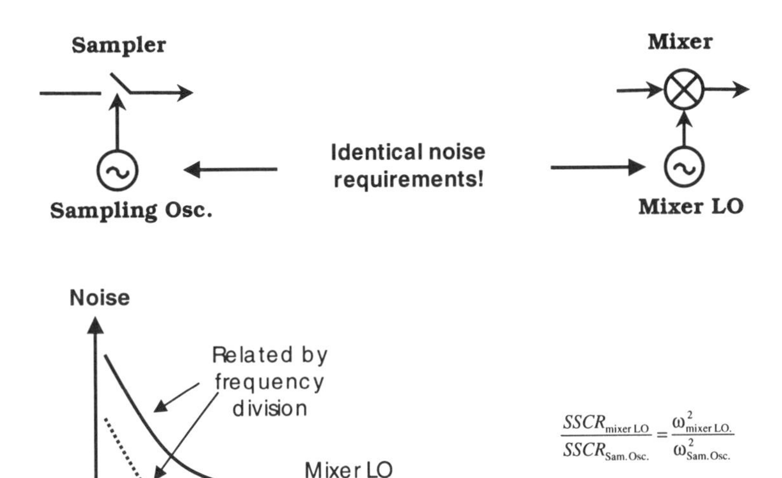

Figure 9.18 Equivalence of blocker sampling and reciprocal mixing.

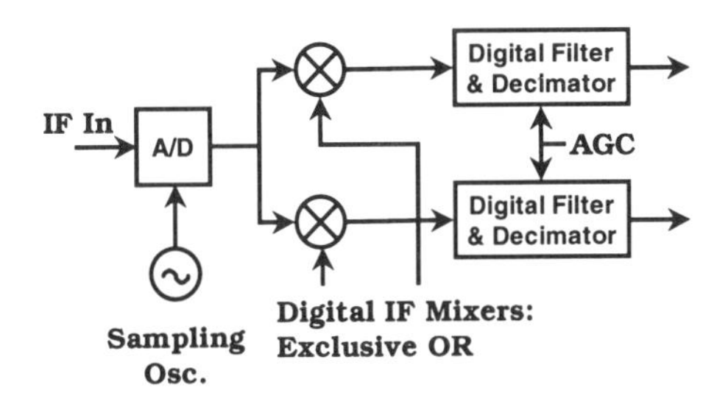

Figure 9.19 Decimator functions.

ments, are shown in Figures 9.21 and 9.22. The circuit behavior in Fig. 9.21 in response to increasingly stronger blockers is easily anticipated. The non-trivial increase in blocker insensitivity at high offset frequencies shown in Fig. 9.22 is unexpected but welcome. Further investigations are necessary to fully explain this property.

9.5 CONCLUSIONS

Designing IF-sampled transceivers is no simpler than designing conventional architectures, despite the promise and advantages of a more digital implementation. Careful tradeoffs between radio-system and circuit issues have to be made in the context of cost and power dissipation budgets. For narrow-band digital wireless systems, such as GSM, it is likely that IF-sampled portable transceivers will become practical. Based

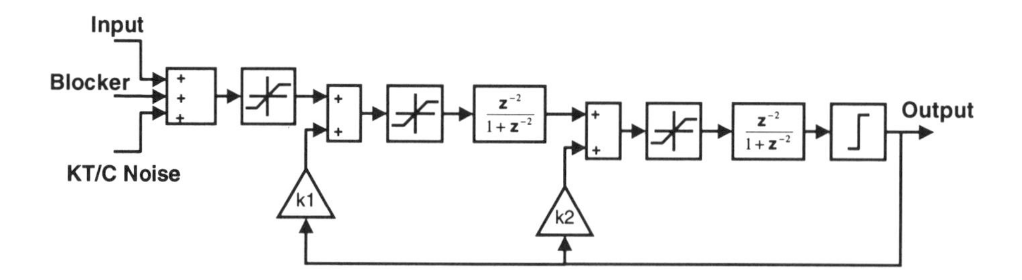

Figure 9.20 Conventional band-pass sigma-delta modulator.

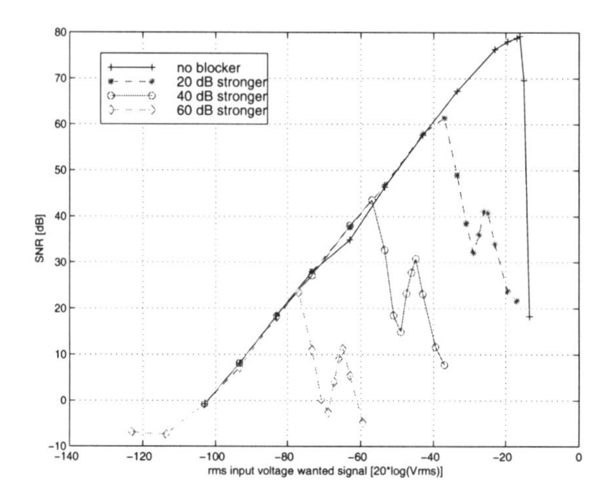

Figure 9.21 S/N ratio of A/D converter with 1.6 MHz-offset blockers.

on the considerations in this paper, it is reasonable to expect that such a solution will use a moderate level of under-sampling and passive IF filters.

References

[1] R. Gregorian and G. Temes, *Analog Integrated Circuits for Signal Processing*, John Wiley & Sons, 1986;

[2] F. W. Singor and W. M. Snelgrove, "Switched-Capacitor Bandpass Delta-Sigma A/D Modulation at 10.7 MHz", *IEEE J. Solid-State Circuits*, vol. 30, No. 3, March 1995, pp. 184–192;

[3] A. Hairapetian, "An 81-MHz IF Receiver in CMOS", *IEEE J. Solid-State Circuits*, vol. 31, No. 12, December 1996, pp. 1981–1986.

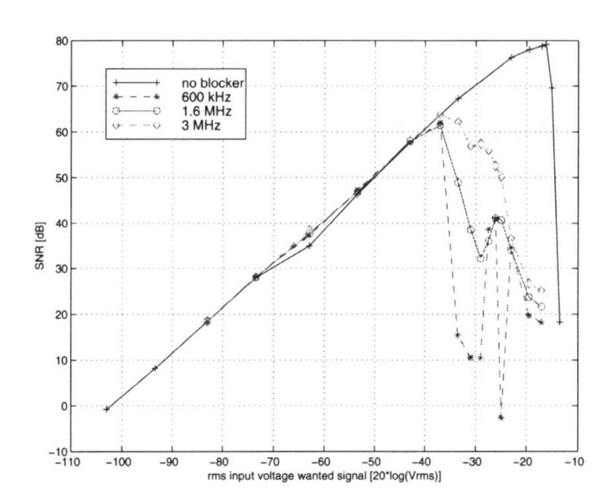

Figure 9.22 S/N ratio of A/D converter with 20 dB blockers operating at different offset frequencies.

10 TRANSMITTER CONCEPTS, INTEGRATION AND DESIGN TRADE-OFFS

Stefan Heinen and Stefan Herzinger

Infineon Technologies AG i. Gr.
RF IC Design Center
Düsseldorf/Munich, Germany

Abstract: An overview of different transmit architectures is given. The major scope are the concepts used in the GSM system. A comparison with respect to integration level and system cost is given. The modulation loop concept commonly used today is discussed in detail, with respect to implementation details and trade-offs.

10.1 INTRODUCTION

The rapidly growing mobile communication market requires increasing the efforts to achieve higher integration levels. The ultimate goal is the integration of a complete mobile on a single die — preferably on a CMOS die. The example given in Fig. 10.1 is still a dream. Today, we are far away from this goal. RF CMOS is more or less a research topic, although there are already a lot of publications dealing with this subject [1–6]. A demonstration of a fully functional GSM-like high-end system is not yet in sight. This would require a circuit which achieves the system specifications with respect to process, temperature and supply variations. Moreover, this target has to be met in a real environment, including on-chip bias circuitry. For these reasons, we are presently far away from integrating the RF parts of high-end systems on a single CMOS die.

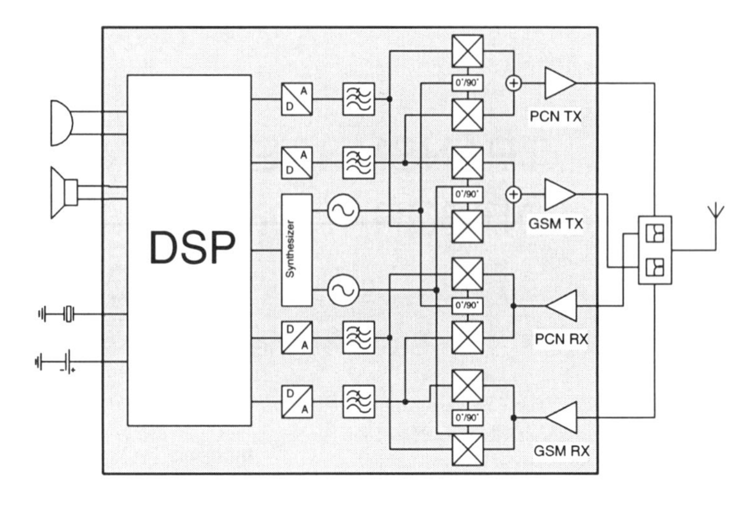

Figure 10.1 The ultimate integration goal.

It has not been demonstrated yet how to integrate large digital blocks along with analog RF blocks. Baseband digital signal processing typically requires a large die area even in advanced technologies. It is therefore desirable to follow the digital shrink path as fast as possible. From this point of view, the single-chip mobile might remain too expensive as long as reasonable cost reduction can be achieved by shrinking the digital functions. The design of the analog RF portion surely is the bottleneck in the product cycle. Therefore research for a single-chip mobile might be postponed until the dies are pad limited, which will provide the die area required for the RF blocks without additional cost. However, the system performance has to be achieved under a lot of constraints:

- reduced supply voltage might result in the need for very high VCO gain and in increased current consumption
- substrate noise
- low substrate resistance
- high process tolerances

Due to the fact that the industry has to deliver products at optimum cost, performance and size, and at a certain point in time, we see an evolutionary development. That means the industry is relying on proven technologies, which are going to allow volume shipping at a low risk with respect to time-to-market. Therefore, architectures as given in Fig. 10.2 will be used in the next integration step.

10.2 GSM TRANSMITTER REQUIREMENTS

In order to derive the requirements of the transmitter building blocks, the GSM system specifications have been reviewed and translated into requirements related to the power spectral density. Fig. 10.3 is generated from [7], Sections 4.2.1 and 4.3.3, which

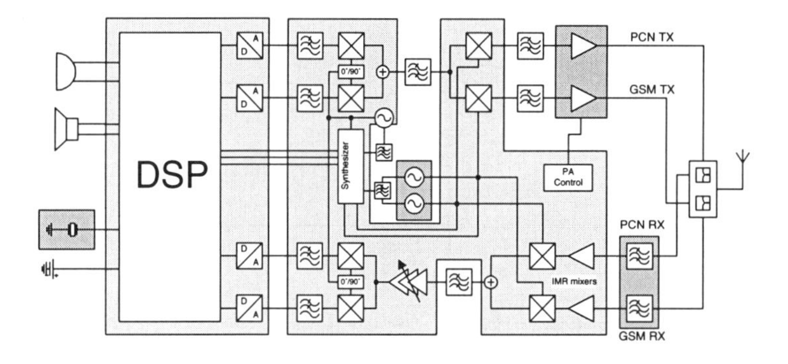

Figure 10.2 The achievable goal.

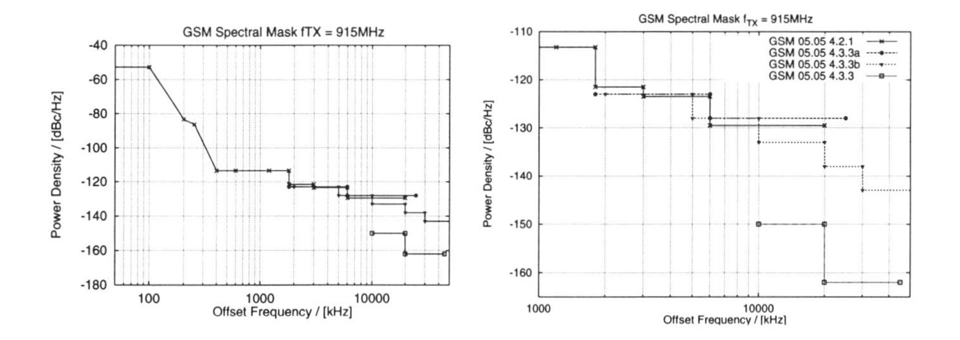

Figure 10.3 GSM unwanted emission requirements.

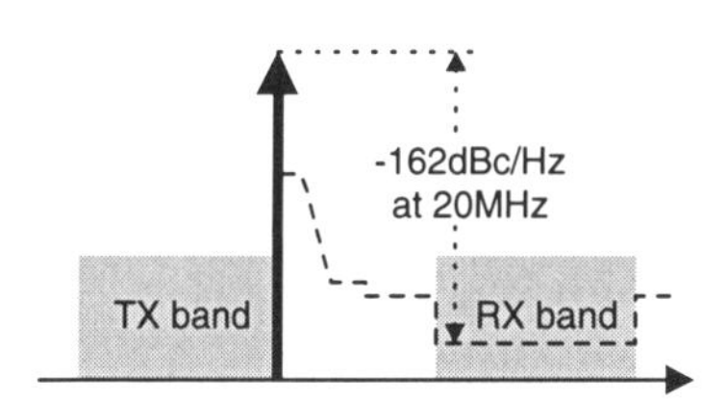

Figure 10.4 Spurious emissions.

are dealing with emissions due to modulation and unwanted emissions. The worst case occurs while transmitting on the highest channel at 915 MHz, due to the spurious emission requirements in the mobile RX band 935 MHz to 960 MHz, given in Section 4.3.3 of [7]. In order to avoid interference with the reception of a nearby mobile, the maximum allowed power in a 100 kHz bandwidth is −79 dBm. Considering power amplifiers of class 4 — which means 2 W or 33 dBm — this results in a requirement of −162 dBc/Hz at 20 MHz offset as shown in Fig. 10.4. For DCS1800 mobiles, this specification is relaxed to −71 dBm.

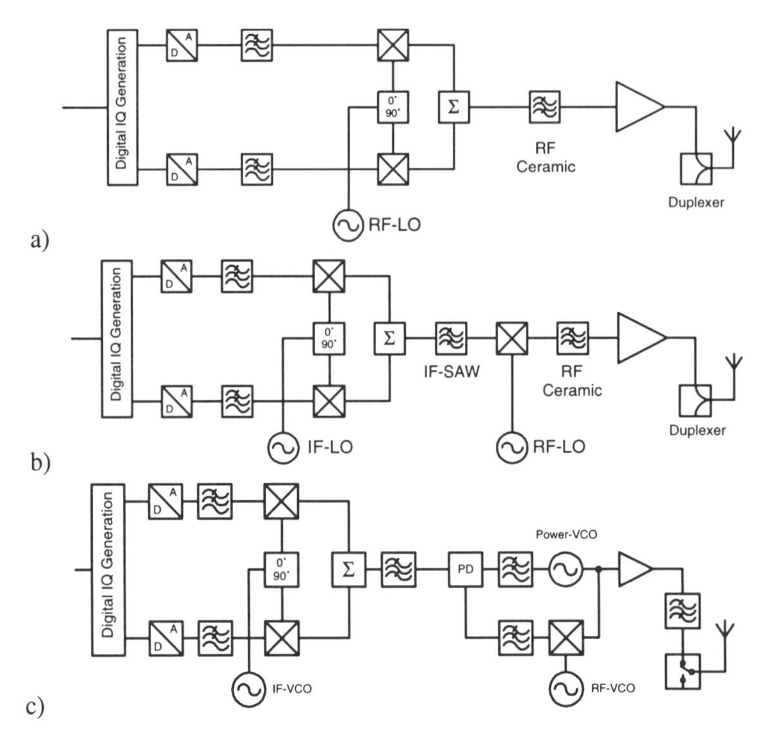

Figure 10.5 Basic GMSK TX Architectures. a) Direct modulation, b) IF modulation with up-conversion mixing, c) modulation loop.

10.3 BASIC GMSK TX ARCHITECTURES

The basic architectures shown in Fig. 10.5 are commonly used today. The aim of this section is to give a comprehensive review of these architectures including a discussion of the advantages and disadvantages in terms of cost and implementation issues. Emphasis will be laid on spurious emissions with respect to nonlinearities. A qualitative study of the effects will be presented, helping to understand the design trade-offs which occur in the implementation of the different architectures. The power amplifier will not be included in this discussion.

10.3.1 Direct modulation

In this section, the basic behaviour of a direct-modulation system as shown in Fig. 10.6 will be studied. Most of the basic characteristics discussed here will also apply to the other architectures using the basic vector modulator at a different frequency. The first target of the modulator is to generate a GMSK modulation which is compliant with the system specification [7]. Moreover, the given modulation mask sub-clause 4.2.1 of [7]

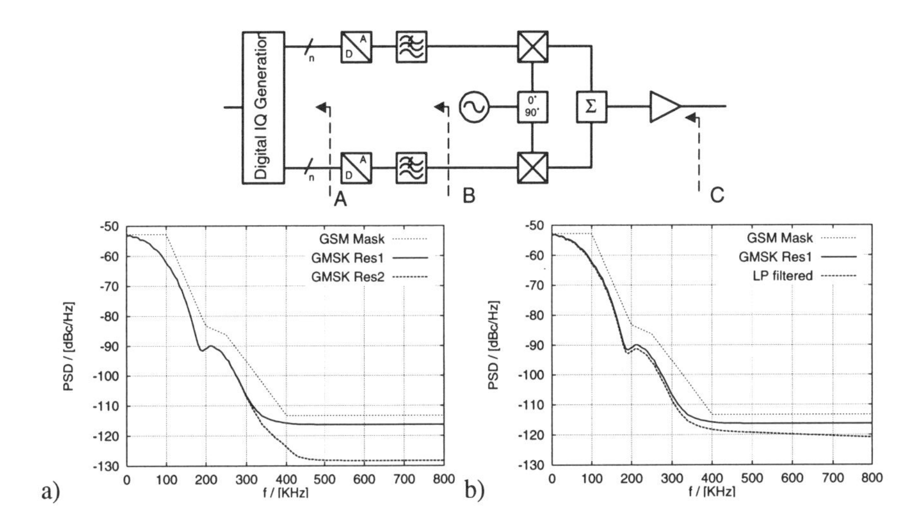

Figure 10.6 Direct-modulation system: Spectrum due to modulation and wideband noise. a) Influence of digital resolution. b) Influence of lowpass filter.

has to be fulfilled. This means that the unwanted power of a mobile unit measured at 30 kHz offsets between 400 kHz and 1800 kHz has to be lower than 60 dBc, where the carrier power is also measured inside a bandwidth of 30 kHz. Including a reduction of 8.5 dB to take into account the reduced bandwidth used to measure the carrier, the requirement will be -113.3 dBc/Hz. The digital part has to deliver a resolution which is sufficient to achieve this requirement, because the anti-aliasing lowpass filters after the DACs cannot provide an additional attenuation below 1 MHz (see Fig. 10.6). The lower limit for the lowpass cut-off frequency is given by the required phase accuracy. In order to improve the performance or to reduce the ADC resolution, compensation in the digital domain might be used. The LO-synthesiser causes additional unwanted emissions at point C of Fig. 10.6 in the form of reference spurs, which have to be considered as well.

The modulator, as shown in the grey box of Fig. 10.7, will have a wideband noise floor of about -140 dBc/Hz. In the TX chain after the modulator, no further narrowband filtering is possible, which is due to the fact that the RF-LO is used to adjust the TX channel. Thus, the noise floor of the modulator itself will dominate the unwanted emissions above 2 MHz.

Using a vector modulator as single-side-band modulator, a typical measurement result is given in Fig. 10.8a). Part b) of the figure shows simulation results of SSB modulation and a GMSK modulation. An ideal modulator would generate only a single tone for the SSB case, whereas the real modulator will generate a lot of different components. The image rejection will be limited by the accuracy of the phase shifter. There might be LO leakage as well due to DC offsets. The other tones are generated by the nonlinearities of the modulator. They will result in a wider spectrum as shown in the simulation part b) of Fig. 10.8.

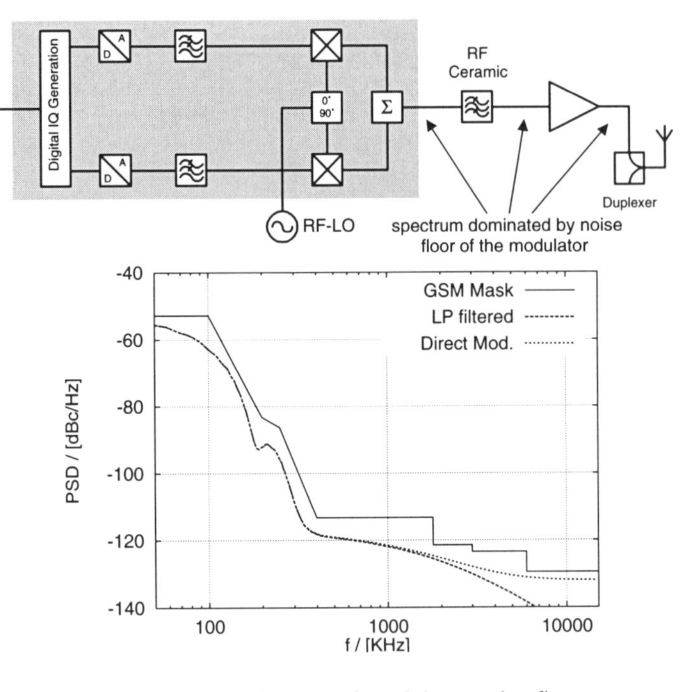

Figure 10.7 Influence of modulator noise floor.

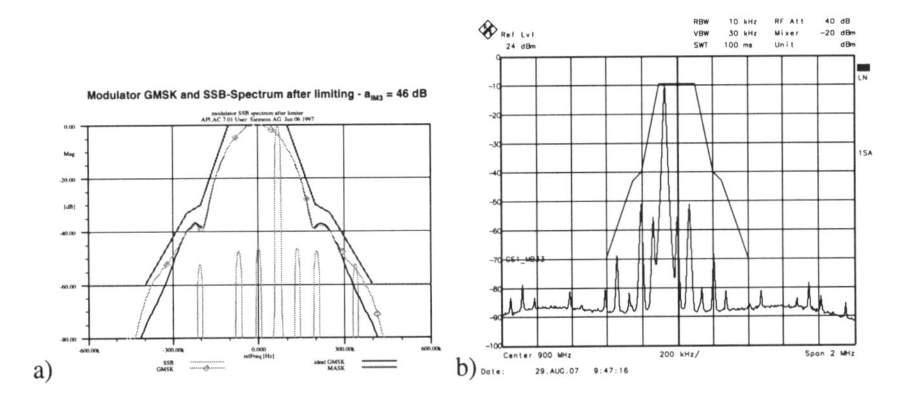

Figure 10.8 Influence of nonlinearity within the modulator: a) SSB measurement, b) simulation SSB and GMSK.

Therefore, nonlinearity is of major concern in the implementation of a vector modulator. In order to understand the basic influence of the different nonlinearities, the first step is to separate them.

Considering only limiting amplifiers in the input stage of the modulator as shown in Fig. 10.9 at first results in the tones given in the same figure. Assuming that the

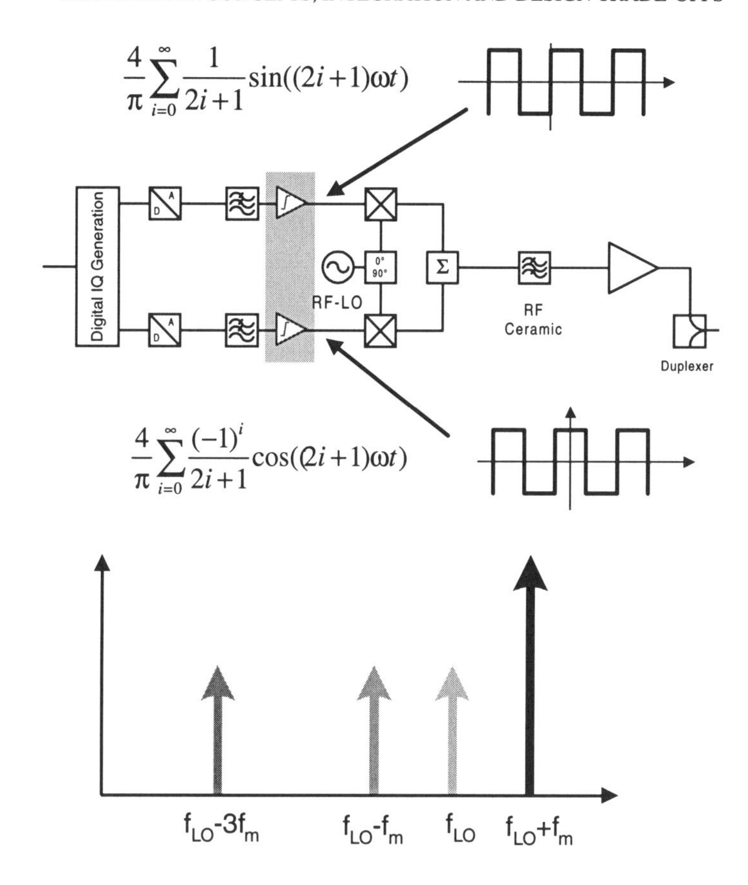

Figure 10.9 Influence of nonlinearity within the modulator input.

input nonlinearity is hard limiting, the input signals are square waves with a phase shift of 90 degree. It can be seen from the Fourier coefficients that the phase of the harmonics in one of the input paths will change by 180 degree, e.g., the 3rd harmonics compared to the fundamental. This means for the 3rd harmonics the upper sideband will be rejected and the lower sideband located at $f_{LO} - 3f_m$ will occur. In Fig. 10.10, only nonlinearities after the LO phase shifter and in the output stage are considered.

For reasons similar to the ones discussed previously, the input signal will be modulated onto the lower sideband of the 3rd harmonics of the LO. That means a tone will occur at $3f_{LO} - f_m$. Because of the output nonlinearity, an intermodulation product will be generated at $-f_{LO} + 3f_m$, which is mirrored to the positive frequency $f_{LO} - 3f_m$.

Finally, taking the nonlinearity of the power amplifier into account, additional intermodulation products will be generated as shown in Fig. 10.11.

What does this really mean for the system? The nonlinearities will generate a wider GMSK spectrum. There are two ways to meet the system specifications: The first is to make the modulator as linear as possible, thereby avoiding the wider spectrum. This

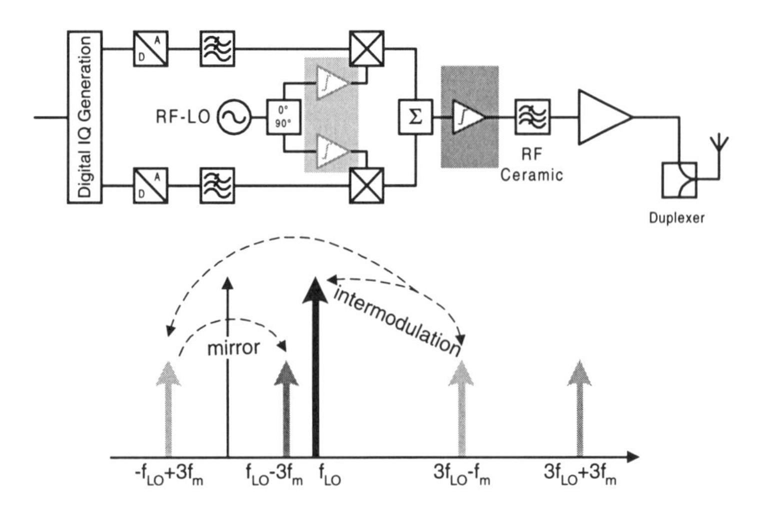

Figure 10.10 Influence of nonlinearity within the modulator LO path and output stage: generated SSB tones.

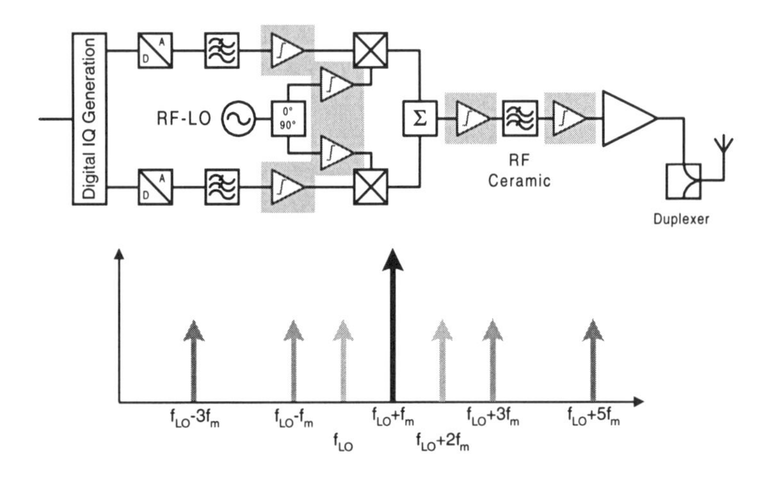

Figure 10.11 Influence of nonlinearity within the power amplifier.

results in an increase in power consumption. The second way is to make the power amplifier more linear in order to reduce the intermodulation. However, the result with respect to power is even worse. Accordingly, this is surely not the way to go for a battery-powered mobile.

Another quite important requirement is the wideband noise floor at the modulator output. As shown in Fig. 10.3, the TX in band requirements are −133 dBc/Hz below 20 MHz offset and −138 dBc/Hz above. In order to achieve these specifications, an output power of about 0 dBm is required, which is hard to achieve when using a CMOS implementation.

The direct-up-conversion architecture shown in Fig. 10.6 has the drawback that the RF-VCO operates at the output frequency. Therefore, depending on the quality of shielding, a re-modulation may occur. This will result in a phase error which might be too high. In order to avoid this problem, almost all commercially available implementations use an offset mixing scheme in order to generate the desired output frequency.

10.3.2 IF modulation

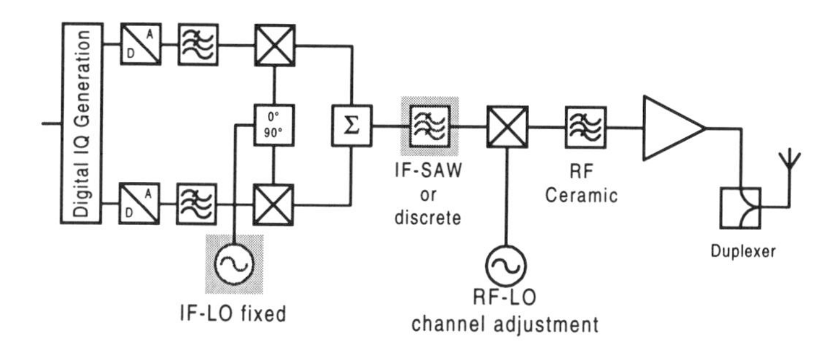

Figure 10.12 IF modulation.

In the last section, the basic characteristics of the vector modulator have been discussed intensively in combination with direct up-conversion architecture. Most of the issues apply as well to the IF modulation architecture given in Fig. 10.12. The IF modulation has a major advantage compared to the direct up-conversion. First, the re-modulation problem does not occur and the isolation requirements are relaxed. Second, a narrow IF-SAW filter might be used to reduce the unwanted emissions. Moreover, the architecture is well suited for dual-band applications (GSM in combination with DCS1800). The IF-modulator will be used for both systems at a common IF, whereas two up-conversion mixers are used to generated the different TX frequency bands.

As described in Section 10.2, a noise floor of -162 dBc/Hz has to be realized in a GSM system. The direct-up-conversion architecture requires a duplexer in the front-end because the noise floor generated by the modulator is quite high. The IF-modulation system comes up with a lower noise floor, which might be in the range of -160 dBc/Hz. In order to guarantee the performance in high-volume production, something like -165 dBc/Hz would be needed. Therefore, the IF-modulation architecture cannot avoid the expensive duplexer. During the last few years, the modulation loop architecture has reached maturity and is the only system approach which avoids the duplexer.

10.3.3 Modulation loop

As already mentioned, the modulation-loop-based transmitter of Fig. 10.13 has become quite popular. The first reason is the fact that no duplexer is needed. A second

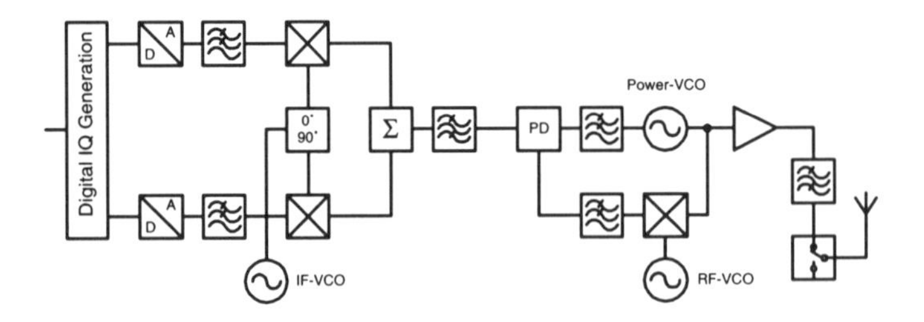

Figure 10.13 Modulation-loop-based transmitter.

major advantage is the dual-band capability of the architecture. The modulation loop is more or less a PLL in which the reference of the phase detector is modulated. In the feedback path of the loop, a down-conversion mixer is used to provide the down-converted VCO signal to the phase detector. The RF-VCO provides the LO for the down-conversion mixer, synthesised by a conventional PLL in order to adjust the desired channel.

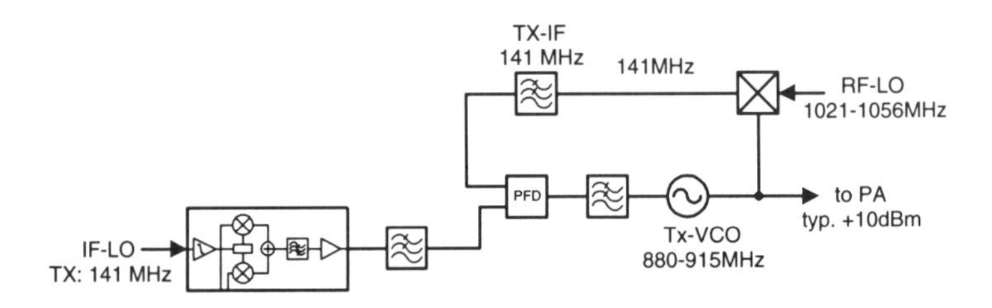

Figure 10.14 Modulation loop: modulate reference path.

Due to the high reference frequency of 100 MHz or higher (refer to Fig. 10.14) there are several hundred phase comparisons for one transmitted bit. Therefore the loop is able to copy the phase modulation of the reference path onto the output of the power VCO, which is within the loop. The GMSK modulation used in GSM-like systems is a pure phase modulation and does not exhibit any amplitude modulation. The power-VCO provides a very low noise floor at 20 MHz offset — typically less than −165 dBc/Hz. This VCO is realized as a module and is capable of driving the power amplifier directly without additional noisy buffer stages.

The configuration can be seen as a transformation of an RF bandpass into a lowpass filtcr.

In Fig. 10.15, the principal block diagram of an up-conversion loop with modulation in the feedback loop is given along with its frequency plan. This configuration requires that the output of the vector modulator within the feedback path does not exhibit any phase modulation. Accordingly, the VCO must have phase modulation which is cancelled by the vector modulator in order to fulfill the requirements of the phase-locked loop.

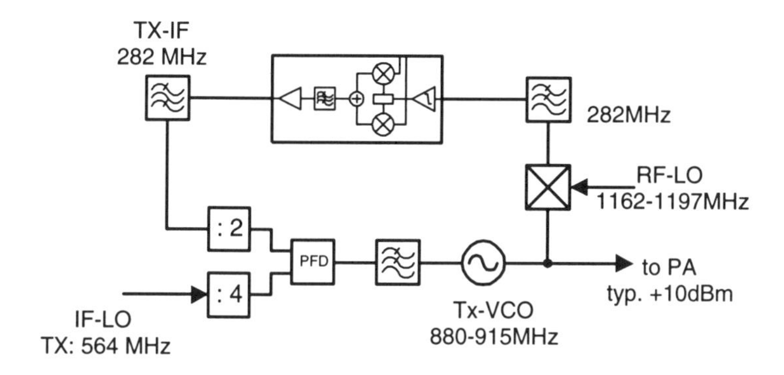

Figure 10.15 Modulation loop: modulate feedback path.

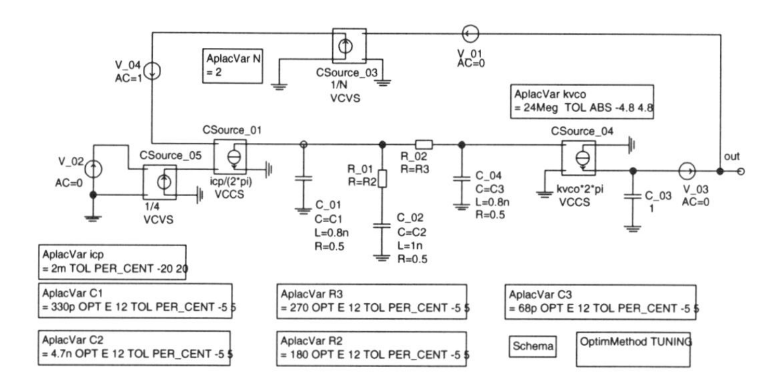

Figure 10.16 Modulation loop: Simulation schematic for loop dynamics.

The GSM system requires an RMS phase accuracy of under 5 degrees, which calls for a high cut-off frequency of the loop filter. On the other hand, only the VCO wide-band noise should be present at 20 MHz frequency offset. In Fig. 10.16, an APLAC simulation setup [8] is given which is used to study the different noise contributions within the modulation loop. The different noise sources, the corresponding transfer functions and the resulting output noise contributions with respect to the major building blocks of Fig. 10.15 are given in Fig. 10.17 and Fig. 10.18.

Outside the modulation bandwidth and below 5 MHz, the vector modulator is the dominant noise source, where the noise of the modulation itself is dominated by the digital noise of the I-Q signal generation. Above 10 MHz, the noise is dominated by the noise of the power VCO. The phase-noise requirements of the IF-LO and the RF-LO are relaxed due to the attenuation of the transfer function within the loop.

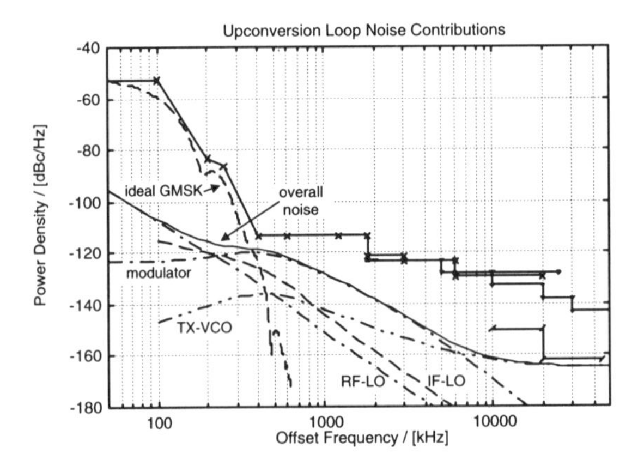

Figure 10.17 Overall noise and different contributions within the modulation loop.

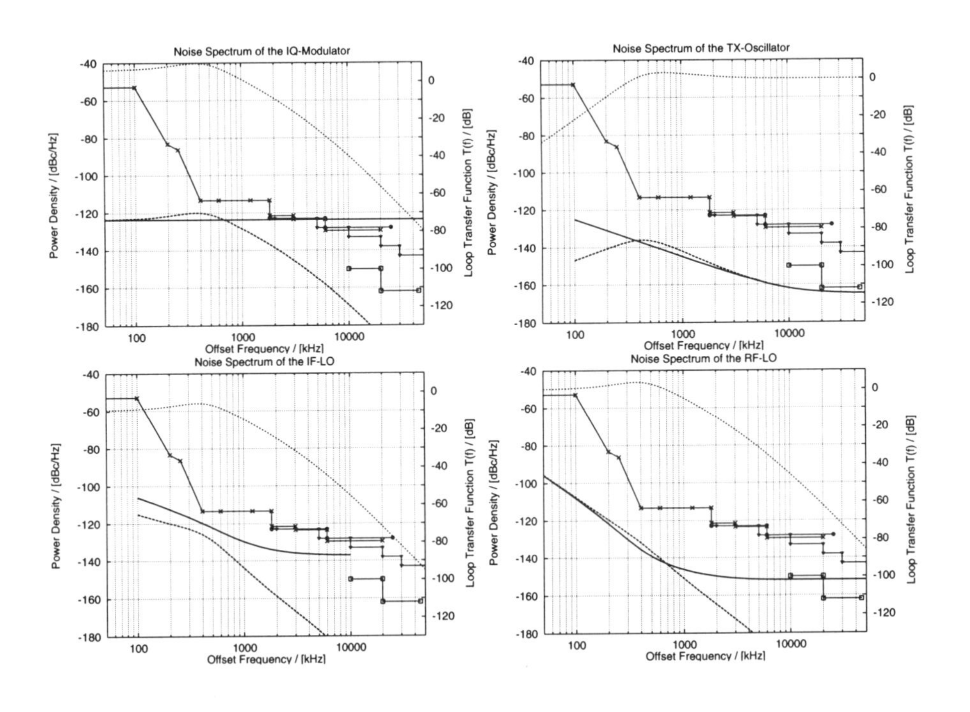

Figure 10.18 Noise contributions and transfer functions of the major building blocks.

10.4 IMPLEMENTATION RESULTS

A major trade-off in designing the loop transfer function is to keep the bandwidth high enough to keep the phase error low while achieving the -162 dBc/Hz at 20 MHz offset. As shown in Fig. 10.19 and Fig. 10.20, a transmitter based on the PMB 2251

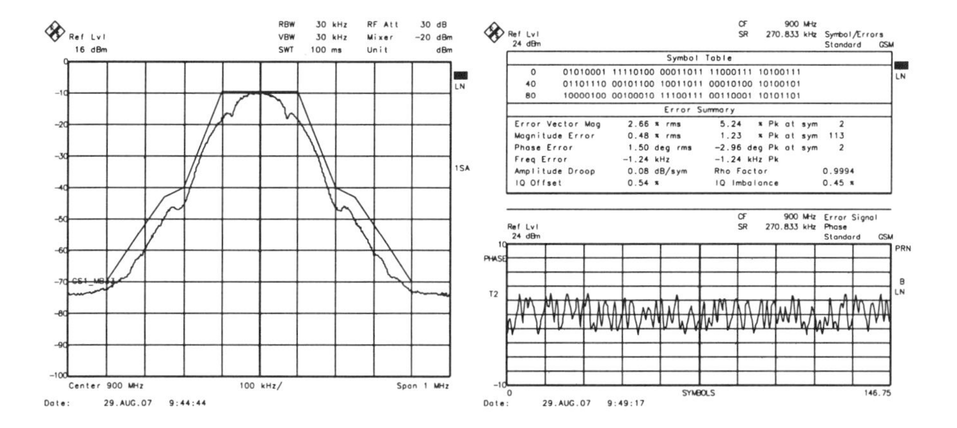

Figure 10.19 Measured output spectrum of a PMB 2251 modulation loop.

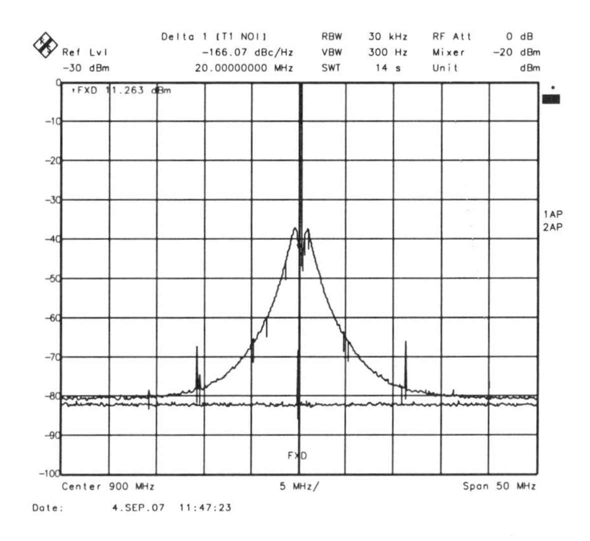

Figure 10.20 Measured phase noise within the RX band at 20 MHz offset.

modulation is able to achieve this goal. Detailed information on the application circuit and the measurement setup can be found in [9].

Fig. 10.21 shows the results of an SSB-modulation generated by the PMB 2251 modulation loop. Due to the limiting behaviour of the type-4 PFD used in this application, the unwanted tones are symmetrical with respect to the wanted signal. The PFD removes any AM components, meaning that only a PM component remains. The IM3 components are quite low at -60 dBc. Therefore, the modulation mask can be achieved easily by using a power amplifier with relaxed specifications. The CMOS

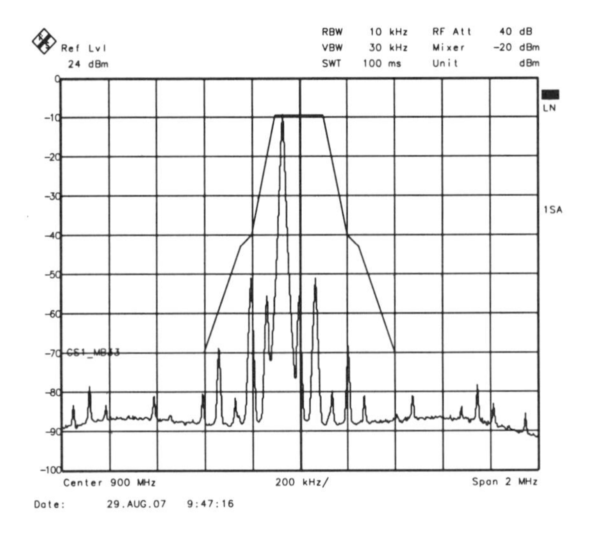

Figure 10.21 Measured single-side-band spectrum of a PMB 2251 modulation loop.

DCS1800 direct-conversion modulator reported in [1] exhibits an IM3 suppression of 35 dB.

10.5 CONCLUSIONS AND FUTURE REQUIREMENTS

The increasing number of cellular subscribers requires the implementation of optimised architectures of multi-band systems combining at least two standards out of GSM, DCS1800 and DCS1900. Therefore, problems like choosing the best possible frequency plan or reducing the number of VCOs may dominate the R&D for at least the next generation. Another challenge will be the implementation of multi-mode system like DCS1800/DECT or GSM/UMTS. In these combinations, we have to cope with different data rates and modulation methods. This will have a significant influence on the implementation and may require a step back in the integration level.

Another issue is the reduction of size and weight of the cellular phone. This will require improved batteries along with reduced power consumption. Concerning the talk time, the efficiency of the PA has to be improved in some way. The standby time is important to the user as well, meaning that the power consumption of the receive path has to be reduced.

As shown for the TX implementation in this Chapter, the overall system design requires an optimum balance between a lot of different effects. Taking the presented problems into account, an increased integration level may result in higher system cost. However, from the commercial point of view, system cost is the major driving force. It really does not matter which technology might be used: digital CMOS only or digital CMOS, analog CMOS, BiCMOS or GaAs. A minimum system cost with respect to a certain point in time has to be achieved. Therefore, issues like yield, design cycles and availability of deep sub-micron technologies have to be considered. The final conclusion from the authors' point of view is that reasonable integration steps are required depending on the system requirements, time schedule and the available technology.

Acknowledgments

We would like to thank Georg Lipperer, Stefan Beyer, and the other colleagues form the application group for helpful discussions and for their providing some of the material used in this chapter.

References

[1] Steyaert *et al.*, "A single-chip CMOS transceiver for DCS-1800 wireless communications," in *Proceedings ISSCC98*. ISSCC, Feb. 1998, IEEE, TP3.3.

[2] Rudell *et al.*, "A 1.9 GHz wide-band IF double conversion CMOS integrated receiver for cordless telephone applications," in *Proceedings ISSCC97*. ISSCC, Feb. 1997, IEEE, SA 18.3.

[3] Behzad Razavi, *RF Microelectonics*, Prentice Hall, 1998.

[4] J. Craninckx and M. Steyaert, *Wireless CMOS Frequency Synthesizer Design*, Kluwer Academic Publishers, 1998.

[5] Thomas H. Lee, *The Design of CMOS Radio-Frequency Integrated Circuits*, Cambridge University Press, 1998.

[6] J. Crols, *CMOS Wireless Transceiver Design*, Kluwer Academic Publishers, 1997.

[7] GSM05.05, "GSM digital cellular telecommunication system (phase 2+): Radio transmission and reception (GSM 05.05)," ETSI, July 1996, Version 5.2.0.

[8] Valtonen *et al.*, "Aplac: An object-oriented analog circuit simulator and design tools," Helsinki University of Technology, Circuit Theory Laboratory, March 1994, Version 6.2.

[9] Siemens Semiconductor Group, "Upconversion modulation loop PMB 2251 v1.2," Application Note, July 1997, Version 1.0.

11 RF CHALLENGES FOR TOMORROW'S WIRELESS TERMINALS

Petteri Alinikula

Nokia Research Center
Itämerenkatu 11-13, 00045 Helsinki, Finland

Abstract: Wireless telecommunications will enter a new era in the coming years. New systems and new customer needs are already introducing vast challenges, in particular for the implementation of the terminals. The RF section is a major contributor to the size and cost of the terminal. Consequently, the winning terminal implementations will be those with the best RF section.

None of the implementation technologies alone seem to provide the full answer to the RF challenges. Instead, the winning RF will be achieved by careful joint optimisation of the overall technology portfolio with the proper radio architecture.

11.1 INTRODUCTION

From the early 1980s when the first generation of analog systems NMT, AMPS, and TACS were first introduced, the cellular phone market has exploded to be one of the major consumer product markets. By the end of 1997 there were close to 207 million cellular subscribers worldwide. There was an increase of 70 million during 1997 alone. According to some predictions the number of cellular subscribers will exceed one billion by 2005 [Nokia Press Release, September 1998]. As a world record for the cellular phone penetration, Finland has just exceeded 50 % of the population (August 1998). Currently, digital narrow-band systems led by GSM continue to conquer the world. In the near future, however, the voice-based second-generation cellular systems will be complemented by new systems. The main interest in Europe is toward the

WCDMA radio access technology that ETSI selected in January 1998 to be the basis for the Universal Mobile Telecommunication Service (UMTS), also referred to as the third-generation cellular system. The driving force behind the WCDMA system is the increased need for data capacity, particularly for mobile multimedia applications.

Furthermore, there is a growing interest toward wireless data-communication applications. The "Bluetooth" concept,[1] working at 2.4GHz, will provide low capacity short range wireless data transfer for various applications and accessories. The Wireless LAN systems at 2.4 and 5GHz will increase the wireless data transmission capacity in office environments.

Although many of the new systems are targeted to be global, the trend is toward higher diversity of co-existing systems. As a result, terminals offering access to several systems are needed, and the integration becomes extremely challenging. In particular, the RF section — being a major contributor to the wireless terminal size and cost — faces requirements that push the technologies to their limits.

The objective of this article is to discuss the main trends affecting the RF implementation of tomorrow's wireless terminal: the emerging wide-band systems, upcoming multi-system terminals, continuous miniaturisation, and cost-reduction. The first two items are emerging new trends, whereas the latter items are more evolutionary ones.

11.2 CAPACITY FOR WIRELESS MULTIMEDIA

The wide-band wireless communication systems are finally moving forward from feasibility studies to serious development programs. The main advantage of the new systems, as advertised, are the high bit-rates ranging from hundreds of kbit/s up to 2Mbit/s for cellular wide-band. In WLAN systems peak bit-rates could even exceed 20Mbit/s. The result of the high capacity is the increasing need for DSP computational power. In terminals, the requirements for the wide-band data flow are increasing up to hundreds or thousands of MIPS. Fortunately, the increasingly complex DSP becomes feasible as the evolution of the sub-micron CMOS technology continues.

One reason for the capacity being so high is the use of spectrum-efficient modulation methods. In order to meet the demanding specifications of the transmitted signal's spectrum, the transmitter has to be highly linear. For example, the adjacent-channel power level must be in the order of −45dBc for the wide-band cellular terminals. With linear power amplifiers, it is extremely difficult to achieve a sufficient battery efficiency; typically, the battery efficiency falls into the region of 5 to 20 %. This is not satisfactory for cellular phone users who are used to relatively long talk times resulting from battery efficiencies above 60 %.

The efficiencies can be boosted up with different transmitter linearisation schemes. Some of the most promising methods are local negative feedback, pre-distortion, and feed-forward linearisation, illustrated in Fig. 11.1. Using the local negative feedback is highly desirable, since it does not affect the radio architecture, and since it also makes it possible to integrate the linearisation circuitry with the power amplifier.

For the pre-distortion, there are several alternative approaches: analog, digital fixed, and digital adaptive pre-distortion. The analog pre-distortion scheme is difficult to con-

[1] see http://www.bluetooth.com

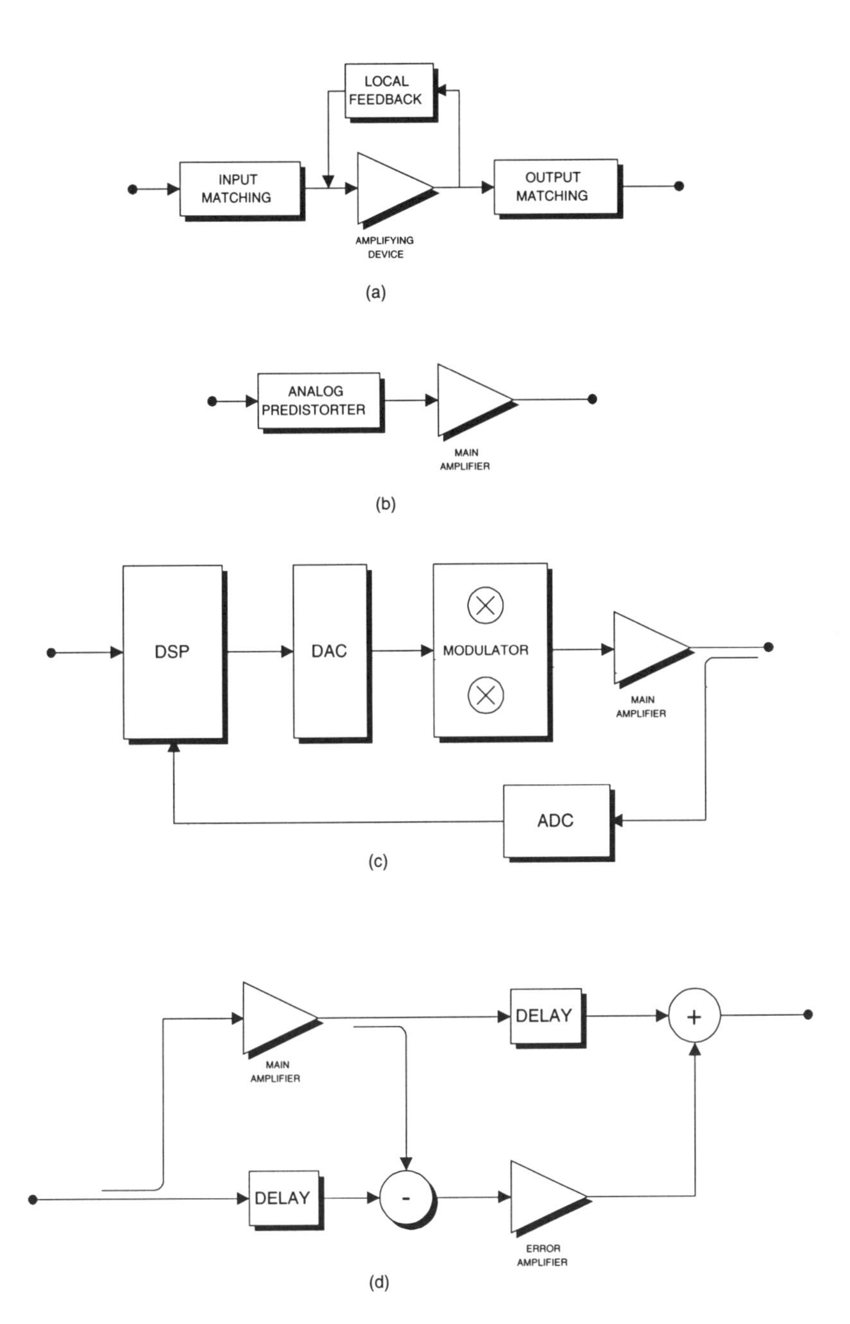

Figure 11.1 Power amplifier linearisation concepts: (a) local negative feedback, (b) fixed analog pre-distortion, (c) adaptive digital pre-distortion, (d) feed-forward.

trol over the variations of temperature, supply voltage, and circuit parameters. With the fixed digital scheme, the system can be taught for different conditions; still, major variations of the nonlinear characteristics of the transistors are extremely difficult to control. An adaptive concept would be flexible for varying conditions, but it suffers from the overall complexity. The digital pre-distortion schemes introduce modifications to the whole transmitter chain: wider bandwidths of the filters and better resolution of the D-to-A converters.

For achieving the ultimately linear operation, the feed-forward linearisation has the highest potential. The feed-forward linearisation is truly adaptive to the different conditions. Still, the method suffers from the complicated control of the phase and amplitude characteristics of the two loops, and the power consumption can be unsatisfactorily high for hand-held terminals.

An additional difficulty, in particular for the CDMA systems, is the large power control range, being in the order of 60 to 90 dB. Maintaining the high linearity over a wide range of power levels is a challenge for the power amplifier design. CDMA systems do not often operate at the maximum power level. Maintaining an adequate efficiency at moderate output powers is therefore an extremely important objective.

The low efficiency combined with the increasing packaging density can lead to heat transfer problems. To overcome these problems, mechanical and thermal design have to be integrated with the electrical design flow of future terminals.

Other new requirements of high bit-rate systems include fast-settling synthesisers for variable-capacity TDMA systems and high-speed A-to-D and D-to-A converters for the baseband section.

11.3 MULTIPLE RADIOS IN ONE UNIT

As the wide-band systems are coming, it is important to notice that the R&D for the second generation terminals (GSM, GSM1800/1900, IS-136, IS-95, PDC ...) continues actively; the market is not showing any signs of saturation. Multiple systems will co-exist in the same geographical areas. Accordingly, there is a market need for terminals offering access to several systems. The non-compatibility of the different systems causes difficulties for the implementation. In the baseband section, core platforms for several radio access protocols can be implemented, whereas in the RF section it is more difficult to design core solutions cost-efficiently.

The key task in implementing the RF section for a multi-system multi-band terminal is to minimise the number of parallel functions. Many of the active functions, such as variable-gain amplifiers and mixers, can be designed to be adaptive to different specifications, whereas passive circuits, such as filters and matching circuits, need to be selectable for each designated frequency band. High-frequency filter technologies are clearly the key for finding implementation solutions for future miniaturised multi-mode terminals. An example RF architecture of a dual-mode terminal is shown in Fig. 11.2.

None of the semiconductor technologies alone will provide an optimum platform for the multi-mode transceivers. The best and only way to find the winning RF solution is to optimise the whole RF technology portfolio including the ICs, packaging, substrates, passives, and filters with the RF architectures for the systems selected for a multi-system product. BJT, BiCMOS, CMOS, SiGe, GaAs, SOI, and others will all

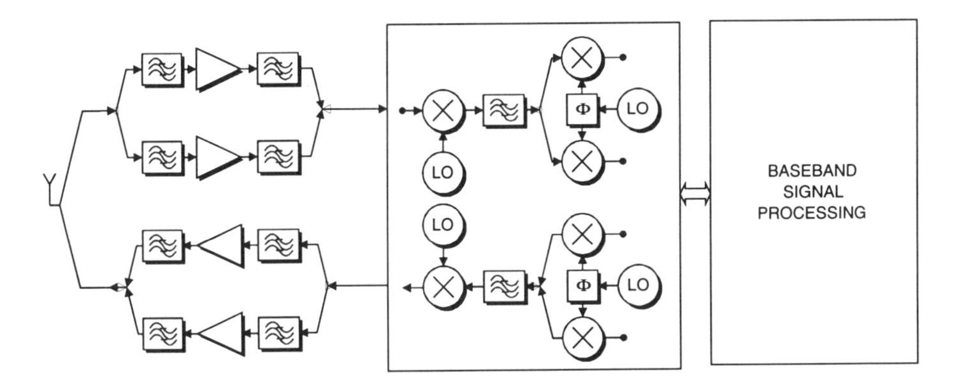

Figure 11.2 Example RF architecture of a dual-mode terminal.

remain possible enabling technologies for the optimum RF solution, but not targets by themselves. The micro-electro-mechanical systems (MEMS) will also eventually find their way into wireless terminals. Possibly, MEMS could offer the needed tunability for the passive functions: filters and matching circuits.

In addition to the component technologies, radio architectures more suitable for multi-mode operation can be applied. As a part of the overall architecture, the selection of reference clock frequencies is of utmost importance.

11.4 THE SMALLER THE BETTER

There seems to be more market demand on ultimately small phones than first predicted. Moreover, miniature phone engines are vital for the multi-mode terminals in which the electronics of two or more terminals are integrated into one product, since the new multi-mode terminals cannot be larger than current high-end single-mode cellular phones.

Figure 11.3 shows the development of the volume of a typical GSM hand-portable. The corresponding component count in the RF section is shown in Fig. 11.4. The significant improvements in size and component count of the hand-held digital cellular phones have been achieved by increasing the level of integration in the phone. Currently, all active functions are integrated, even in the RF section. On average, there are currently three RF ICs in a terminal. Simultaneously, there are approximately 100 passive discrete components in the RF section, as can be seen in Fig. 11.4. These discrete components dominate the size; consequently, it is obvious that reducing the number of discrete components is much more important than integrating the remaining ICs into a single chip.

In the evolutionary prediction illustrated in Fig. 11.3 and 11.4, the main improvements are made by applying new packaging and interconnection technologies. The old microwave hybrid technology is reborn with a slightly new appearance: custom ICs are assembled using chip-scale miniature packages, e.g. flip-chip, on a high-per-

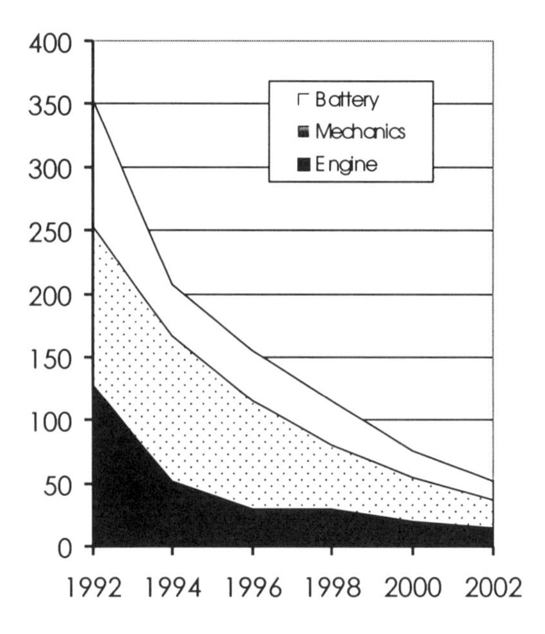

Figure 11.3 Development of volume (in cubic centimetres) of a typical GSM hand-portable terminal.

formance substrate with passive filters, interconnections, and external discrete components.

More revolutionary improvements in the miniaturisation might be achieved with advanced transceiver architectures shown in Fig. 11.5. The direct conversion architecture can be applied to some systems to remove the bulky IF filters. An ultimate goal could be an all-digital transceiver adaptive to all major wireless systems. In the all-digital transceiver, the sampling is carried out at the signal frequency. For several reasons, the approach is not yet feasible. The power consumption of the required Gbit/s signal processing and the high-resolution high-speed A-to-D conversion exceeds many times the acceptable level for portable devices. Furthermore, the stopband rejection specification for the remaining RF filter becomes extremely demanding: approximately 100 dB for narrow-band TDMA systems. With today's technology, this would lead to a filter having a physical size too large for hand-portable terminals.

11.5 ... FOR THE LOWEST COST

The prices of the cellular phones have dropped by tens of per cent annually over the last years. In order to maintain the competitive edge, the total cost of the product needs to be reduced. This trend will continue in the near future, and there is even more demand for cost reduction since the more complex — and hence more expensive — wide-band and multi-mode terminals are coming. It seems evident that customers are not willing to pay more for the new advanced terminals.

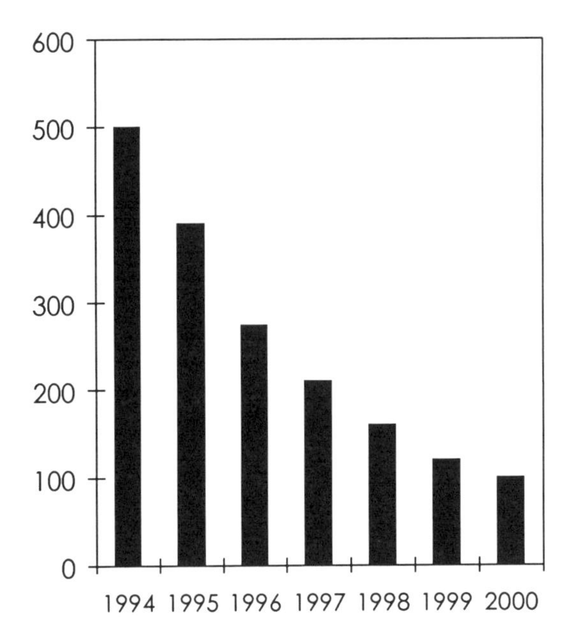

Figure 11.4 Development of component count in the RF section of a typical GSM hand-portable terminal.

Fortunately, the technologies that enable the miniaturisation provide, for the most part, also cost-reduction. The key issue is to optimise the radio architecture and the technology portfolio for the lowest total cost and still meet the specified performance. The production volumes of cellular phones have reached such high figures that in some technology areas, e.g. analog and RF ICs, the cellular phones have become the driver application. As a result, large terminal manufacturers can have a significant influence on the price and the development of the technology. The cost breakdown of the typical RF section is shown in Fig. 11.6. The cost implication of the direct-conversion approach versus the super-heterodyne is also shown.

The duration of the design cycle is a major contributor to the total cost of the terminal's RF section. A key goal, in particular in RF IC design with its long turnaround time, must be the first-time success. In this respect, the quality of the design environment and the device models play a critical role. The RF design environment must be highly integrated to offer seamless system- and transistor-level simulation capabilities, access to electro-magnetic simulation tools, and capability to extract parasitics and substrate couplings from the layout. With the increasing packaging densities, the electro-thermal simulation becomes a necessity as well.

11.6 CONCLUSIONS

The overall goal, namely to be able to design the world's best miniature multi-mode wide-band terminals with the ultimately lowest cost, is so challenging that it requires RF designers to do miracles in the coming years. New technologies introduce new

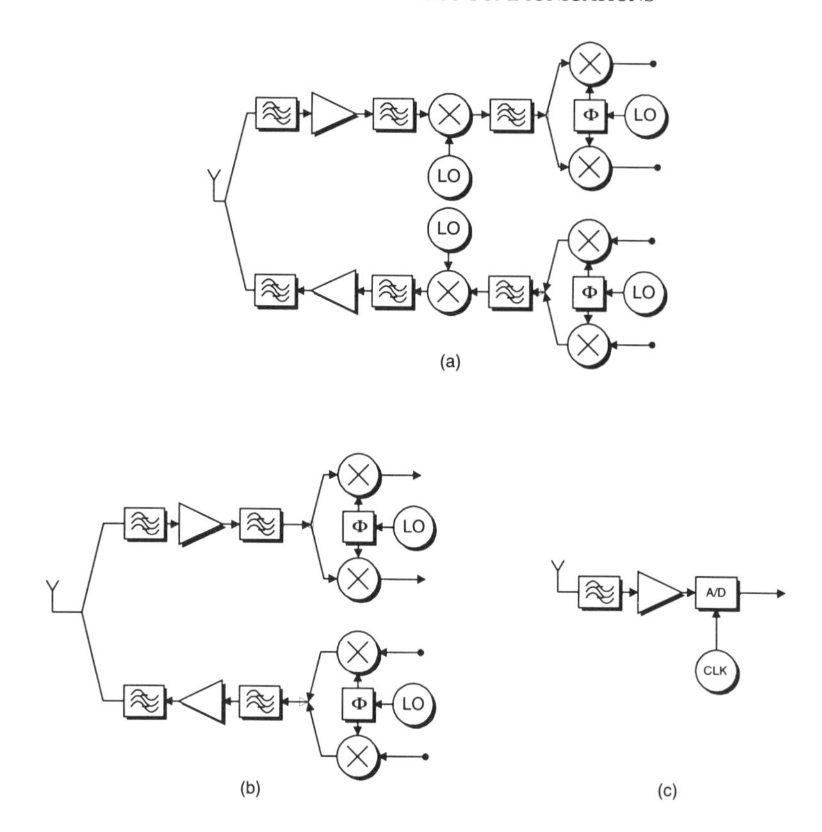

Figure 11.5 Radio front-end architectures: (a) super-heterodyne, (b) direct-conversion, (c) all-digital receiver.

solutions but, at the same time, also new problems. None of the implementation technologies will, however, offer a clear winning RF solution. Only by improving all factors — technologies, architectures, and design environment — the optimum leading edge RF section can be developed. For innovative RF development, the designers must master larger systems and there is a clear need for multi-discipline experts, e.g., in RF, DSP, and materials areas. With hard work, good vision, and committed design teams we will see that these miracles will take place — as so many other technological breakthroughs have in the past.

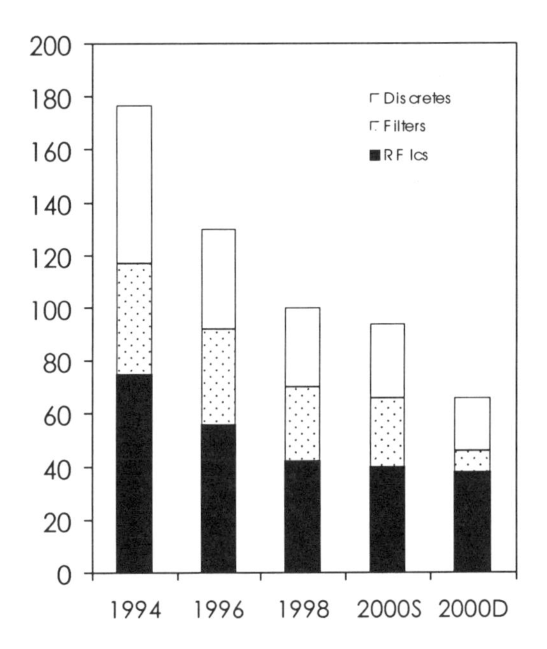

Figure 11.6 Relative cost breakdown in the typical RF section: 2000S is an evolutionary prediction for super-heterodyne transceiver and 2000D for direct conversion, respectively. The scale is normalised to be 100 in 1998.

III Wideband Conversion for Software Radio

12 WIDEBAND CONVERSION FOR SOFTWARE RADIO

José E. Franca

IST Centre for Microsystems
Instituto Superior Tecnico Av. Rovisco Pais, 1
1096 Lisboa Codex, Portugal

12.1 INTRODUCTION TO THE FOLLOWING PAPERS

Radio-frequency (RF) circuit design used to be realized using discrete bipolar devices, and later integrated bipolar technology was used as well. However, because of the dramatic growth of mobile communications and the need for providing highly miniaturised, cheaper equipment, there has been a recent surge of interest in realizing RF circuits in CMOS technology.

One of the challenges in RF design concerns the design of the front-end receiver section, whose main role is to interface the radio frequency bands allocated for communication with the baseband unit, which performs the bulk of the information processing functions needed in modern radio communications. Conceptually, such an RF front-end receiver can be viewed as a mixed-signal processing and conversion chain which combines a tunable bandpass filtering function, an analog–to–digital converter (ADC), and a digital signal processor (DSP), as schematically illustrated in Fig. 12.1.

Whilst previous sections of this book have been concerned with the practical realization of circuit and system solutions that provide the equivalent functionality of the tunable bandpass filter, this section is concerned with the design and integrated-circuit implementation of the analog–to–digital conversion function. At a later stage, the de-

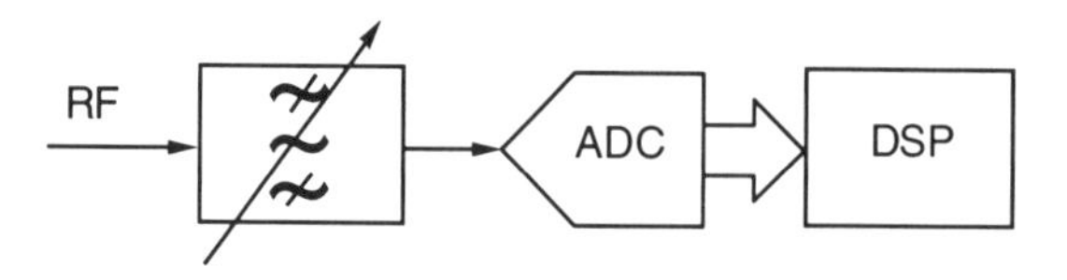

Figure 12.1 Simple conceptual representation of an RF receiver.

sign and implementation of dedicated DSP units for communication applications will also be considered.

Most of the work and design solutions reported to date have addressed the problem of narrow-band receivers, whereby only one channel is selected for baseband processing from the available multi-channel RF band [1]. Therefore, as indicated in Table 12.1, the effective channel bandwidth to be digitised by the ADC has varied typically from about 25 kHz for narrow-band frequency modulated receivers to approximately 1250 kHz for receivers based on spread-spectrum processing techniques. Neither represents a particularly stringent requirement for the design of integrated ADCs [2–6].

Table 12.1 Typical radio bandwidths and channel spacings for popular mobile radio systems.

Standard	*Radio Bandwidth [MHz]*	*Channel Spacing [kHz]*
PDC		25
GSM	25	200
DCS-1800	75	200
DECT	20	1728
IS-95 (CDMA)	25	1250
AMPS	25	30

The present conservative situation, however, is changing very rapidly, due to the insatiable demand for wider-band mobile communication systems capable of supporting high-speed data transmission as well as multimedia applications. Not too far in the future, the ADCs in radio receiver front-ends are expected to digitise effective signal bandwidths from a few MHz for hand-held terminals of the universal mobile telecommunication system (UMTS) to as much as 20–30 MHz in base stations implementing the software radio concept. High dynamic range, low spurious components, low-voltage operation, and low power dissipation are additional requirements which, besides wide bandwidth, significantly add to the design complexity of such ADCs.

This part of the book addresses the problem of designing high-performance, wide-bandwidth integrated ADCs suitable for future generations of mobile communication systems. First, Roovers introduces the performance metrics that are needed to evaluate the feasibility of receiver architectures with wide-band ADCs and compare the relative performance of alternative forms of realization. It is shown that when the performance of an ADC is mapped onto the requirements of a receiver architecture with multi-channel processing, it should be able to provide not only wide bandwidth but also high dynamic range with respect to noise limitations, distortion, and spurious signals. Since the combination of wide bandwidth and stringent dynamic range demands high power dissipation, such ADCs are not yet adequate for deployment in hand-held terminals, where energy consumption is under tight control. Nevertheless, when such ADCs are used in combination with some form of digitally-controlled signal amplification, they can already find useful applications in base stations, as it is illustrated in a case-study developed for a GSM receiver. It is expected that the continuing research of innovative circuit techniques combined with the availability of higher-performance technologies, including the emerging SiGe, will make such converters also widely used in portable equipment in the not too distant future.

Bang-Sup Song examines the current status of high-performance ADCs and discusses the fundamental techniques, namely pipeline-based and folding architectures, which are expected to enable the realization of at least 14 bit conversion linearity over an effective signal bandwidth higher than 25 MHz. The performance of pipeline ADCs is fundamentally limited by the speed and linearity of the interstage residue amplifiers, while the offset of the folding amplifier and overall circuit complexity appear to be the major limitations of folding ADCs in achieving high-performance operation. For both architectures, several improvements are discussed in order to increase effectively both the number of conversion bits and the conversion speed while dissipating significantly lower power than current solutions. Predictions based on recent work and trends indicate that high-performance pipeline ADCs will be capable of achieving 15 bit to 16 bit linearity with a spurious-free dynamic range over 90 dB at a 100 MHz sampling rate, and which will, therefore, make the widespread use of software radios feasible.

In the last contribution to this part of the book, Azeredo-Leme *et al.* begin by examining the contributions and requirements of different building blocks in a traditional receiver front-end architecture, with the aim of allowing a more relaxed realization of filters and synthesisers at the expense of a more complex ADC. For this purpose, they introduce the concept of an image-rejection sub-sampling ADC, which is a combination of a highly oversampled ADC together with a sub-sampling block implementing a double-quadrature mixer. It is shown that, for present mobile communication standards, this allows the digitisation of multiple radio frequency carriers and hence enabling the final channel tuning operation to be carried out fully in the digital domain. An additional advantage offered by this approach is the digital control over selectivity and tuning resolution, which lead to simplified implementations of multi-mode terminals. It is further shown that no penalty to power dissipation arises when compared with classical implementations.

References

[1] J. Rapeli, "IC Solutions for Mobile Telephones", *book chapter in Design of VLSI Circuits for Telecommunications and Signal Processing*, edited by J. Franca and Y. Tsividis, Prentice Hall 1994.

[2] Rabaey, J. Sevenhans, "The Challenges for Analog Circuit Design in Mobile Radio VLSI Chips", *in Advances in Analog Circuit Design*, edited by R. J. van de Plassche, W. M. Sansen, and J. H. Huijsing, Kluwer 1994.

[3] A. Abidi, "Low-Power Radio-Frequency ICs and System Architectures for Portable Communications", *book chapter in Low-power HF microelectronics: a unified approach*, edited by G. A. Machado.

[4] Sheng et al., "Analog and Digital CMOS Design for Spread-Spectrum Wireless Communications", *book chapter in Low-power HF microelectronics: a unified approach*, edited by G. A. Machado.

[5] "MOST RF Circuit Design", contributed by several authors, *Part I of Analog Circuit*, edited by W. Sansen, J. Huijsing, R. van de Plassche, Kluwer 1996.

[6] C. Rudell, *et al.*, "A 1.9 GHz wide-band IF double conversion CMOS integrated receiver for cordless telephone applications", *in Digest Technical Papers ISSCC 97*, February 1997.

13 WIDE-BAND SUB-SAMPLING A/D CONVERSION WITH IMAGE REJECTION

C. Azeredo-Leme, Ricardo Reis, and Eduardo Viegas

Instituto Superior Técnico, Microssystems Centre
Av. Rovisco Pais 1, 1096 Lisboa Codex, Portugal

13.1 INTRODUCTION

Mobile communications have enjoyed a very fast growth and acceptance virtually across all continents. However, today we observe a multitude of standards with little or no compatibility with each other. They differ both in the allocated bands and in the modes of operation (FDMA, TDMA, CDMA to mention only the multiple access techniques) as shown in Tab. 13.1. This has created the need for multi-band and multi-mode terminals.

However, the classical super-heterodyne architecture, with the large number of accurate frequency selective filters it requires, makes it difficult to build such a multi-standard radio in a way compatible with today's demand for very high level of integration and long stand-by times.

This led to the development of the concept of "Software Radio" where the functions of channel selection and filtering would be done digitally by software. For example, the telephone would have a memory module with the available standards at the purchase date, and updates to new standards could be loaded later from a computer connection.

Table 13.1 Standards parameters for popular mobile radio systems.

New Standards	*Centre Band*	*Channel Spacing*	*Multiple Access*	*Modulation*
PDC	810–826 MHz (Rx) 940–956 MHz (Tx)	25 kHz	TDMA	$\pi/4$ DQPSK
GSM	935–960 MHz (Rx) 890–915 MHz (Tx)	200 kHz	TDMA	GMSK
DCS-1800	1805–1880 MHz (Rx) 1710–1785 MHz (Tx)	200 kHz	TDMA	GMSK
DECT	1880–1900 MHz	1728 kHz	TDMA	GMSK
IS-95	869–894 MHz (Rx) 824–849 MHz (Tx)	1250 kHz	CDMA	QPSK
AMPS	869–894 MHz (Rx) 824–849 MHz (Tx)	30 kHz	FDMA	FM

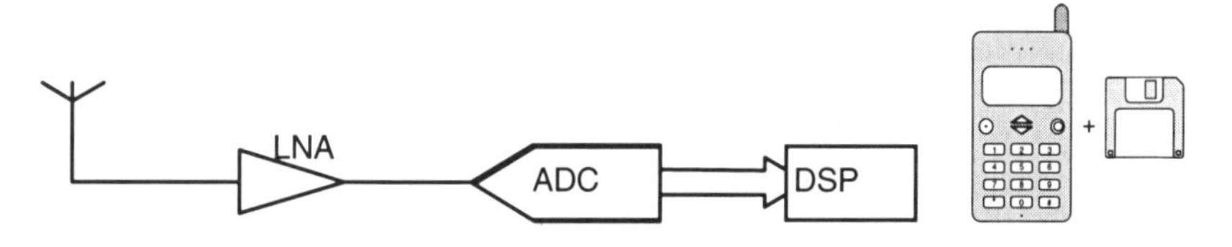

Figure 13.1 Unrealizable software radio concept where the radio is totally software adaptable.

Consider the receiver architecture. A broad-band LNA followed by an ADC digitise the received antenna signal, and all processing would continue in a DSP, as shown in Fig. 13.1.

Obviously, this solution is completely unrealizable. Even if the DSP processing power were to ever become available, the specifications on the ADC would be prohibitive. The sampling rate would have to be of the order of 1 GHz in order to avoid tunable filters, and the dynamic range would have to be well in excess of 100 dB since no blocking filter is present.

However, the search for a receiver architecture requiring as few high-frequency filters as possible is very active at present.

Practical implementations of receivers basically rely on the sequencing of down-converting mixers and IF filters until the desired signal is at baseband. That is the case with the classical super-heterodyne architecture, see Fig. 13.2. The zero-IF receiver (which is also classical) eliminates the IF but also places all channel selection and fil-

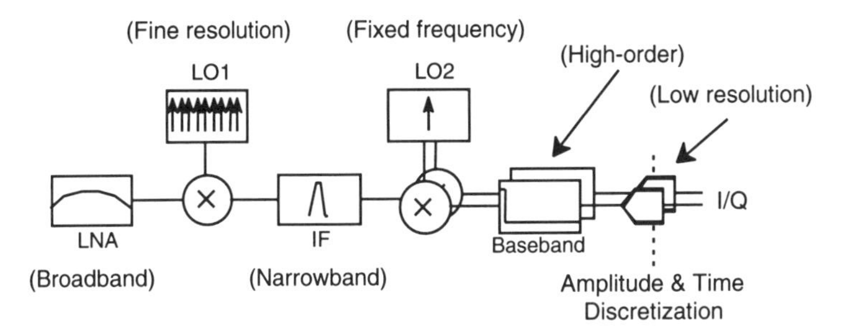

Figure 13.2 Classical super-heterodyne architecture. Tuning is performed by the RF synthesiser LO1 while the IF synthesiser is fixed. Channel filtering is performed at IF and in the baseband. The ADC requirements are very relaxed.

tering in front of the ADC. More sophisticated architectures, like low-IF and quasi-IF, relax the requirements placed on the IF filters by careful use of image rejection mixers, but still place the ADC after channel selection and filtering.

The result is that the ADC is the component with the most relaxed specifications in the receiver, while the filters need to accommodate very large dynamic ranges and are of high order (and high Q in super-heterodyne architectures).

On the other hand, the need to perform the channel selection requires that the frequency synthesisers have a very fine frequency resolution equivalent to the channel spacing. Depending on the standard, the filters' dynamic range can be above 100 dB, and the synthesiser's frequency resolution can be as low as 30 kHz. The question we want to answer here is this: "Is it possible to relax the requirements placed on the filters and frequency synthesisers at the expense of a more complex ADC, but with no penalty on power dissipation or monolithic compatibility?"

13.2 OVERSAMPLED ARCHITECTURES

It is better to place the ADC before the channel filters, as shown in Fig. 13.3. The ADC must then be highly oversampled in order not to alias high frequency interferers into the baseband.

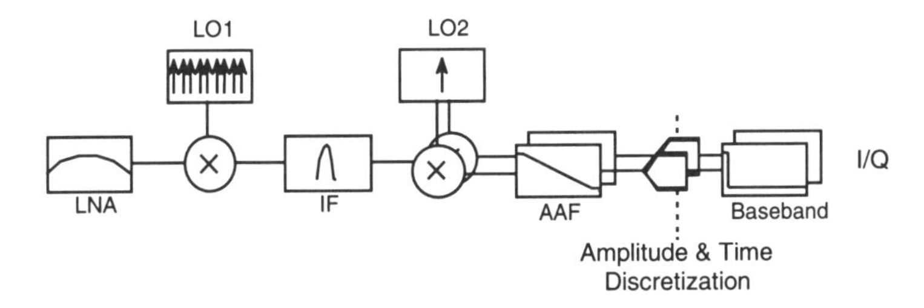

Figure 13.3 Super-heterodyne architecture with oversampling.

Irrespective of the receiver architecture, the baseband signal produced by the down-conversion mixer contains the channel signal at a weak level in addition to out-of-band interferers that can be much stronger that the desired channel itself. Depending on the IF filters' selectivity and on the standard, these interferers can be as high as 70 dB above the desired signal. Taking into account fast-fading effects and AGC margins, the ADC will have to support a large dynamic range of 60 to 90 dB.

The most convenient type of ADC to place at this point is a delta-sigma modulator (two are required, since the signal is in I and Q representation). Strong adjacent channels are combined with quantisation noise and actually act as a dither source and help to randomise the bit stream. The digital decimating filter will then serve the dual function of attenuating both the quantisation noise and the adjacent channels interference.

In practice, the order of the baseband filter and the ADC has been interchanged. The filter is now implemented digitally, reusing existing hardware — the decimating filter. Only a simple anti-aliasing filter remains before the ADC, and the ADC has been highly oversampled to avoid aliasing by unfiltered high-frequency interferers and noise. As a result, the dynamic range, which was supported by the filter in the classical architecture, is now supported by the ADC.

The advantages of this architecture are numerous. The I/Q matching accuracy is very good, depending basically on the matching of the two ADC input stages. The portability between different technologies is facilitated since more of the signal processing is done digitally. The filters can have a very accurate frequency response and a linear phase characteristic important for digital modulation techniques. Besides, the channel bandwidth can be easily programmed.

But the ADC is required to operate at a very high sampling rate and must achieve a large dynamic range. We now compare the power dissipation of this ADC and that of the original filter it replaced.

13.3 POWER DISSIPATION ISSUES

The power dissipation of an analog circuit depends on many considerations, most of them related to the limitations of the technology being used. But it is instructive to consider the fundamental limitations of analog signal processing.

The basic function of an analog circuit is to process a voltage (or current) signal, identified by its bandwidth B and amplitude, with a maximum degradation characterised by distortion and noise. Regarding distortion, there are no relevant physical limits determining a minimum power dissipation level. But in the case of noise, there is a clear relationship involving temperature, signal bandwidth and circuit dynamic range. For the case of voltage signals and optimal supply voltage corresponding to the signal amplitude, it is given by the following expression [1]:

$$P_{\text{diss}} = 8kT \cdot B \cdot DR\,, \tag{13.1}$$

where k is the Boltzmann constant and T is the absolute temperature. The dynamic range, DR, is defined as the power ratio between the maximum signal amplitude (can be an out of band interferer) to the thermal noise integrated over the signal bandwidth B. The power associated to (13.1) is dissipated by the necessary charging and discharging of the capacitance on each circuit node carrying the signal. In a filter there are, as a minimum, N such nodes, where N is the filter order. In an ADC, only one such node is necessary as a minimum: the sampling node.

This analysis is valid irrespective of continuous-time or discrete-time circuits. The power dissipation is not affected by the actual sampling frequency of the ADC for a fixed signal bandwidth and dynamic range. This leads to two conclusions:

1. The power dissipation of the ADC is not necessarily higher than that of the analog filter it replaced.

2. The ADC can be oversampled by a large factor without penalty on the power dissipation.

This second conclusion is important because it allows to use a sufficiently large sampling frequency so that the ADC images will be far enough from the desired channel for the anti-aliasing filters to be of simple construction. It means that for a fixed signal bandwidth (the channel bandwidth) and fixed dynamic range, the sampling frequency can be increased without necessarily increasing the power consumption.

There are practical limits to this, however. In switched-capacitor (SC) circuits, as the sampling frequency increases, the sampling capacitors can be decreased by the same factor, to keep the thermal noise $k \cdot (T/C)$ constant within the signal band. The reduced load capacitors allow to keep the same amplifier slew rate at the original bias currents, but the amplifier needs to have a higher gain-bandwidth product. That, in turn, cannot be made higher than the internal non-dominant poles of the amplifier, otherwise the phase margin will be reduced to unacceptable values.

Considering CMOS technology, the most power-efficient operating point of the transconductance defining MOS transistors in an amplifier is in moderate inversion with an over-drive voltage $V_{OV} = V_{GS} - V_{\mathrm{T}} \approx 50\,\mathrm{mV}$. At this point the ratio G_M/I_{D} is maximum. However, the transit frequency, F_{T}, of the transistors in the signal path must be above the sampling frequency by a factor of typically 20, at least, to preserve phase margin.

In CMOS, F_{T} of the transistors is a function of the operating point:

$$F_{\mathrm{T}} = \frac{\mu \cdot V_{\mathrm{OV}}}{L^2} \,. \qquad (13.2)$$

An NMOS transistor in a 0.35 μm technology has an F_{T} of about 2 GHz, allowing sampling frequencies approaching 100 MHz at optimum power dissipation, in the sense that the transistors responsible for poles are biased at the maximum G_M/I_{D} operating point. In fact, there are already several reports of delta-sigma modulators validated for sampling frequencies in excess of 50 MHz [2–4].

Above that frequency, the over-drive voltage of the transistors will have to be increased at least proportionally to the sampling frequency in order to keep their transit frequency high enough. Since the MOS transconductance is of the form

$$G_M = 2\frac{I_{\mathrm{D}}}{V_{\mathrm{OV}}} \,, \qquad (13.3)$$

the bias current will have to be increased as well in order to keep the transconductance values fixed. This makes the power consumption increase linearly with the sampling frequency, until the over-drive voltage is large enough for the transistors to enter the velocity saturation region. Then the power dissipation will increase very fast with increases in sampling frequency. This is illustrated in Fig. 13.4.

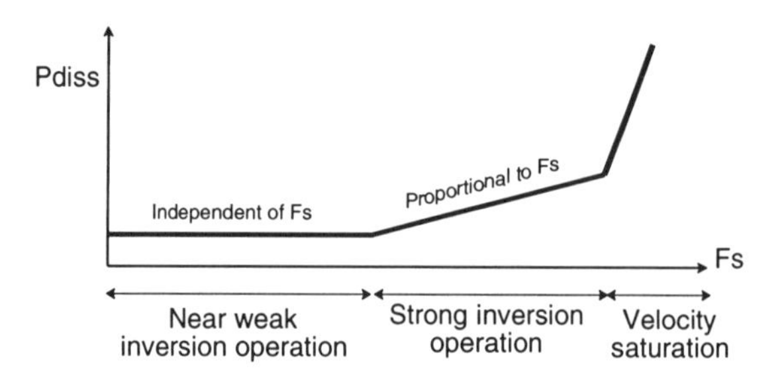

Figure 13.4 Typical power dissipation of a CMOS amplifier for constant $(F_s \cdot C_L)$.

13.4 IF SAMPLING ARCHITECTURES

IF sampling architectures directly interface the ADC to a signal band centred at IF. This leads to a convenient system partitioning where the IF is the interface between the baseband signal processing chip and the RF front-end.

The analysis in the previous section showed that it is possible to efficiently increase the sampling frequency of an SC circuit up to a frequency of 100 MHz. However, the IF frequencies are often above this frequency, requiring more advanced architectures to allow direct sampling of the IF signal.

The solution is to operate in a sub-sampling mode. In this mode, the amplifiers operate at a reduced rate, and only the input SC branches need to settle accurately with IF input signals. Since these are simple passive branches, their operating speed can be much higher than that of the active charge transfers performed by the amplifiers, and can easily allow sampling signals well above 100 MHz.

There are several possible approaches to integrate the complete IF-sampling receiver in CMOS.

One is to keep exactly the architecture of Fig. 13.3. The mixer could either be implemented as a classical Gilbert cell, using triode-region-biased transistors [6], or using a sub-sampling SC circuit. Double-quadrature architectures can also be considered for improved image rejection in the RF front-end providing I and Q IF.

13.5 IMAGE-REJECTION SUB-SAMPLING ADC

A more ambitious approach is to embed the sub-sampling SC mixer in the ADC, as shown in Fig. 13.5. The anti-aliasing filter is eliminated and its function is performed by the IF filters. This approach results in a significant reduction of receiver complexity and a potential power dissipation reduction since fewer blocks are required.

The IF filters can be strongly relaxed since they only need to attenuate the aliasing bands of the ADC. The sub-sampling SC mixer can be implemented by appropriate phasing of the input SC branches of the first integrator of the delta-sigma modulator.

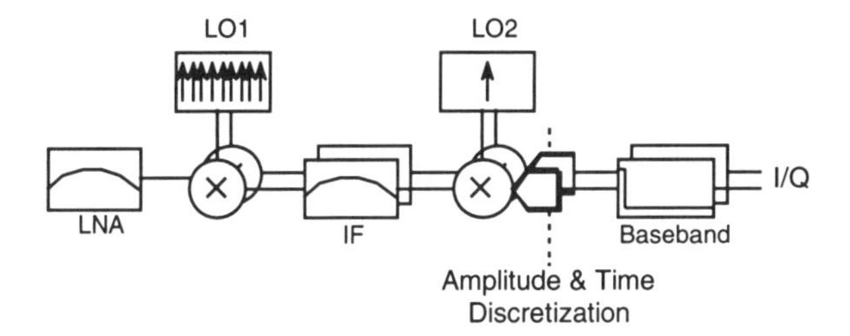

Figure 13.5 Receiver architecture using an image-rejection sub-sampling ADC.

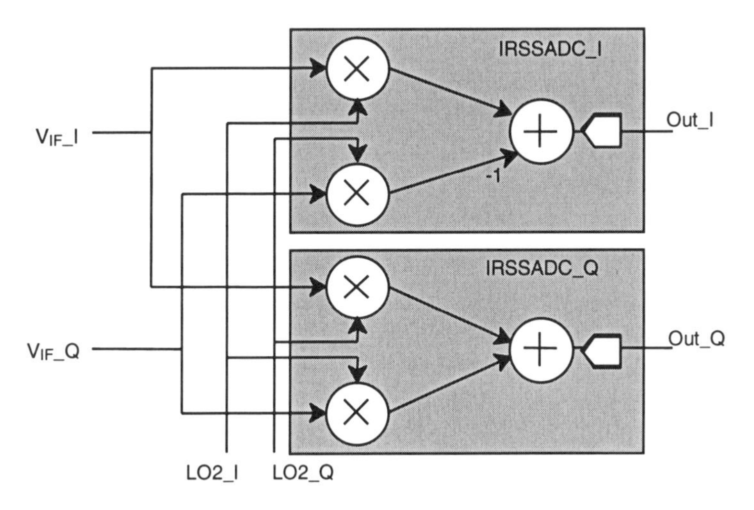

Figure 13.6 Block diagram of IF sampling block using two IRSSADCs.

We denote the component combining the functions of sub-sampling mixing, image rejection (due to the double-quadrature topology) and analog-to-digital conversion as Image-Rejection Sub-Sampling ADC (IRSSADC). The application of IRSSADC for the realization of the IF sampling block is shown in Fig. 13.6, where $LO2_I$ and $LO2_Q$ are the in-phase and quadrature signals of the IF local oscillator. Two IRSSADCs are required, realizing a double-quadrature down-conversion from an I and Q IF.

The mixing and addition operations are realized directly at the input SC branches of the ADC. These are exemplified in Fig. 13.7 where the output is connected to the virtual ground of the first delta-sigma integrator. In a practical realization, the circuit would be fully differential and would use parasitic-insensitive SC branches.

The clock phases in Fig. 13.7 realize a LO frequency at half the clock rate of the delta-sigma modulator, F_s. Since the switches implement impulse sampling, all odd harmonics of the LO exist, thereby allowing sub-sampling operation.

Since the LO is a complex-domain circuit, the LO frequency should actually be analysed also in the complex domain, where the frequency axis is not symmetrical. Assuming a model for the mixer in the form of a multiplier, the time waveforms repre-

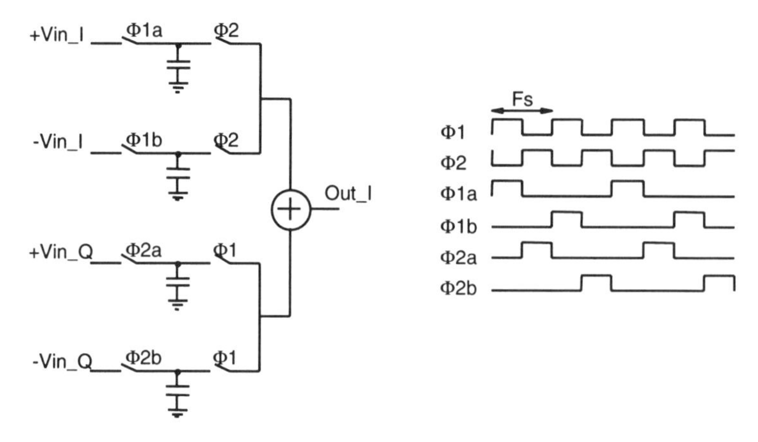

Figure 13.7 Input SC branches with corresponding clock phases for implementing the IRSSADC.

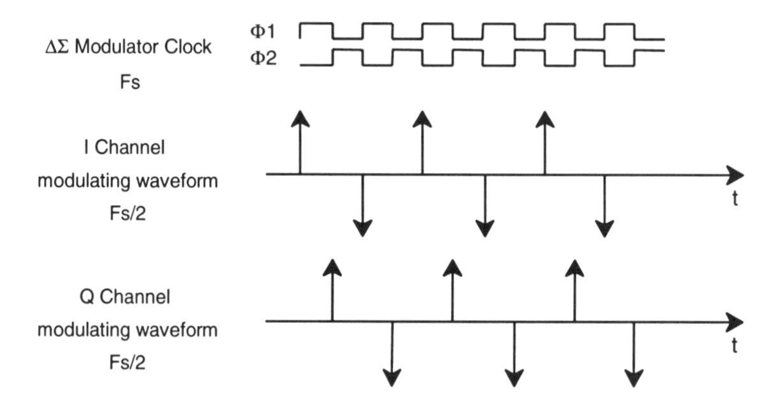

Figure 13.8 Time domain equivalent waveforms representing the I and Q LO.

senting the I and Q LO signal are as shown in Fig. 13.8. The arrows denote the Dirac function corresponding to ideal impulse sampling.

Applying the Fourier transform, we obtain the frequency representation shown in Fig. 13.9. As expected, the spectrum is not symmetrical. Since the IF signal is also in a complex representation, the relevant band is located in the positive frequency axis. The frequency band in the negative frequency axis corresponds to the image band.

The operation of this mixer is now clear: Consider the bands around $F_S/2$, $5F_S/2$, $9F_S/2, \ldots$ Since the local oscillator frequency, LO2, is $F_S/2$, these bands correspond to the fundamental, 5th harmonic, 9th harmonic, etc., of LO2. Therefore, a sub-sampling operation by a factor of 1, 5, 9, ... , is possible. If we eliminate the inversion in Fig. 13.6, it can be similarly shown that the sub-sampling factors become 3, 7, 11, ...

The presence of multiple input bands is typical of sampled systems and the desired band should be selected by an anti-aliasing filter in front of the mixer. This filter can

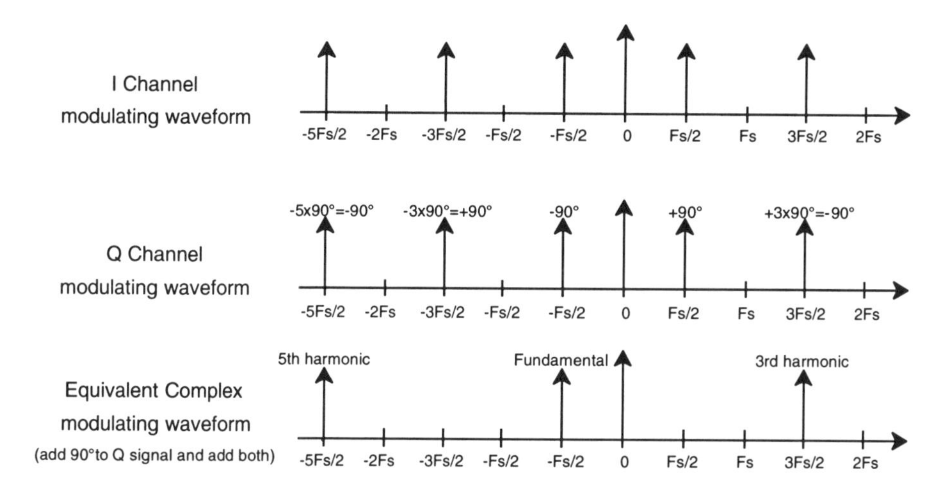

Figure 13.9 Frequency domain equivalent spectrum representing the I and Q LO.

be the IF filter itself and should reject the bands ±4 times LO2 away from the centre IF band.

13.5.1 Case study

As a practical example, for an IF centre frequency of 150 MHz, and when operating with a sub-sampling factor of 5, the local oscillator frequency is 30 MHz and the delta-sigma modulator sampling frequency is 60 MHz. The IF filter should reject the aliasing bands around 30 MHz and 270 MHz.

Assuming a typical application such as GSM, where the baseband channel bandwidth is 100 kHz, the delta-sigma modulator would be operating with an oversampling ratio of $60000/200 = 300$. This would easily allow to achieve a dynamic range in excess of 90 dB [4]. This dynamic range is higher than the ratio of maximum blocking levels (within GSM band) to the minimum signal level, which is about 80 dB.

The previous example confirms that the IF filters can be substantially relaxed since their function is basically to reject aliasing bands.

13.6 DIGITAL TUNING

As already mentioned, the dynamic range of the ADC is increased by proper sizing of the sampling capacitance, which, for a large oversampling frequency, can be very small. This concerns the level of thermal noise. In delta-sigma modulators one must also consider the quantisation noise. In these circuits, the signal-to-quantisation-noise ratio (SQR) is a strong function of the oversampling ratio. By proper use of techniques such as

- high order modulator loops,
- multi-bit quantising,
- data-weighted averaging linearisation,

an SQR above 90 dB can be easily achieved at a low oversampling ratio such as 16. This means that in the present application where the oversampling ratio is of the order of 100, the quantisation noise is negligible compared to thermal noise within the channel bandwidth.

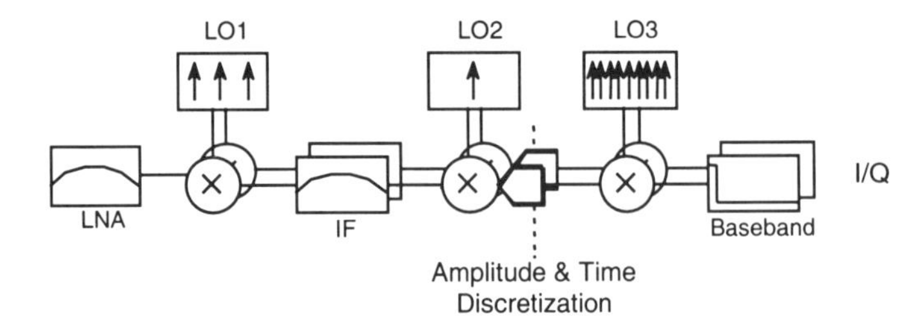

Figure 13.10 Receiver architecture using an image-rejection sub-sampling ADC and digital tuning.

This motivates a more ambitious application of the IRSSADC. The idea is to use broadband digitalisation where a group of channels is down-converted to a very low IF. The channel selection is then performed by a third mixer, as shown in Fig. 13.10. At this point, signal processing is already digital, which allows much flexibility in defining channel spacing and channel filtering.

For example, in the case study above, the IRSSADC, at an oversampling ratio of 16, can digitise a band of 60 MHz / (2x16) = 1875 kHz. This allows to include 9 GSM channels at once (the channel spacing is 200 kHz).

The high flexibility due to easy digital programmability of channel centre frequency and spacing is very advantageous for implementing a multi-mode/multi-band terminal.

Furthermore, the RF frequency synthesiser can now have a coarse frequency resolution, which significantly relaxes its design specifications.

13.7 ARCHITECTURE SIMULATION

A Matlab/Simulink model has been developed to represent the architecture shown in Fig. 13.11 consisting of an IF sub-sampling quadrature mixer, I and Q low-pass filters, I and Q Delta-Sigma modulators, digital decimating filters and digital tuning. The IF filter is not included.

The Delta-Sigma modulators are of third order with 3-bit quantisation. They use dynamic data averaging linearisation of feedback multi-bit DACs. The GSM case-study presented above is used here again.

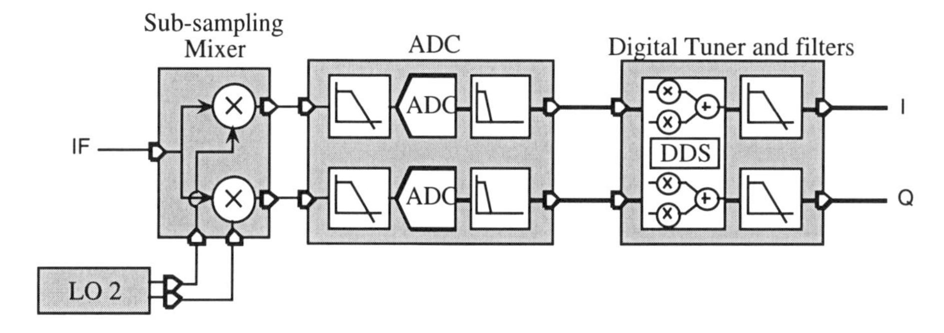

Figure 13.11 Architecture considered for simulation.

The sub-sampling mixer is a switched-capacitor circuit that samples the signal at an F_s of 60 MHz and mixes it with two sampled sine waves in quadrature with a frequency of $F_s/2 = 30$ MHz. This mixer generates I and Q sampled signals, which are filtered by two anti-aliasing filters. The first one is a first-order switched-capacitor low-pass filter (actually it is embedded in the mixer), and the second is a continuous-time second-order Butterworth low-pass filter. The signal is then converted to the digital domain, using a third-order delta-sigma converter sampling at F_s and decimated with a sinc^4 filter into $60/8 = 7.5$ MHz. Afterwards, the signal is mixed digitally with a DDS-generated sine wave, converting the desired channel to base-band. Finally the signal is low-pass filtered and decimated by 16 to obtain the desired I and Q components of the signal.

The signals applied are:

- Desired signal at 150.8 MHz with an amplitude 100 μV (0 dBref)
- Interferer 1 at 150.6 MHz (1 channel offset), amplitude 300 μV (+9.5 dBref)
- Interferer 2 at 150.4 MHz (2 channels offset), amplitude 12 mV (+41 dBref)
- Interferer 3 at 149.2 MHz (8 channels offset), amplitude 12 mV (+41 dBref)
- Interferer 4 at 147.8 MHz (15 channels offset), amplitude 630 mV (+76 dBref)
- Interferer 5 at 89.25 MHz (near 3/5 of IF), amplitude 25 μV (−12 dBref)
- Interferer 6 at 119.8 MHz (near 4/5 of IF), amplitude 100 mV (+60 dBref)

The spectrum at the input of the Delta-Sigma modulator is shown in Fig. 13.12a, where the signals around IF were shifted to DC by the sub-sampling mixer. The interferer I6 is down-converted but strongly attenuated by the anti-aliasing filters (not shown). The interferer 5 is down-converted without attenuation, because it lies around the third harmonic of the IF local oscillator (90 MHz). If a complex IF were used, as in Fig. 13.5, this image would be strongly attenuated. Fig. 13.12b shows the spectrum at the output of the delta-sigma modulator, together with the transfer functions of the anti-aliasing filters (curve 1) and of the sinc^4 decimating filter (curve 2). The signal dynamic range is still 76 dB (ratio between the strongest interferer, I4, and the desired signal).

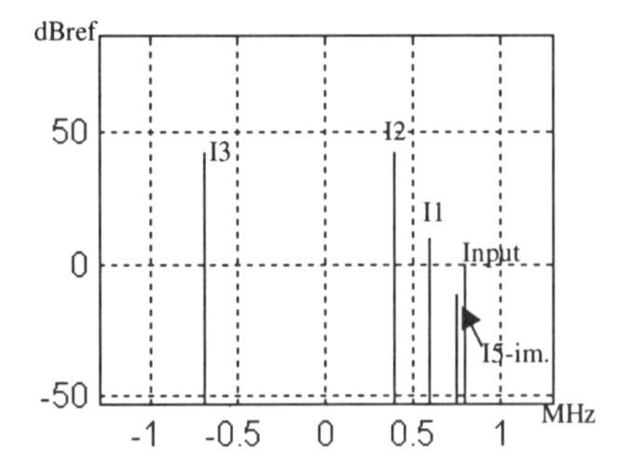

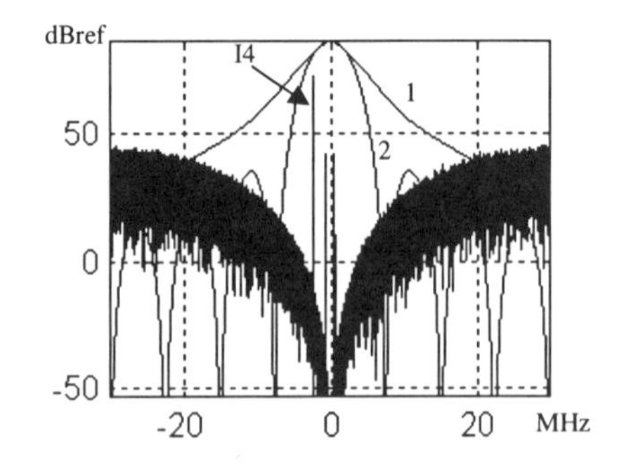

Figure 13.12 ADC $\Delta\Sigma$ converter input (a), ADC ($\Delta\Sigma$) converter output (b).

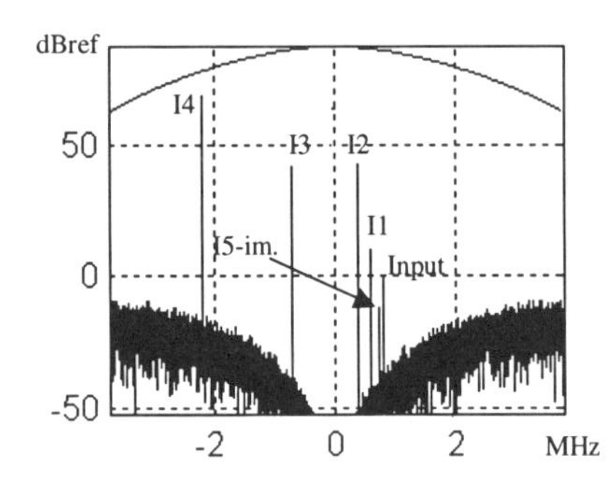

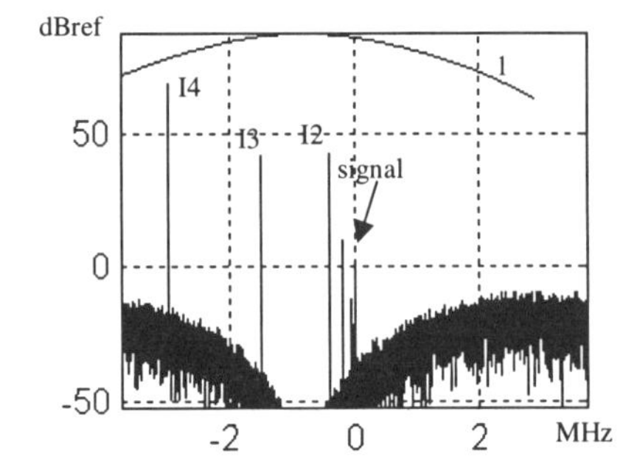

Figure 13.13 Decimator filter output (a), Digital mixer output (b).

Fig. 13.13a shows the resulting spectrum at the output of the decimating filter, before the digital mixer, together with the transfer function of the anti-aliasing filter plus sinc. The Delta-Sigma converter quantisation noise is well below the desired signal.

Fig. 13.13b represents the digital mixer output. Curve 1 corresponds to the combined transfer function of the anti-aliasing filter plus sinc^4. Now the desired signal is centred at DC and all interferers are in the rejection band of the digital low-pass decimation filters (not shown for clarity).

In Fig. 13.14, the spectrum at the output of the system is shown with the interferers strongly attenuated and the desired signal well above the noise level. The last digital low-pass filter transfer function is also represented. The image of interferer 5 can also be seen, generated in the sub-sampling mixer, and the image of interferer 2 generated in the last decimation by two in the digital filter. Since this filter has a stop-band attenuation of 65 dB and the interferer is 41 dB above the desired signal, the image is attenuated to 23 dB below the desired signal.

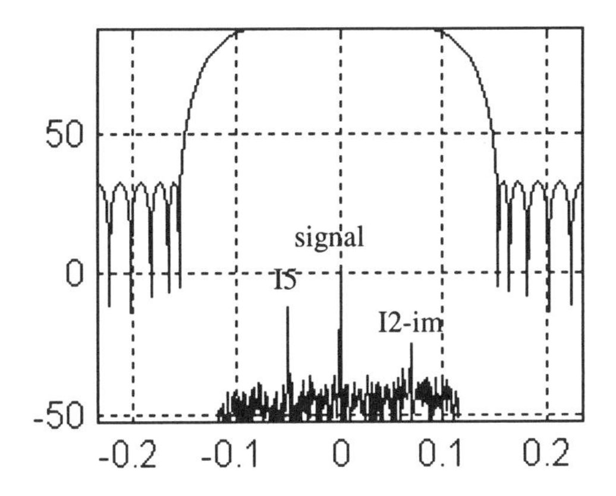

Figure 13.14 System output.

13.8 CONCLUSIONS

We introduced the concept of the image-rejection sub-sampling ADC, which is a combination of a highly oversampled ADC associated with a sub-sampling block implementing a double-quadrature mixer.

This allows to implement wide-band conversion for radio receivers with associated advantages. On the one hand, software tuning and selectivity becomes possible, simplifying implementation of multi-mode capability. On the other hand, the specifications of the IF filters and of the RF frequency synthesiser can be relaxed, the former due to the large dynamic range of the ADC that accepts a high level of blocking signals, the latter due to the software tuning that allows coarse frequency resolution in the RF frequency synthesiser.

An analysis was presented that shows that these advantages should be possible with no penalty on power dissipation, and simulation results demonstrate the efficacy of the principle.

Acknowledgments

The authors wish to acknowledge the financial support of the European Commission under ESPRIT project n. 25476, PAPRICA.

References

[1] R. Castello and P. Gray, “Performance Limitations in Switched-Capacitor Filters”, *IEEE Transactions on Circuit and Systems*, Vol. CAS 32, No. 9, Sep. 1985, pp. 865–876.

[2] B. Brandt and Bruce Wooley, “A 50-MHz Multibit Sigma-Delta Modulator for 12-b 2-MHz A/D Conversion”, *IEEE J. Solid-State Circuits*, Vol. 26, No. 12, Dec. 1991, pp. 1746–1756.

[3] G. Yin and W. Sansen, "A High-Frequency and High-Resolution Fourth-Order $\Sigma\Delta$ A/D Converter in BiCMOS Technology", *IEEE J. Solid-State Circuits*, Vol. 29, No. 8, Aug. 1994, pp. 857–865.

[4] A. Marques, V. Peluso, M. Steyaert and W. Sansen, "A 15-b Resolution 2-MHz Nyquist Rate $\Sigma\Delta$ ADC in a 1-μm CMOS Technology", *IEEE J. Solid-State Circuits*, Vol. 33, No. 7, Jul. 1998, pp. 1065–1075.

[5] A. Feldman, B. Boser and P. Gray, "A 13-Bit, 1.4-MS/s Sigma-Delta Modulator for RF Baseband Channel Applications", *IEEE J. Solid-State Circuits*, Vol. 33, No. 10, Oct. 1998, pp. 1462–1469.

[6] J. Crols and M. Steyaert, "A Highly Linear CMOS Downconversion Mixer", *IEEE J. Solid-State Circuits*, Vol. 30, No. 7, Jul. 1995, pp. 736–742.

14 WIDE-BAND A/D CONVERSION FOR BASE STATIONS

Raf L. J. Roovers

Mixed-Signal Circuits & Systems group
Philips Research Laboratories
Prof. Holstlaan 4, 5656 AA Eindhoven, The Netherlands

14.1 INTRODUCTION

In order to realise a multi-channel receiver system, a wide-band A/D converter must be available which can convert a complete band into the digital domain. The basic schematic of a multi-channel receiver is shown in Fig. 14.1.

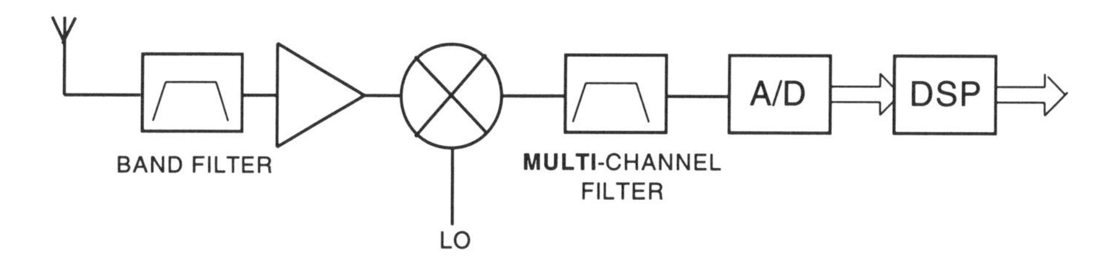

Figure 14.1 Multi-channel receiver schematic.

The multi-channel receiver has an antenna input (RF) followed by a band filter and a low noise amplifier (LNA). The RF signal is down-converted to an IF frequency and again some filtering can be applied. The IF signal containing the complete signal band is then converted to the digital domain by a wide-band A/D converter. The channel selectivity is implemented in the digital domain. The advantages of such a system are obvious: increased flexibility and reduced analog hardware when more than one channel is to be received simultaneously. These advantages of multi-channel receiver are particularly interesting for cellular base stations, where multiple channels are received. The signal at the input of the A/D converter combines a wide bandwidth with a large dynamic range. This large dynamic range is due to the fact that neither channel filtering nor variable gain can be used before the A/D conversion in a multi-channel receiver. This means that the converter has to handle at the same time maximum blocking and interfering signal levels while still receiving channels with signals at minimum power levels. This puts severe performance requirements on the linearity and noise contributions of the A/D converter,

14.2 PERFORMANCE METRICS FOR A/D CONVERTERS

There are two reasons to introduce performance metrics for A/D converters:

- to evaluate the feasibility of receiver architecture with a wide-band A/D converter,
- to compare different A/D converter realisations (benchmarking).

The performance metrics used here are based on the effective performance in terms of dynamic range, bandwidth, and power dissipation. These three parameters are combined into a global figure of merit. It should be clear to the reader that other parameters such as latency, power supply voltage, input range, input impedance, etc. are important as well, but these are not included in the figure of merit. First the resolution bandwidth product of an A/D converter is introduced. The A/D converter can be regarded as a black box: analog in, digital out as shown in Fig. 14.2.

Assuming that a full-scale, single-sine-wave input signal is applied to the A/D converter, the digital output can be analysed using its Fast Fourier Transform (FFT) spectrum. This FFT spectrum contains the input signal together with noise and harmonic components. From this FFT spectrum the *SINAD* (Signal to Noise And Distortion) is calculated. This *SINAD* can be translated in *ENOB* (Effective Number Of Bits), describing the effective resolution of the A/D converter:

$$ENOB = (SINAD - 1.76\,\text{dB})/6.02 \tag{14.1}$$

This *ENOB* is measured in function of the input signal frequency (c.f. Fig. 14.2). The closer the *ENOB* approaches the nominal number of bits the better. The useful bandwidth of the A/D converter can be limited in two ways: First, the Nyquist criterion gives a maximum bandwidth to be converted equal to half the sampling frequency ($f_s/2$). The second limitation is due to the fact that A/D converters tend to have a reduced *ENOB* for high-frequency input signals. Therefore the effective resolution bandwidth is defined as the minimum of the Nyquist frequency ($f_s/2$) and the input signal frequency for which the *ENOB* is reduced by 0.5 bit ($f_{\text{in, enob}-0,5}$) compared to the *ENOB* at low input frequencies ($ENOB_0$). The effective resolution bandwidth

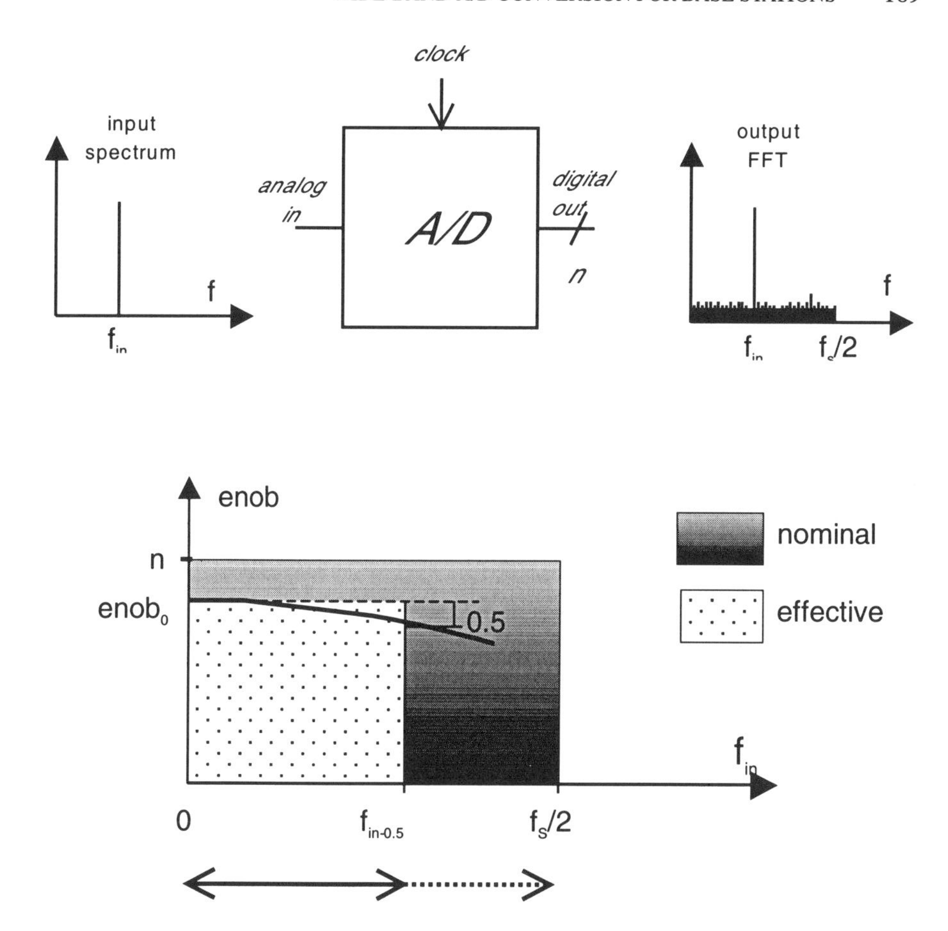

Figure 14.2 A/D converter: Black-box representation and *ENOB* in function of f_{in}.

product (*BWres*) combines both the effective resolution and the effective resolution bandwidth:

$$BWres = 2^{ENOB_0} \cdot \min(f_s/2, f_{\text{in, enob}-0,5}) \tag{14.2}$$

This product is proportional to the area in Fig. 14.2, where the difference to the nominal resolution-bandwidth becomes clear. The effective resolution-bandwidth products for some A/D converters are plotted in Fig. 14.3.

If the effective resolution bandwidth product is combined with the power dissipation of the A/D converter a Figure of Merit (*F.o.M.*) can be defined:

$$F.o.M. = \text{power}/2 \cdot BWres \tag{14.3}$$

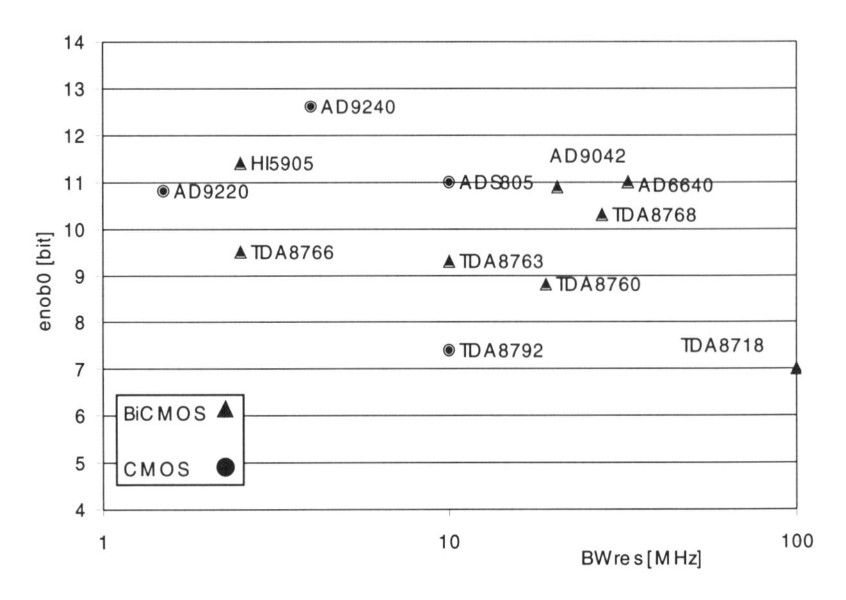

Figure 14.3 *BWres* and $ENOB_0$ for some A/D converters. This data is derived from manufacturers data sheets [2].

The factor of 2 is introduced to be compatible with the existing *F.o.M.* based on nominal performance, i.e. $F.o.M_{\text{nominal}} = \text{power}/(2^n f_s)$. The dimension of the *F.o.M* is Joule per conversion, and for better performance its value should be as small as possible. In Fig. 14.4 the evolution of this *F.o.M.* as a function of time is plotted.

The improvement in A/D converter performance over time is both due to design innovations and technology improvements. At this moment, wide-band A/D converters (10...14 bit) that achieve a *F.o.M.* of 5 pJ per conversion are state of the art. The *F.o.M.* can be used to estimate the A/D converter's power dissipation for various applications in a resolution-bandwidth graph as shown in Fig. 14.5. The lines of constant power dissipation are based on a *F.o.M.* of 5 pJ per conversion.

The wide-band A/D converters to be used in base-station receivers require bandwidths in the order of tens of MHz and resolutions in the order of 12 to 16 bit, depending on the IF frequency used and the amount of filtering in the analog part of the receiver. These A/D converters have a power consumption in the order of 1 Watt and are therefore not suitable for implementation in handsets. For base station application, these levels of power dissipation are acceptable, and the benefits of receivers with wide-band A/D converters can be exploited as multiple channels share the same analog front-end plus A/D converter. The realisation of this A/D converter for multi-channel receivers is the main challenge in A/D converter design.

14.3 RECEIVER ARCHITECTURE AND ADC SPECIFICATION

The resolution and bandwidth required for a multichannel receiver is explained with GSM as an example. The GSM signal spectrum is shown in Fig. 14.6.

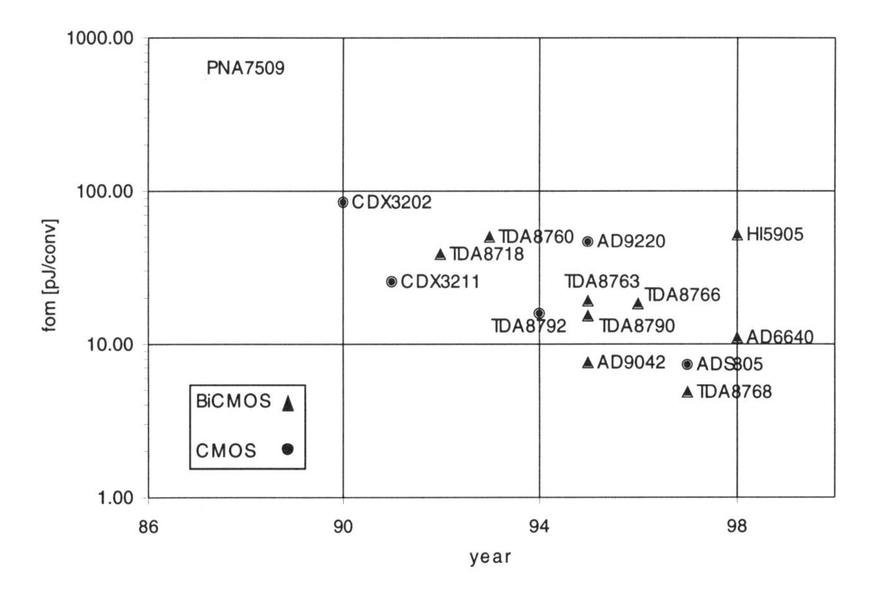

Figure 14.4 *F.o.M.* evolution in function of time [2–6].

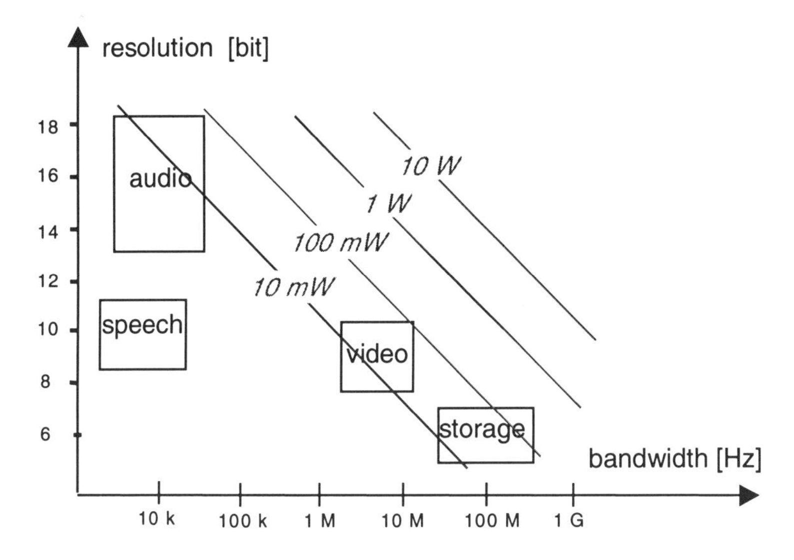

Figure 14.5 ADC power dissipation in resolution-bandwidth graph based on 5 pJ per conversion.

The 25 MHz signal band is divided into 200 kHz-wide channels. The base station has to perform according to conformance test for blocking, interferers and sensitivity. The conversion to the digital domain of a complete band requires a dynamic range in the order of 100 dB and a bandwidth larger than 25 MHz. If only a single channel is to

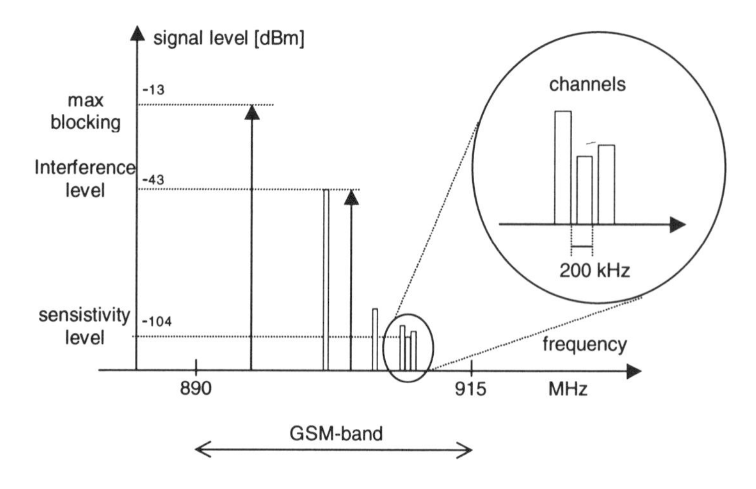

Figure 14.6 GSM signal band.

be converted, channel filtering and variable gain can be introduced in the analog signal path to reduce the dynamic range requirements in the A/D converter [7].

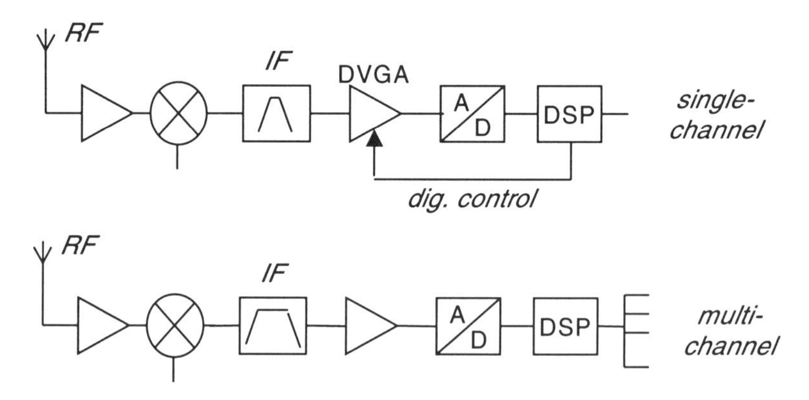

Figure 14.7 Receiver architecture with wide-band A/D converter.

Two architectures containing wide-band A/D converters are shown in Fig. 14.7. The difference between them is the presence of a (single) channel filter in the first one. For this single channel receiver, a variable-gain amplifier can be introduced to reduce the required dynamic range of the A/D converter. The second architecture is a multi-channel architecture that does not allow variable gain in front of the A/D converter to reduce the A/D converter requirements.

When the performance of the ADC is mapped to the system requirements, two kinds of Dynamic Range (DR) should be distinguished: DR due to noise limitation (SNR) and DR due to distortion and spurious signals (SFDR). The theoretical performance of an A/D converter results in an SNR improvement of 6 dB per additional bit and an SFDR improvement of 9 dB/bit (3rd harmonic in ideal quantisation). For the SNR specification we have to take the "processing gain" into account: as the channel bandwidth is smaller than the Nyquist bandwidth, the noise outside the wanted channel will be filtered out, resulting in an SNR improvement of $10\log(BW_{Nyquist}/BW_{channel})$. In Fig. 14.8, the SNR and SFDR of a 12 bit A/D converter is plotted as a function of the input signal level related to the full scale of the A/D converter (dBFS).

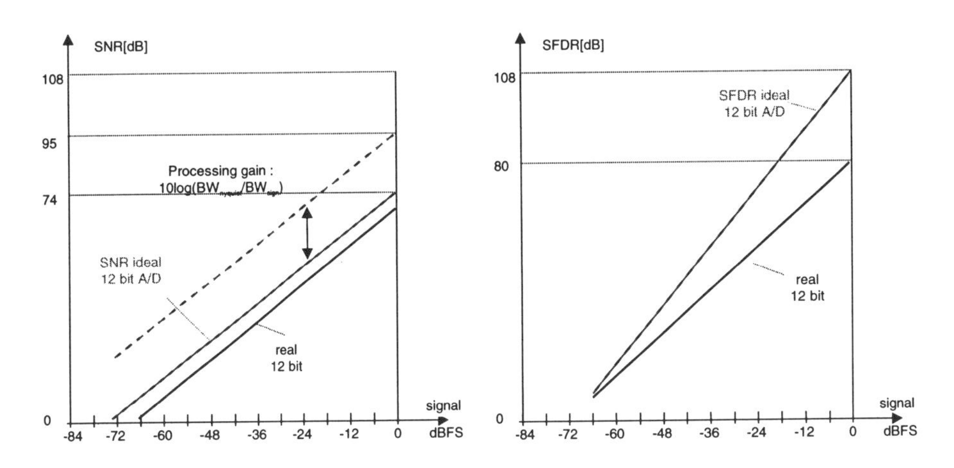

Figure 14.8 SNR and SFDR as a function of signal level.

A real 12-bit converter typically achieves an SFDR of 80 dB (theoretically 108 dB SFDR), which is not sufficient for a multi-channel base station receiver. This requires at least a highly linear 14-bit wide-band A/D converter to achieve an SFDR of about 100 dB.

14.4 CASE STUDY: SINGLE-CHANNEL RECEIVER BASED ON A 12-BIT WIDE-BAND A/D CONVERTER

The difference between real and ideal A/D converters is explained by a 12-bit A/D converter realisation [1]. This BiCMOS A/D converter is based on a cascaded folding-interpolating architecture resulting in a power/area efficient implementation. The folding-interpolating quantiser is preceded by a wide-band Track-and-Hold to obtain a resolution bandwidth equal to $f_s/2$. The limiting factors in *ENOB* are contributed by various sources of noise and distortion: nonlinearity in the Track-and-Hold, INL errors due to mismatching in the quantiser, thermal noise, clock buffer noise, etc. These sources of noise and nonlinearity are often proportional (or quadratically proportional) to the power consumption. As an example, the breakdown of the contributions of noise and distortion for the 12 bit A/D converter realisation is shown in Fig. 14.9.

Despite the state-of-the-art *F.o.M.* of 5 pJ per conversion, this 12-bit A/D converter has not enough dynamic range to be used in a multi-channel receiver for cellular base-stations. However, this wide-band A/D converter can be used to realise a single-chan-

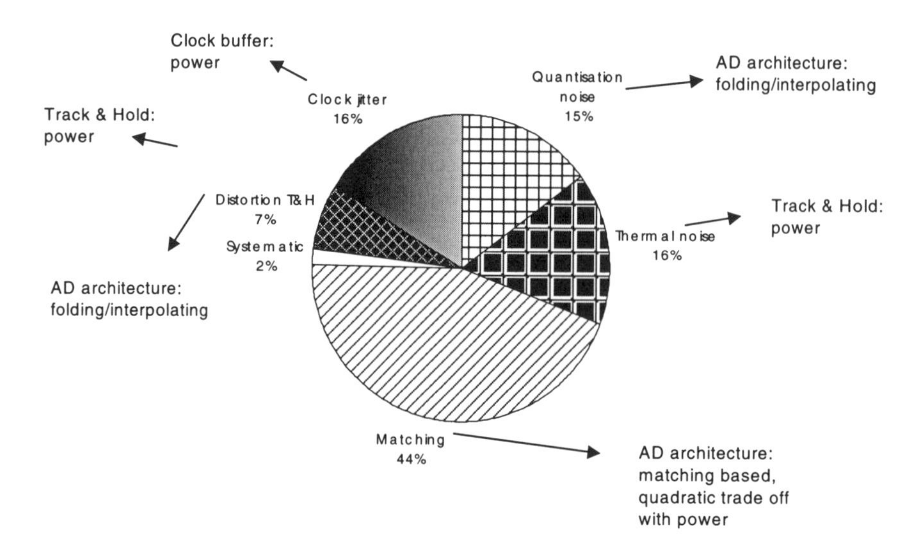

Figure 14.9 Noise and nonlinearity contributions to *ENOB* for a 12-bit A/D converter.

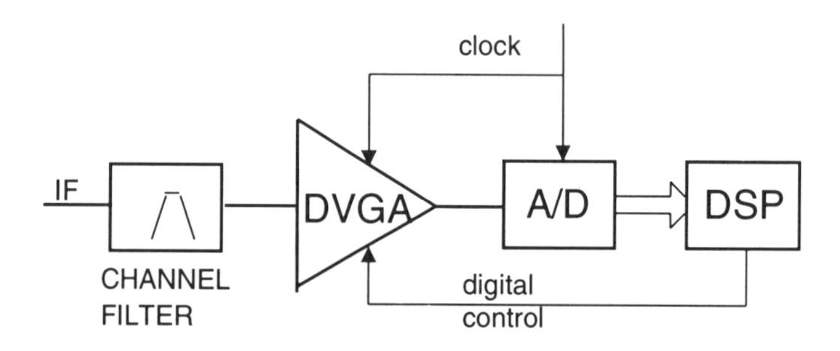

Figure 14.10 Single-channel IF-to-digital conversion system.

nel IF-to-digital conversion system. This system is shown in Fig. 14.10. The channel filter at IF reduces the dynamic range requirements and allows the use of a variable-gain amplifier in front of the ADC. In this case, the variable gain amplifier is digitally controlled and has a gain variable from 6 to 30 dB in 6-dB steps. The basic schematic of the Digitally Controlled Variable Gain (DVGA) is shown in Fig. 14.11. This additional gain range extends the dynamic range for noise of the single channel receiver to about 100 dB for a 200 kHz channel.

This DVGA was designed for low noise, low distortion, and fast settling, and it replaces an input buffer amplifier to drive an A/D converter. The SFDR of the combined DVGA and ADC is shown in Fig. 14.12.

While the SFDR of the A/D converter decreases with higher input signal levels, the distortion of the DVGA increases. Since a channel filter is introduced in front of the converter, the global SFDR of DVGA and ADC is sufficient to realise a single-channel IF-to-digital converter.

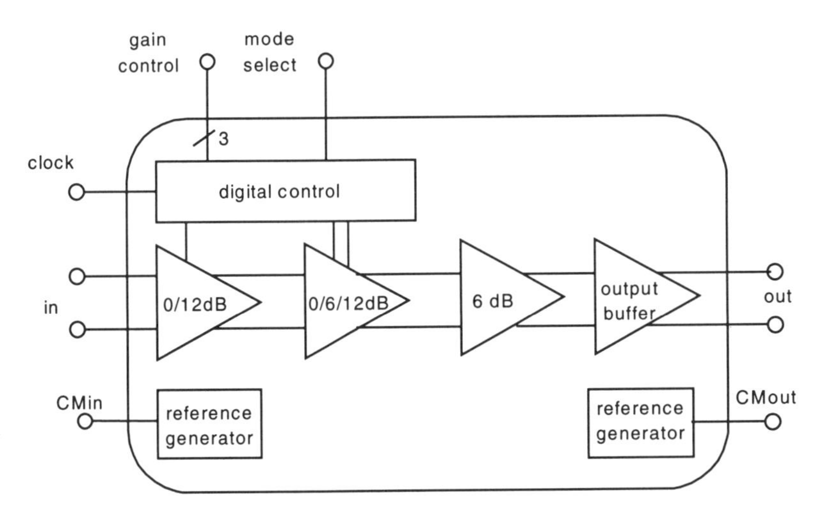

Figure 14.11 Digital Controlled Variable Gain Amplifier (DVGA) schematic.

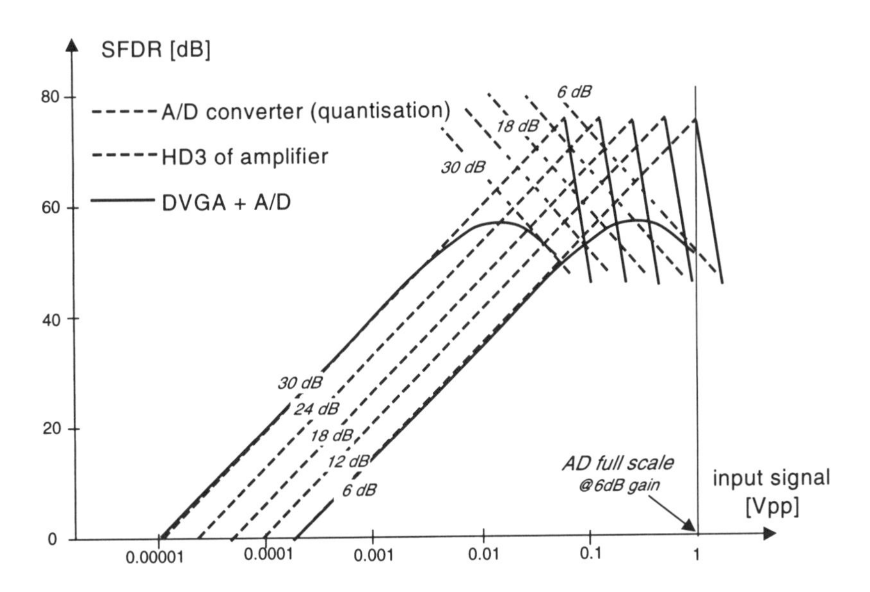

Figure 14.12 SFDR for DVGA and ADC.

14.5 CONCLUSIONS

A/D converters for multi-channel receiver base-stations is *THE* challenge in ADC design. Based on performance metrics and observations of state-of-the-art A/D converters, it can be concluded that the combined dynamic range and bandwidth requirements for multi-channel receivers will not be low power: these A/D converters will be *hot*. As ADCs for multi-channel receivers are not yet available, wide-band A/D converters can be used in combination with DVGA to build IF-to-digital conversion for single-channel receivers.

References

[1] P. Vorenkamp and R. Roovers, "A 12-b, 60Msample/s Cascaded Folding and Interpolating ADC", *IEEE J. Solid-State Circuits*, vol. 32, pp. 1876–1886, Dec. 1997.

[2] *Datasheets Philips Semiconductors, TDA8760, TDA8768, TDA8718, TDA8790, TDA8792.*

[3] *Datasheets Analog Devices, AD6640 AD9042, AD9220.*

[4] *Datasheets Harris Semiconductors, HI5905.*

[5] *Datasheets Sony, CXD3211, CXD3202.*

[6] *Datasheets Burr Brown, ADS805.*

[7] R.Roovers, "Digital Controlled Variable Gain Amplifier (DVGA) for IF to Digital Conversion System" *proc. ESSCIRC98*, pp. 152–155, Sep. 1998.

15 LOW-SPURIOUS ADC ARCHITECTURES FOR SOFTWARE RADIO

Bang-Sup Song

Department of Electrical and Computer Engineering
University of Illinois, Urbana, IL 61801, USA

15.1 INTRODUCTION

In digital wireless systems, a need to quantize a block of spectrum with low intermodulation has become one of many technically challenging problems. Implementing IF filters digitally has already become a necessity in cell sites and base stations as explained in Fig. 15.1. Even in hand-held units, placing data conversion blocks closer to the RF front end has many benefits. Substantial savings in system cost and complexity can be attained by implementing highly selective IF filters digitally. Digital IF filtering can increase immunity to adjacent and alternate channel interferences, and the RF transceiver architecture can be made independent of the system and adapted to different standards using software. However, all these predictions for the software radio rely on the availability of low-spurious ADCs and DACs that consume low power at low supply voltages.

Table 15.1 summarizes various RF communication standards. The fundamental limit in quantizing the IF spectrum is crosstalk and overload, and the system performance heavily depends on the SFDR (spurious-free dynamic range) of the sampling ADC. Spurious components result from two non-linearity sources during the quantization process. One is the differential non-linearity (DNL), and the other is the

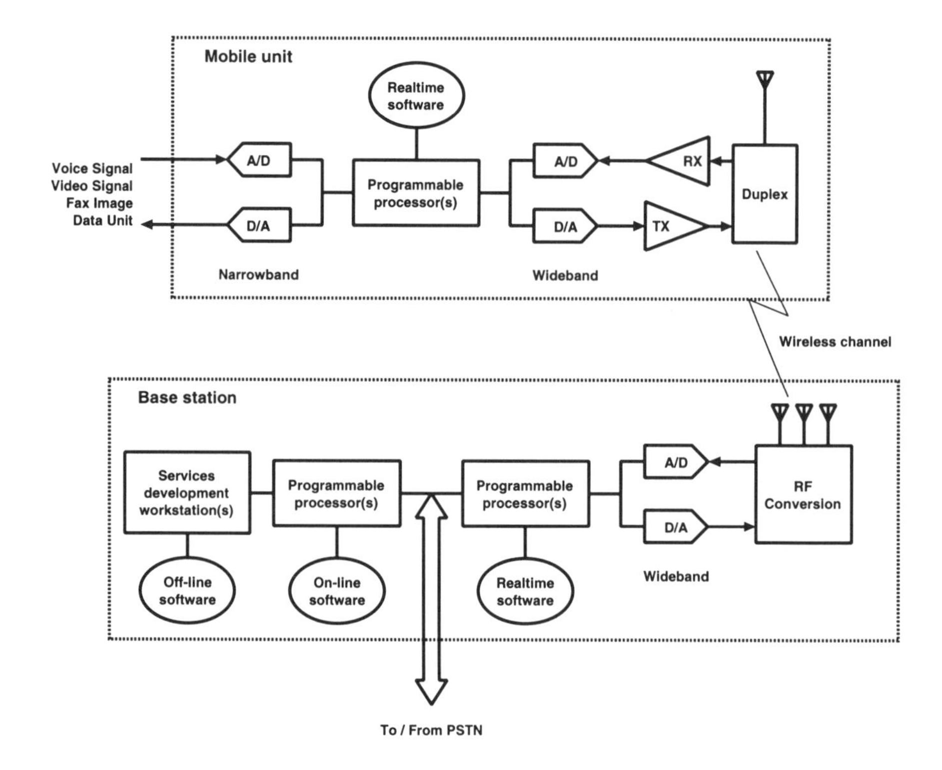

Figure 15.1 Software programmable multi-standard radio transceivers.

Table 15.1 Parameters of various RF communications.

Applications	*IF BW*	*Modulation*	*Baseband*
FM mobile	30 kHz	FM	4 kHz
GSM	200 kHz	GMSK	13 kb/s
CDMA	2 MHz	M-PSK	8 kb/s
Military FH	30 kHz	FH-QPSK	16 kb/s
JTIDS	3 MHz	MSK	150 kb/s
Microwave	20 MHz	64 QAM	90 Mb/s

integral non-linearity (INL). The former produces harmonically unrelated spurious components in the quantized spectrum while the latter produces intermodulation components. A possible solution to reduce spurious components resulting from the limited DNL is a dithering method to spread the spurious components randomly [1]. On the other hand, the limited INL can only be improved by straightening the ADC transfer characteristic. In general, only high-resolution techniques using oversampling or calibration are known to improve INL.

15.1.1 Technical challenges in digital wireless

Fig. 15.2 explains why wide dynamic range is needed to quantize a weak signal embedded in strong interferences. To meet this growing demand, low-spurious ADCs are being actively developed in ever increasing numbers. Current commercial products for this purpose are only a few, such as Analog Devices's AD6640 (12 b, 65 MHz) and National's CLC952 (12 b, 41 MHz), but more are being announced this year. Typical performance currently possible with bipolar or BiCMOS technologies ranges from 12 b to 14 b sampling at 65 MHz with a typical SFDR of 70 dB to 80 dB, although numbers higher than 80 dB are often quoted for converters using a dithering method [1]- [3]. Typical power consumption of such parts is well over half a watt. For a 14b-level dynamic range while sampling at over 50 MHz, it is necessary to control the sampling jitter to below 0.3 ps. The low-jitter requirement is on the border line of the bipolar chip performance, but very challenging for scaled CMOS technologies. However, unlike non-linearity that causes inter-channel mixing, the random jitter in IF sampling increases only the random noise floor. As a result, the random jitter is not considered as a fundamental problem in this application.

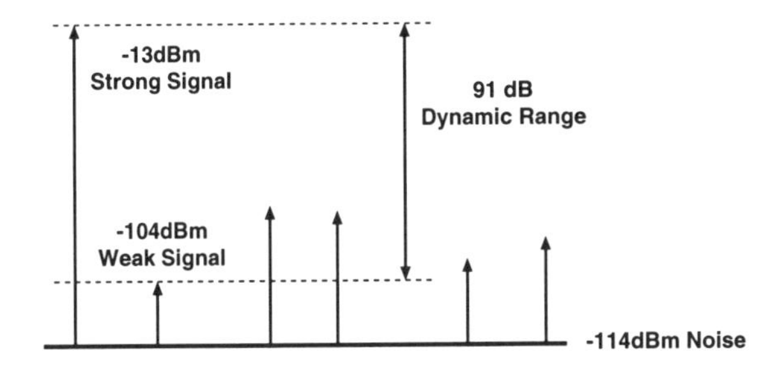

Figure 15.2 Spectral density of the RF spectrum.

Considering the current state of the art in CMOS ADCs, most architectures known to date are unlikely to achieve a high sampling rate of over 50 MHz even with 0.2 to 0.3 μm technologies [4–13], but the situation looks better for bipolar or BiCMOS [1–3, 14–16]. The ADC architecture varies depending on system requirements.

15.1.2 ADC state of the art

Fig. 15.3 is a resolution-versus-speed plot showing the trend of ADC-development works recently reported at the ISSCC. Three architectures stand out for three important areas of applications. For example, the oversampling converter is exclusively used to achieve high resolution above 12 b level at low frequencies. The difficulty in achieving better than 12 b matching in conventional techniques gives a fair advantage to the oversampling technique. For medium speed with high resolution, pipeline ADCs are promising. At extremely high frequencies, only flash and folding ADCs survive, but with low resolution. As both semiconductor process and design technology advance, the performance envelope will be pushed further. Table 15.2 lists the state of the art of >12 b ADCs in different technologies.

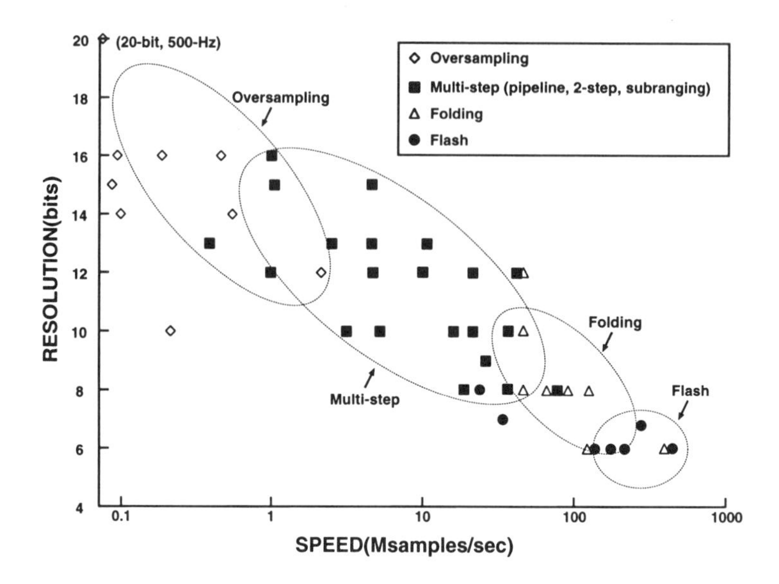

Figure 15.3 ADC performance, resolution versus speed.

The ADC performance is often represented by a figure of merit L as shown in Fig. 15.4. The higher L is, the more bits are obtained at higher speed with lower power. The plot of L versus year shows the low-power and high-speed trend both for integrated CMOS and bipolar/BiCMOS ADCs published at the ISSCC during the last decade. Overall, the low-power trend is evident though not dramatic.

15.2 TECHNIQUES FOR HIGH-RESOLUTION ADCS

There exist many high-speed ADC techniques for a wide range of resolution and speed. Table 15.3 summarizes some of them and their features. In general, the main criteria of choosing ADC architectures are resolution and speed, but additional requirements such as power, chip area, supply voltage, latency, operating environment, or technology often limit the choices. The current trend is toward low-cost integration without using expensive discrete technologies. Among many high-speed data conversion architectures, two high-speed candidate architectures, pipeline and folding, and their variations with new system approaches are potential candidates to challenge the limit set by the software radio. This paper is focused on these two architectures and presents two possible ways to improve their linearities.

In general, ADC design examples are so diverse that it is difficult to fit all of them into a few fixed frames. While pipeline ADCs are limited by the DAC resolution and residue amplifier gain, folding ADCs suffer from poor offset of the folding amplifier. Therefore, high-resolution techniques should be tailored for each architecture.

Table 15.2 State-of-the-art ADCs.

Author *Note*	*Speed* *Bits*	*Architecture* *Technology*	*Power* *Area (mm^2)*
I. Opris ISSCC98	20 M 12 b	Pipeline w/ self calib. 0.7 μm CMOS	250 mW 5 V 3.2 × 3.1
L. Singer VLSI96	10 M 14 b	Pipeline 0.8 μm CMOS	210 mW 5 V 19
S. Kwak ISSCC97	5 M 15 b	Pipeline w/ calibration 1.4 μm CMOS	60 mW 5 V 4.4 × 6.2
P. Vorenkamp ISSCC97	50 M 12 b	Cascaded folding 13 GHz 1 μm BiCMOS	300 mW 5 V 2.5 × 3.3
T. Shu VLSI94	10 M 13 b	Pipeline w/ calibration 1.4 μm BiCMOS	350 mW 5 V 6.7 × 5.0
M. Mayes ISSCC96	1 M 16 b	Pipeline w/ calibration 1 μm BiCMOS	200 mW 5 V/3.3 V 5.7 × 6.1
R. Jewett ISSCC97	128 M 12 b	Two-step w/ dithering 25 GHz Bipolar	5.7 W 5/-5.2 V 5.6 × 7.3

Analog high-resolution techniques reported to date to improve ADC linearity are as follows:

- Dynamic matching (Plassche, JSSC, June 1979)
- Ratio-independent (Li, JSSC, Dec. 1984)
- Reference-feedforward (Sutarja, JSSC, Dec. 1988)
- Capacitor-error averaging (Song, JSSC, Dec. 1988)
- Current sampling (Schouwenaars, JSSC, Dec. 1988)
- Voltage sampling (Kerth, JSSC, April 1989)
- Inherently monotonic (Wu, ISCAS, 1994)

The dynamic matching technique is efficient, but it generates high-frequency noise. The ratio-independent, reference-feedforward, and inherently monotonic techniques either require more than two clock cycles or are limited to improvement in DNL only. The capacitor-error averaging technique requires three clock cycles, and the voltage/current sampling techniques are sensitive to clock sampling errors.

The ADC self-calibration concepts have originated from the direct code-mapping concepts using memory. The code-mapping technique is to predistort the digital input of the DAC so that the DAC output can match the ideal level from the calibration equipment. Due to the precision equipment needed, this method has limited use. The first self-calibration concept applied to the binary-ratioed successive-approximation ADC

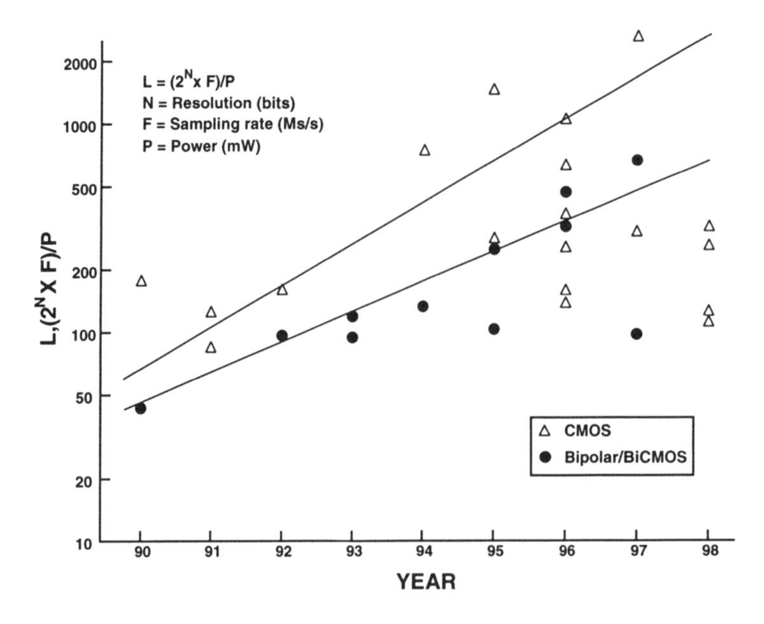

Figure 15.4 Figure of merit (L) versus year.

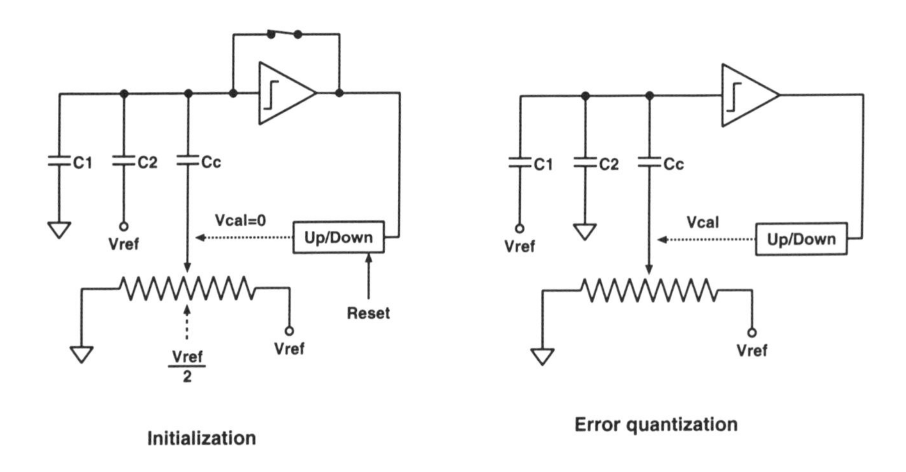

Figure 15.5 Capacitor ratio error measurement.

is to internally measure capacitor DAC ratio errors using a resistor-string calibration DAC as shown in Fig. 15.5 [17]. Later, an improved concept of the digital-domain calibration was developed for the multi-step pipeline ADC [6], and the trend continues to calibrate ADCs in the background while ADCs are in operation [11, 15].

Table 15.3 High-speed ADC techniques and their features.

Techniques	*Features*
Flash	Fastest, but complexity increases exponentially.
Interpolation/averaging	Reduces the number of preamplifiers and improves DNL.
Pipeline	Fast, but limited by DAC accuracy and residue amplifier settling.
Folding	Smaller number of comparators than flash, but limited by speed and offset of the folders.
Time interleaving	Parallel multi-path system, but limited by fixed pattern noise and clock accuracy.
Subranging	Smaller number of comparators than flash and monotonic, but limited by comparator accuracy.
Double sampling	Have more time, but need extra interstage S/H.
Current domain	Fast, but no precision linear current source exists.

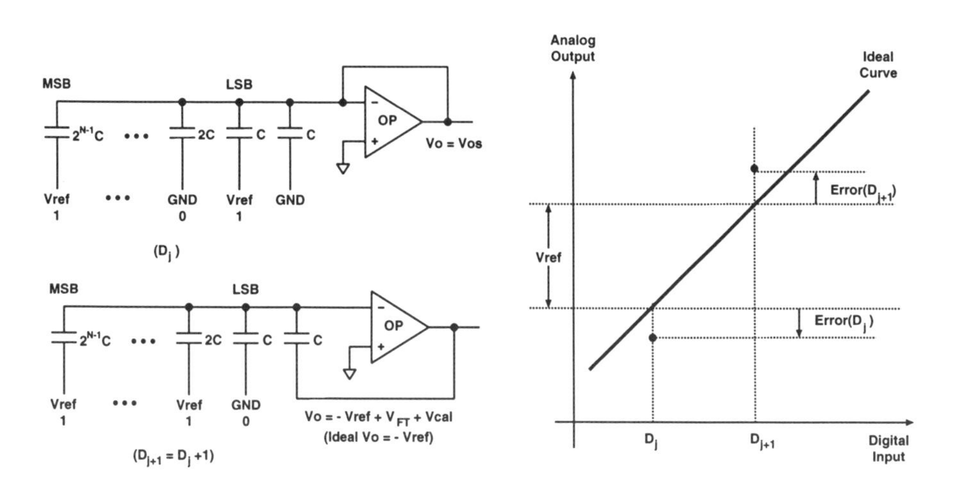

Figure 15.6 Digital code-error calibration technique.

The general concept of the digital-domain calibration is to measure code or bit errors, to store them in the memory, and to subtract them during the normal operation.

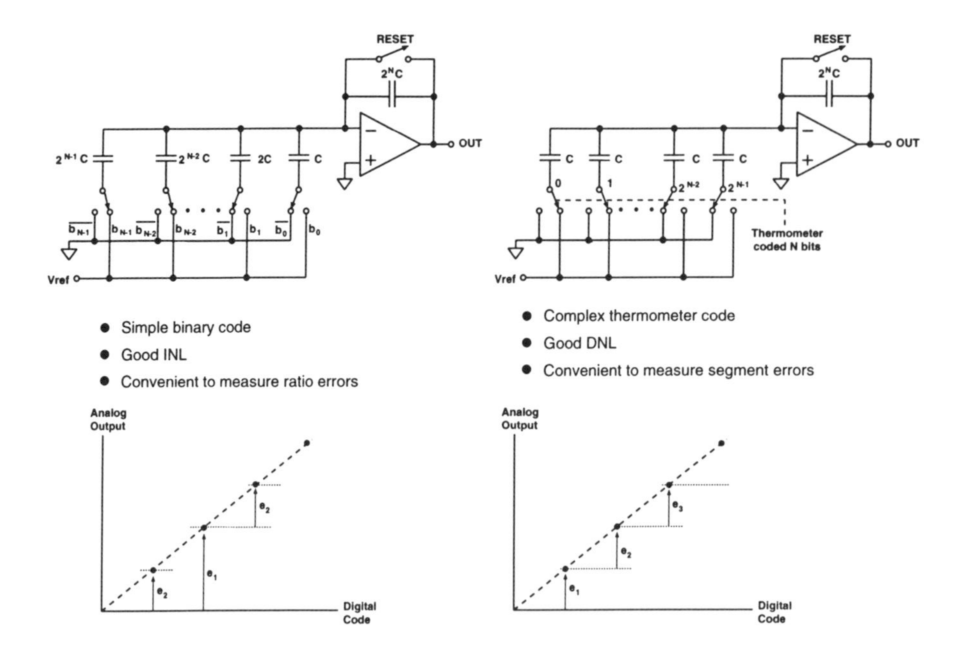

Figure 15.7 Comparison of binary and thermometer code-error measurements.

The concept is explained in Fig. 15.6 using a generalized N-bit MDAC using a capacitor array [6]. If the DAC code increases by 1, the MDAC output should increase by V_{ref} or $V_{\mathrm{ref}}/2$ if redundancy is used for digital correction. Any deviation from this ideal step is defined as a segment error. The code errors are obtained by accumulating segment errors. This segment-error measurement needs two cycles. The first cycle is to measure the switch-feedthrough error, and the next cycle is to measure the segment error. The segment error is simply measured using the LSB-side of the ADC by increasing the digital code by 1. In the case that N= 1, the segment error is the bit error. If binary bit errors are measured and stored, the code error should be calculated for subtraction during the normal operation. How to store DAC errors is a trade-off issue. Ways to measure binary and thermometer code errors are explained in Fig. 15.7. Examples of the digital calibration are well documented for the cases of the segment error [11] and bit error [14] measurements.

The static linearity can be improved using many techniques as discussed, but the speed improvement is not straightforward. Although digital calibration can improve the resolution of the pipeline ADC to over 15 b, the settling time of the interstage residue amplifier has been the bottleneck in achieving high conversion rate.

15.3 OUTLOOK

New circuit techniques and architectures will lead to the implementation of data converters for digital IF applications. Assume the goal is to realize an ADC that meets the 14 b, 50 MHz requirement using low-power CMOS technologies. New high-resolution CMOS ADC architectures need to be explored so that high linearity and speed can be readily obtained with less hardware complexity. The following are identified as major factors limiting performance in two candidate architectures.

- Pipeline ADC
  - Speed of the residue amplifier
  - Linearity and calibration
- Folding ADC
  - Offset of the folding amplifier
  - Complexity for high resolution
  - Linearity and calibration

In the pipeline ADC, the settling time of the interstage residue amplifier has been the bottleneck in achieving high conversion rate. The system can be reconfigured so that the interstage gain may be obtained using a transresistance amplifier to speed up. Also the resolution of the folding ADC can be improved using a subranging concept combined with cascaded folding. Subranging will greatly simplify the folding architecture, and the idea of calibrating the folding ADC is feasible with manageable complexity. These two approaches have never been applied to ADCs with more than 14 bits. To achieve such a linearity at over 50 MSample/s rate using CMOS, we have to overcome numerous technical barriers. The ADC of this caliber has yet to be demonstrated. An oversampling delta-sigma calibrator will enable all calibrations in the background.

Fig. 15.8 illustrates a residue amplifier that can be used as a front-end of the pipeline ADC. The system uses a current-domain residue amplifier, which has an added advantage of using a current DAC which can be controlled more accurately at high speed than voltage DACs. In the proposed architecture, linearity is limited only by the matching between current sources and resistors. Since the matching in IC processing is not enough, we need to calibrate current sources in the background using an oversampling delta-sigma calibrator [15]. The ultimate limit of the conversion rate is set by the settling time of the residue amplifier. For example, if the amplifier settles within 10 ns with a 12 b resolution, the closed-loop time constant is about 1.1 ns. This in turn requires a closed-loop bandwidth of about 145 MHz. This closed-loop bandwidth is believed to be well within the reach of the current CMOS technologies. The system differs from others having the following distinct features.

- Current-domain residue amplifier for fast settling
- Background calibration of current-domain MDAC
- Oversampling delta-sigma calibrator

The overall system works as follows. The input sample-and-hold amplifier makes a new sample ready for the residue amplification. The coarse digital values are also ready to switch differential current DAC elements for the residue amplification. An LSB stage is also assumed to be ideal within its resolution. One left-over phase of the residue amplifier is used to calibrate the current DAC elements, where the output of the residue amplifier corresponding to V_{ref} is compared to the ideal V_{ref} using a delta-sigma modulator. The bitstream output from this modulator is counted and averaged to increment or decrement a small current DAC to trim the current source. Behavioral simulation results of the quantized output show that without calibration, the spurious

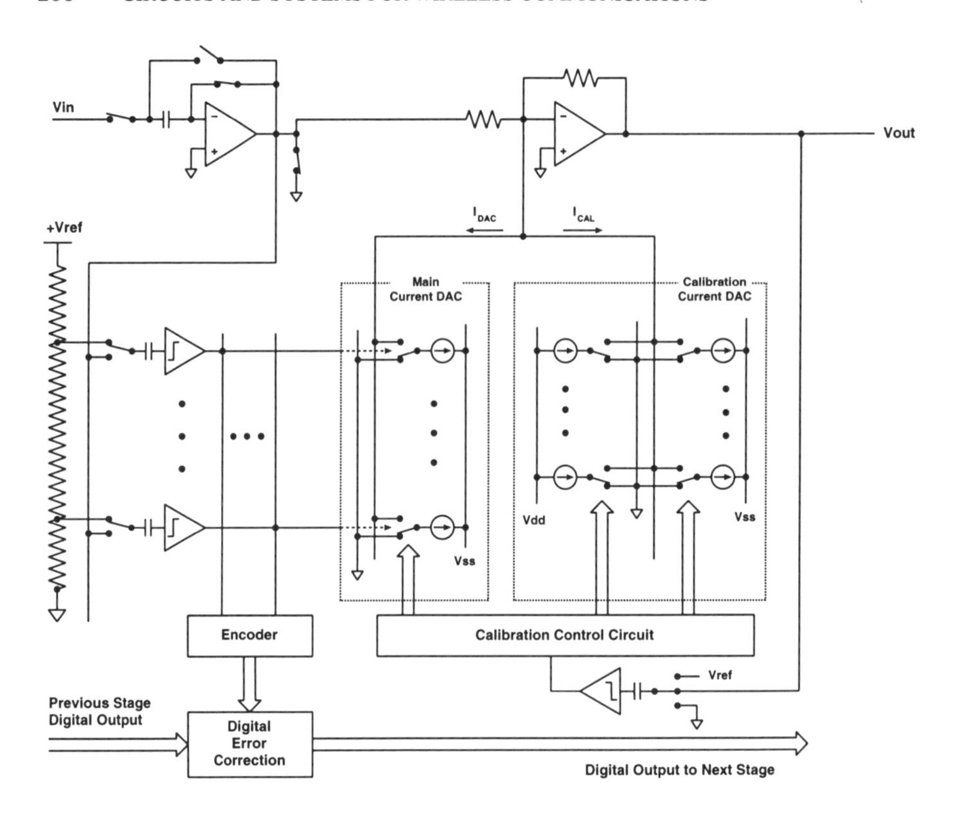

Figure 15.8 Pipeline ADC front-end with background-calibrating current DAC.

level approaches -70 dB assuming 0.2% component mismatch. However, with the calibration engaged, the spurious level is suppressed to below -100 dB.

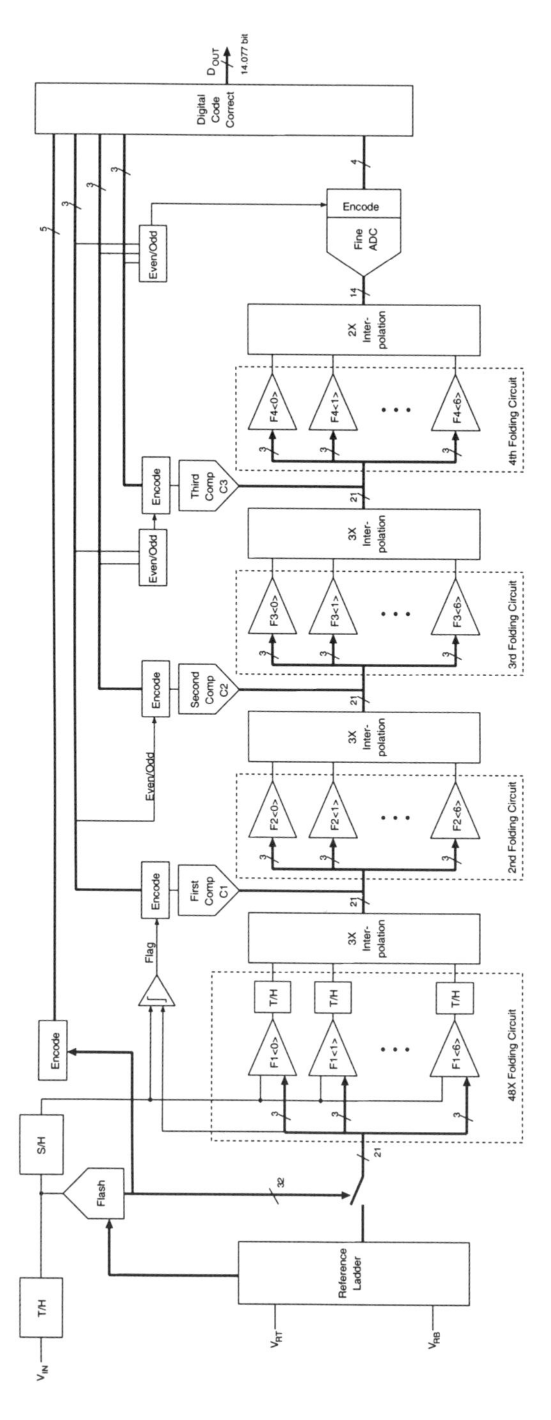

Figure 15.9 A background-calibrating pipeline folding ADC with subranging and interpolation.

The folding architecture has been most frequently adopted for high-speed ADCs. The folding circuit simultaneously produces many zero crossing points, but they need to be as accurate as the resolution of the ADC. Two limiting factors are the accuracy of the folding thresholds and the linearity of the folding amplifier. Due to the difficulty of controlling these, folding architectures have been known for high speed but for resolution typically lower than 8 to 10 bits [18–20]. However, at a 14 b level, the system also becomes prohibitively complex, and a new efficient folding architecture is needed. In addition, the new architecture should be calibratable for high linearity.

Several improvements can be made on the basic folding and interpolation architecture to overcome its limitations. The following features can increase the number of bits effectively.

- Cascaded folding with an odd number of folding blocks
- Subranging concept in folding
- Background calibration of the reference and folding amplifier offsets
- Oversampling delta-sigma calibrator

For cascading, it is necessary to generate folded signals from the first folding amplifier, and a Gilbert-type multiplier circuit has been used [3]. However, the multiplier requires level shifters and cascode circuits that make the implementation more complex and introduce errors in the folding process. This difficulty can be overcome by using the same current switch as in the first folding amplifier for the intermediate stage. For this, the folding scheme needs to be modified so that the number of folding blocks becomes an odd number. If the number is a multiple of an odd number, the second folding outputs can be generated by summing the first folding outputs. Since the summation of such outputs can be implemented using current switches, the same circuitry used for the first folding amplifier can be used repeatedly. Using this and the subranging concept, one solution to a calibratable cascaded folding architecture can be configured as shown in Fig. 15.9. The number of output codes is not a power of two, but the number of folding current switches is reduced to only 84. The first-stage folding amplifiers' offsets are of interest and subject to calibration.

The linearity of the resistor ladder network can be calibrated without interfering with the normal conversion using a delta-sigma calibrator [15]. An offset of the folding amplifier has the same effect as the reference error, and it is the main source of errors. To achieve 50 MSample/s conversion speed, it is necessary to have the reference plus folding outputs settle within 10 ns. To guarantee high resolution, a calibration scheme to cancel the folding amplifier offsets should be incorporated. Fig. 15.10 shows such an implementation of the folding current switch. When one folding current switch is under calibration, its output is separated from the second folding block and fed into an error measurement block. The input to the folding amplifier under calibration is shorted to null the output. The output necessary for the second folding block is generated by interpolation. The error measurement block measures the folding amplifier output, and the calibration logic adjusts the bias current of the folding current source until the output of the folding amplifier is nulled. Like most pipeline ADCs, redundancy can be used in the folding architecture for digital correction. A prototype of the LSB 10 b ADC to be used in this architecture is implemented separately to prove the speed of this architecture. Fig. 15.11 shows the 19 MHz input sampled at 157 MHz, and its reconstructed waveform shows distinct steps.

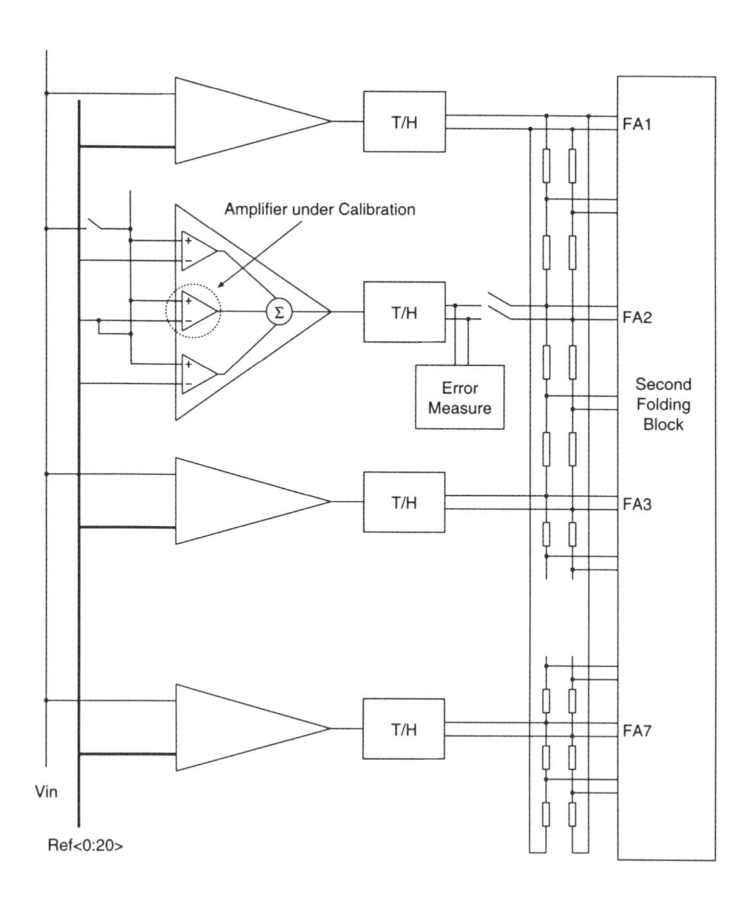

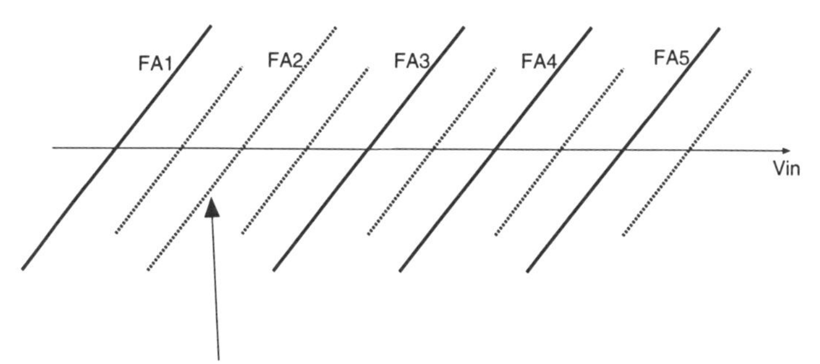

Figure 15.10 Folding amplifier offset calibration scheme.

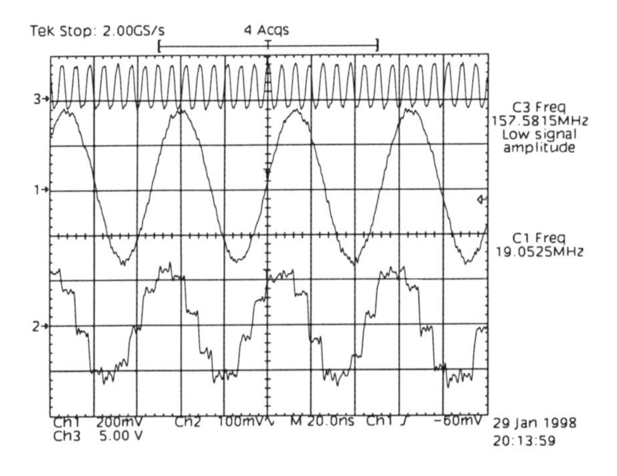

Figure 15.11 A/D-D/A outputs of a 10 b pipeline folding ADC.

15.4 CONCLUSIONS

Digitization is now about to move into IF for wireless systems. The software-programmable radio offers the technically most challenging tasks to the data acquisition design community. Only high-performance ADCs and DACs, implementable in low-cost VLSI technologies and consuming one order of magnitude less power than current circuits, will make the wide-spread use of software radios feasible. CMOS data converters are not yet in this arena. However, two examples of high-resolution ADCs shown here prove that even CMOS ADCs can move into the 100 MHz range with high resolution if new robust architectures are developed. The current ADC art, estimated by recent works, is projected to achieve a linearity of 15 b to 16 b with a SFDR of over 90 dB at a 100 MHz range in the near future.

References

[1] R. Jewett, K. Poulton, K.-C. Hsieh, and J. Doernberg, "A 12b 128MS/s ADC with 0.05LSB DNL," in *ISSCC Digest of Technical Papers*, pp. 138–139, Feb. 1997.

[2] F. Murden and R. Gosser, "12b 50MSample/s Two-Stage A/D Converter," in *ISSCC Digest of Technical Papers*, pp. 278–279, Feb. 1995.

[3] P. Vorenkamp and R. Roovers, "A 12-bits, 50 MSPS Cascaded Folding & Interpolation ADC," *IEEE Journal of Solid-State Circuits*, pp. 1876–1886, Dec. 1997.

[4] Y.-M. Lin, B. Kim, and P. R. Gray, "A 13-b 2.5-MHz Self-Calibrated Pipelined A/D Converter in 3-μm CMOS," *IEEE Journal of Solid-State Circuits*, vol. SC-26, pp. 628–636, Apr. 1991.

[5] B. Razavi and B. A. Wooley, "A 12-b 5-MSample/s Two-Step CMOS A/D Converter," *IEEE Journal of Solid-State Circuits*, vol. 27, pp. 1667–1678, Dec. 1992.

[6] S.-H. Lee and B.-S. Song, "Digital-Domain Calibration of Multi-Step Analog-to-Digital Converters," *IEEE Journal of Solid-State Circuits*, vol. 27, pp. 1679–1688, Dec. 1992.

[7] S.-I. Lim, S.-H. Lee, and S.-Y. Hwang, "A 12b 10MHz 250mW CMOS A/D Converter," in *ISSCC Digest of Technical Papers*, pp. 316–317, Feb. 1996.

[8] D. W. Cline and P. R. Gray, "A Power Optimized 13-b 5 Msamples/s Pipelined Analog to Digital Converter in 1.2 μm CMOS," *IEEE Journal of Solid-State Circuits*, vol. 31, pp. 294–303, Mar. 1996.

[9] L. A. Singer and T. L. Brooks, "A 14-Bit 10-MHz Calibration-Free CMOS Pipelined A/D Converter," *1996 Symposium on VLSI Circuis, Digest of Technical Papers*, pp. 94–95, June 1996.

[10] P. C. Yu and H.-S. Lee, "A 2.5-V, 12-b, 5-MSample/s Pipelined CMOS ADC," *IEEE Journal of Solid-State Circuits*, vol. 31, pp. 1854–1861, Dec. 1996.

[11] S.-U. Kwak, B.-S. Song, and K. Bacrania, "A 15b 5-MSample/s Low-Spurious CMOS ADC," *IEEE Journal of Solid-State Circuits*, pp. 1866–1875, Dec. 1997.

[12] J. Ingino, Jr. and B. Wooley, "A Continuously-Calibrated 10MSample/s 12b 3.3V ADC," in *ISSCC Digest of Technical Papers*, pp. 144–145, Feb. 1998.

[13] I. Opris, L. Lweicki, and B. Wong, "A Single-Ended 12b 20MSample/s Self-Calibrating Pipelined A/D Converter," in *ISSCC Digest of Technical Papers*, pp. 138–139, Feb. 1998.

[14] A. N. Karanicolas, H.-S. Lee, and K. L. Bacrania, "A 15-b 1-Msample/s Digitally Self-Calibrated Pipeline ADC," *IEEE Journal of Solid-State Circuits*, vol. 28, pp. 1207–1215, Dec. 1993.

[15] T.-H. Shu, B.-S. Song, and K. Bacrania, "A 13-b 10-Msample/s ADC Digitally Calibrated with Oversampling Delta-Sigma Converer," *IEEE Journal of Solid-State Circuits*, vol. 30, pp. 443–451, Apr. 1995.

[16] M. K. Mayes and S. W. Chin, "A 200 mW, 1 Msample/s, 16-b Pipelined A/D Converter with On-Chip 32-b Microcontroller," *IEEE Journal of Solid-State Circuits*, vol. 31, pp. 1862–1872, Dec. 1996.

[17] H.-S. Lee, D. A. Hodges, and P. R. Gray, "A Self-Calibrating 15 Bit CMOS A/D Converter," *IEEE Journal of Solid-State Circuits*, vol. SC-19, pp. 813–819, Dec. 1984.

[18] J. van Valburg and R. J. van de Plassche, "An 8-b 650-MHz Folding ADC," *IEEE Journal of Solid-State Circuits*, vol. 27, pp. 1662–1666, Dec. 1992.

[19] B. Nauta and A. G. W. Venes, "A 70-MS/s 110-mW 8-b CMOS Folding and Interpolating A/D Converter," *IEEE Journal of Solid-State Circuits*, vol. 30, pp. 1302–1308, Dec. 1995.

[20] M. P. Flynn and D. J. Allstot, "CMOS Folding A/D Converters with Current-Mode Interpolation," *IEEE Journal of Solid-State Circuits*, vol. 31, pp. 1248–1257, Sept. 1996.

IV Process Technologies for Future RF Systems

16 PROCESS TECHNOLOGIES FOR FUTURE RF SYTSEMS

Urs Lott

Lab. for Electromagnetic Fields and Microwave Electronics
Swiss Federal Institue of Technology (ETH)
Zurich, Switzerland

16.1 INTRODUCTION TO THE FOLLOWING PAPERS

Today, silicon CMOS, bipolar and BiCMOS technologies dominate the field of digital and low-frequency analog monolithic integrated circuits. At RF frequencies of 500 MHz and above, the choice of technology is no longer so straightforward, as other technologies based on III-V semiconductors (GaAs, InP) or SiGe heterojunction devices show distinct advantages at least for certain building blocks like RF power amplifiers. The advantages and disadvantages of each of these technologies with respect to performance, cost, ease of design, etc. are currently a subject of hot debate, both in industrial companies and academic circles. Furthermore, the discussion with regard to the optimum choice of integrated circuit technology for a certain part of an RF radio system partly overlaps with the — no less controversial — debate on the desirability and usefulness of the so-called "single-chip radio." The overlap of both controversies about technology and the "right level of integration" also shows up in the three contributions of this part.

Some contributions in this volume deal with aspects of the single-chip radio mainly from a CMOS point of view. This seems natural because most of the push towards the "single chip radio" currently comes from the people working their way up from CMOS baseband processing to include more and more RF functions. A glance at the current publications on single-chip RF radios, which are too numerous to be cited here, reveals, however, that most RF single chips presented still lack very important parts of a complete transceiver, such as the final power stages, RF switches, selective filters, and, sometimes, oscillators.

The extent to which all these functions can be integrated in the future depends not only on the advances made by RF circuit designers, but also on the general development of the silicon technology, which is driven by digital applications. It is not clear whether the trends in digital CMOS, e.g. towards ever lower supply voltages (mainly due to reduced gate length) and oxide thickness, do not contradict the requirements of the RF circuit designer.

The first contribution, written by C. A. King, highlights the importance of low-cost processing in advanced BiCMOS technology development. With a "CMOS first" philosophy, the additional processing steps for fabricating the bipolar devices are minimized. Nevertheless, a peak f_T for the bipolar devices of 30 GHz is achieved. Using similar device structures with the same "CMOS first" philosophy, Si/SiGe heterostructure bipolar transistors have been fabricated with a peak cut-off frequency of 54 GHz. The process steps for device fabrication are shown and some properties of the Si/SiGe material system are discussed. The evaluation of the RF device performance and the extracted delay elements show that further improvements in f_T should be possible by reducing the junction capacitances.

R. Christ's contribution compares GaAs-Metal-Semiconductor-Field-Effect-Transistor (MESFET) RF IC technology with competing technologies seen from the viewpoint of a commercial GaAs IC supplier. The transmitter linearity requirements in advanced code-division multiple-access (CDMA) systems such as IS-95 are considerably higher than in the GSM (global system for mobile communications) system. Today, the IS-95 specifications can be met more easily with power amplifiers based on the GaAs MESFET or GaAs Heterojunction Bipolar Transistor (HBT) than with silicon devices. However, it is not clear whether the HBT or MESFET will be more economical, as HBT technology generally has a higher cost per wafer, but achieves smaller die size for a given power output. In the following discussion of the single chip radio, it is stated that full integration for its own sake makes no sense commercially. Aspects like development time, product life time, and design flexibility often call for less-than-maximum integration levels in commercial RF chip sets. Finally, the key parameters of a state-of-the-art GaAs RF MESFET process featuring enhancement and depletion MESFETs, high-Q inductors ($>$ 17 at 1.9 GHz), MIM capacitors, and NiCr resistors are described.

In the third paper by U. Lott and W. Bächtold, different RF IC technologies are compared regarding their frequency range of application and suitability for RF and baseband integration. Thereafter, examples of RF ICs working at 2 GHz and above are described. Ultra-low-power, low-noise amplifiers (LNAs) have been realised in both a silicon bipolar and a GaAs MESFET technology at 1.9 GHz. Another LNA including the switching function for a three-input-antenna diversity system at 4.5 GHz has demonstrated low noise and low power performance. Finally, a 200 mW power amplifier for 17 GHz using High-Electron-Mobility-Transistor (HEMT) technology is shown. These examples highlight the potential of advanced semiconductor technologies for realising integrated circuits at multi-GHz frequencies, where RF CMOS will lag behind GaAs MESFET technology for several years to come.

17 LOW-COST Si AND $Si/Si_{1-x}Ge_x$ HETEROSTRUCTURE BiCMOS TECHNOLOGIES FOR WIRELESS APPLICATIONS

Clifford A. King

Bell Laboratories
Lucent Technologies
600 Mountain Avenue
Murray Hill, NJ, 07974, USA

Abstract: BiCMOS for analog circuits is finding applications for which the high transconductance, low noise, high output resistance, and additional speed of the bipolar devices yield a performance not easily obtained through a CMOS technology alone. However, for BiCMOS to remain a viable alternative for these applications, costs must be kept low. Our BiCMOS approach, MBiC (Modular BiCMOS), starts with a state-of-the-art CMOS technology and adds a bipolar transistor in a modular fashion while minimizing additional process steps. Using process steps such as high-energy implantation in place of epitaxy over a heavily doped sub-collector, and amorphous-Si refill techniques under the extrinsic base region, we produce Si homo-junction devices with cutoff frequencies of about 30 GHz with only three additional masking levels. Using similar device structures with the same "CMOS first" philosophy, we also fabricated Si/SiGe hetero-junction bipolar transistors with a cutoff frequency of 54 GHz.

17.1 INTRODUCTION

Si BiCMOS technology is important for wireless applications since it provides low-noise transistors with low power dissipation at microwave frequencies along with the state-of-the-art CMOS needed for logic applications. This capability means that BiCMOS technology provides the most likely opportunity to combine the radio and signal-processing portions of wireless circuitry to make single-chip phones possible. While CMOS alone may provide this capability at the lower bands in the future, markets will not wait for CMOS technology to evolve before striking a demand. Therefore, the development of BiCMOS technologies will continue for current wireless communication applications and for applications in emerging frequency bands.

17.2 SILICON BiCMOS

BiCMOS process development at Lucent Technologies follows a modular approach. The bipolar transistor is integrated into an existing CMOS process without changes to the core process so that re-qualification of the MOS devices is not necessary. In addition, the inclusion of the bipolar transistor is accomplished by adding as few lithography levels and processing steps as possible, in order to keep costs low. Techniques such as high-energy implantation for sub-collector formation and poly-silicon refilling for extrinsic base contacts help to eliminate unnecessary processing steps while preserving device performance.

17.2.1 High-energy-implanted sub-collector

To reduce collector resistance in bipolar transistors, a buried layer, or sub-collector layer, is generally used. This low-resistivity layer is formed by implantation and diffusion into patterned areas on the substrate. These regions are subsequently covered using epitaxial growth by a layer having much higher resistivity to form the active collector region. A high-energy implantation process which eliminates many steps and therefore reduces cost and process complexity can replace this diffusion and epitaxy process. Table 17.1 illustrates the device performance for the 0.25 μm High-Energy Implanted BiCMOS (HEIBiC) process of Lucent Technologies [1]. The change in the cutoff frequency is strongly dependent on the breakdown voltage and thus on the collector depletion layer transit time. A trade-off between the collector–emitter breakdown voltage and the cutoff frequency is clearly evident from Table 17.1.

17.3 LATERAL ETCHING AND AMORPHOUS SI REFILLING PROCESS

Lateral etching of SiO_2 is important for the fabrication of electronic device and micro-machine structures. Micro-machines, which are mechanical surface or bulk devices, use the lateral etching of SiO_2 to free overlying structural layers made of polycrystalline Si or Si_3N_4. Typical thicknesses of the oxide layer to be etched for micro-machine fabrication are in the range from 0.1–0.5 μm, while the lateral etch distance can be hundreds of μm. For the fabrication of high-performance transistor structures, lateral etching to remove an insulating region and replace it with a semiconducting region is a useful process to lower costs and improve performance. In these devices, a cavity is first created by the removal of the insulator, usually SiO_2, using a solution

Table 17.1 HEIBiC device performance.

NPN	*3.3 V*	*2.5 V*	*2.5 V-HS*
A_E	$0.4\times1.2\,\mu m^2$	$0.4\times1.2\,\mu m^2$	$0.4\times1.2\,\mu m^2$
f_T	33 GHz	42 GHz	52 GHz
f_{max}	24 GHz	30 GHz	33 GHz
β	60	112	264
V_A	25	23	10
BV_{CEO}	4.2 V	3.8 V	2.8 V
BV_{CBO}	8.9 V	8.8 V	8.1 V
BV_{EBO}	5.5 V	4.6 V	6.9 V
BV_{CS}	12.2 V	12.6 V	12.6 V

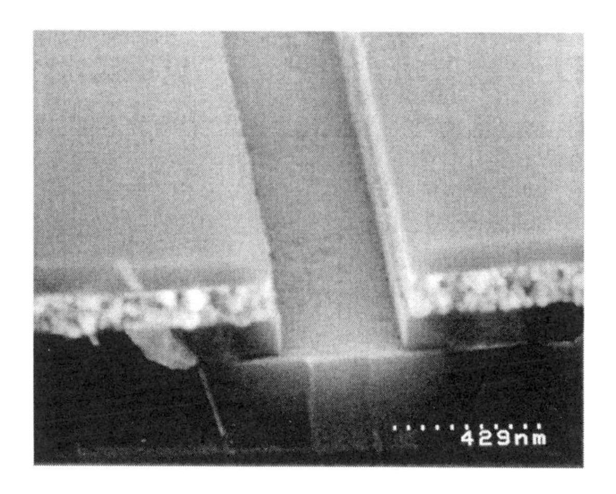

Figure 17.1 An SEM micrograph of a laterally etched trench feature with 6 nm of thermal oxide under the poly-silicon, WSi_x, and nitride layers.

of HF acid. Subsequent to the lateral etching process, the cavity is filled with amorphous Si (α-Si) or epitaxial Si to create an electrical contact to underlying layers. This lateral-etch-and-refill procedure is especially important in Bipolar-CMOS (BiCMOS) processes where integration of a super self-aligned bipolar transistor with the MOS devices must be accomplished with a minimum of steps to keep costs low [2, 3].

In this work, we created extremely small cavity structures (down to only 2.5 nm) and refilled them with α-Si. To the authors' knowledge, these are the smallest structures to be laterally etched and refilled. In addition, we report the etching and refilling characteristics of such small structures and compare the results to previous work. Prior work investigating the etch rate dependence on the cavity height, which is determined by the thickness of the oxide layer, is inconsistent. One group finds a dependence [4] while another does not [5]. In our work with very thin oxide layers, we see a dependence of etch distance on cavity height in the thickness range of 50 nm down to 2.5 nm.

In addition, we checked the dependence of the lateral etch distance on window opening size and trench opening size in the range of 0.18–0.65 μm using scanning electron microscopy (SEM). After creating cavities of various dimensions with lateral etching, we refilled them with an α-Si layer and examined the final structure with SEM and transmission electron microscopy (TEM).

The fabrication of test structures began with conventional 150 mm Si (100) p/p+ epi substrates. We started the process by creating a stack on the wafer resembling a metal-oxide-semiconductor (MOS) gate or a bipolar junction transistor (BJT) extrinsic base. From the substrate, the stack consists of SiO_2, polycrystalline Si, WSi_x, and silicon nitride. The stack formation process began with thermal oxidation. The initial oxidation step was carried out in a conventional vertical oxidation furnace except for the 2.5 nm oxide which was grown with rapid thermal oxidation [6]. Following the oxidation, we deposited α-Si at 550°C to a thickness of 100 nm using low-pressure chemical vapor deposition (LPCVD). Next, WSi_x was sputtered to a thickness of 120 nm on top of the α-Si. To finish the stack formation, we used LPCVD to deposit 100 nm of Si_3N_4. The stack was then patterned using 248 nm deep UV lithography with a mask comprising trench and window sizes ranging from 0.18 to 0.65 μm. The trench patterns have a width equal to the minimum size dimension and a length of at least 1 mm while the window features are circular with a diameter equal to the minimum size dimension. After photo-lithography, we etched the nitride using reactive ion etching and stripped the masking photoresist. Using the nitride as a mask, we continued to etch the WSi_x and polycrystalline Si (converted from α-Si during nitride deposition) layers with an etch that is highly selective to oxide. This selective etch allowed us to stop on the thin underlying oxide (2.5 to 50 nm) [6] and complete the test structure fabrication. With the trench and window structures now formed on the wafer, we immersed the samples for 70 minutes in a recirculating bath of 0.49-weight-percent (wt. %) HF solution controlled to a temperature of 21°C. The etch rate of blanket thermal oxide layers in this bath is about 3 nm/min. Fig. 17.1 shows an SEM micrograph of a trench feature after lateral etching in HF on a sample with a 6 nm thermal oxide. The lateral etch distance is clearly demarcated by the undercut visible in the picture. The polycrystalline Si, WSi_x, and nitride layers are also clearly discernible. In Fig. 17.2, we plot the lateral etch distance in both trenches and windows for samples with underlying oxide thicknesses of 6 and 20 nm. For a given feature (i.e. trench or window), the lateral etch distance is approximately independent of feature size from 0.18 to 0.65 μm. For 20 nm underlying oxide thickness, the average lateral etch distance in trench features is 183 nm, while it is 166 nm for window features. A much larger disparity of the lateral etch distance between trenches and windows is apparent when the underlying oxide thickness is reduced to 6 nm. In this case, the average lateral etch distance for trench features is 119 nm, and for windows it drops to only 84 nm.

Previously published results for lateral etching of much thicker oxide layers are in conflict with each other. Two groups working with nearly identical etching conditions (test structure, etch monitoring scheme, etc.) came to opposite conclusions regarding the dependence of etch rate on oxide thickness. Monk et al. observed almost no difference in the etching characteristics of structures with 134 nm and 2.36 μm oxide layer thicknesses using a concentrated HF solution (49 wt. %) [5]. On the other hand, Liu et al. observed a large dissimilarity in the etching behavior of oxide channels with layer thicknesses between 240 nm and 1.7 μm using a less concentrated 26.5 wt. % HF

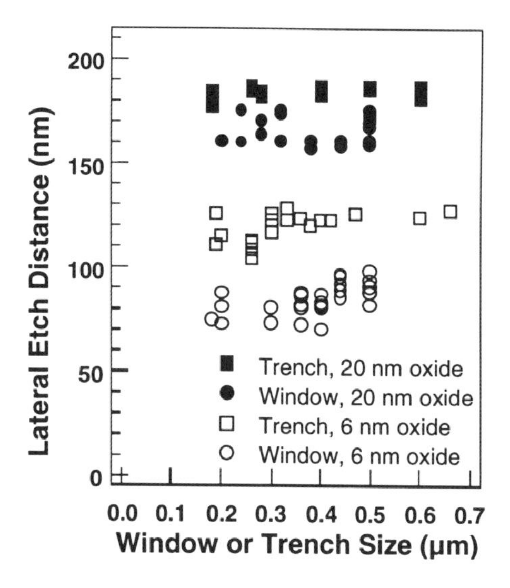

Figure 17.2 The lateral etch distance for two samples with 6 and 20 nm oxide thickness with window and trench features of various sizes. The samples were etched for 70 min. in 100:1 H_2O : HF (0.49 wt. percent).

solution [4]. Liu attributed the thickness dependence of the etching to a liquid-solid interaction between the silicon nitride over-layer and the HF solution. In our very thin oxide layers (less than 20 nm), we also see a pronounced reduction of the etch rate with oxide thickness. It is not readily apparent that the oxide should be etched at all given the extremely small dimension of our thinnest sample (2.5 nm) which is on the order of only 5 Si lattice constants. The etch solution may be forced into the cavity by capillary action or it may be drawn in by the hydrophilic nature of the oxide surface.

The lateral etching of these cavities appears to be roughly described by simple models for capillary flow. The speed of liquid flow into a capillary can be approximated by [7]

$$\frac{dz}{dt} = \frac{R_H(\gamma_{SV} - \gamma_{SL})}{4\eta z}, \tag{17.1}$$

where R_H is the ratio between the volume and surface area of the cavity, γ_{SV} and γ_{SL} are surface free energies between the solid-vapor and solid-liquid interfaces, and η is the viscosity of the liquid. For a rectangular cross-section capillary with a length much greater than its height, the liquid penetration after time t is given by

$$z = \frac{1}{2}\sqrt{\frac{(\gamma_{SV} - \gamma_{SL})th}{\eta}}, \tag{17.2}$$

where h is the cavity height. Therefore the extent of the lateral etch after time t is proportional to the square root of the cavity height. Based on the data of Fig. 17.2, the ratio of lateral etch depths for the two different cavity heights of 20 and 6 nm varies between 1.5 and 2, while the square root of the ratio of the cavity heights is 1.8. This result suggests that capillary action may be important for the etching of these small structures, however, further investigation is necessary.

For transistor applications, the cavity formed by the lateral etch of the oxide must be refilled with a conducting film to facilitate a contact to the crystal region. As the gate length of Si CMOS technology scales downward, the gate oxide thickness of the transistor also scales downward. In order to achieve the maximum cost reduction in a BiCMOS technology, the lateral etch and refill process must be accomplished using the gate oxide to form the cavity. Therefore, we studied oxide thicknesses down to 2.5 nm to cover the low end of the range expected for the next few generations of CMOS technology.

In our test structures, we refilled the cavity with α-Si deposited at 550°C using LPCVD at 250 mTorr. Fig. 17.3 exhibits the TEM results from the refill of α-Si into gaps ranging from 50 nm down to just 2.5 nm. Even in the thinnest sample, the refill of the cavity is complete without the appearance of large voids, which demonstrates that the deposition process is extremely conformal. This result shows that this technique indeed has great promise for future generations of BiCMOS technology. In the samples with cavity thickness greater than 2.5 nm, the convergence of the top and bottom growth fronts is visible. In the 2.5 nm sample, however, the border of the growth fronts is not visible due to taper introduced into the gap before the refill process. This taper may come about from native oxide formation and removal. In all the samples, one can notice that the refilled region is slightly thicker than the oxide thickness, which may be due to native oxide formation along the top and bottom of the cavity walls while awaiting α-Si deposition. The native oxide is removed by a short etch in dilute HF immediately prior to deposition, leaving behind a larger cavity, which is most easily seen in the TEM images of the thinnest samples. We also note that other experiments show that the border created by the coalescence of the top and bottom growth fronts disappears completely in all samples after a rapid thermal anneal (RTA) at 1050°C for 10 seconds.

17.4 $Si/Si_{1-X}Ge_X$ HETEROSTRUCTURE BIPOLAR TRANSISTORS

The gain in a Si bipolar transistor is expressed as

$$\beta \approx \frac{D_B x_E N_E}{D_E W_B N_B}, \tag{17.3}$$

where D is the minority carrier diffusion coefficient, x_E is the emitter thickness, W_B is the base thickness, and N is the doping concentration. Modern Si bipolar devices are quickly approaching vertical scaling limits with base widths in the order of 50–80 nm and base doping levels well above $10^{18}\,\mathrm{cm}^{-3}$. The desire to reduce the base thickness further to increase device speed (f_T) is hampered by two fundamental constraints. First, the base doping level cannot be raised substantially above about $3 \cdot 10^{18}\,\mathrm{cm}^{-3}$ because of the onset of high tunneling currents at the emitter–base junction. A higher base doping level is required to avoid punch-through between the emitter and collector and is desired to minimize base resistance (increase f_{max}). Secondly, the need to maintain sufficient current gain requires that the emitter doping (which is already at the solid solubility) be substantially higher than the base doping.

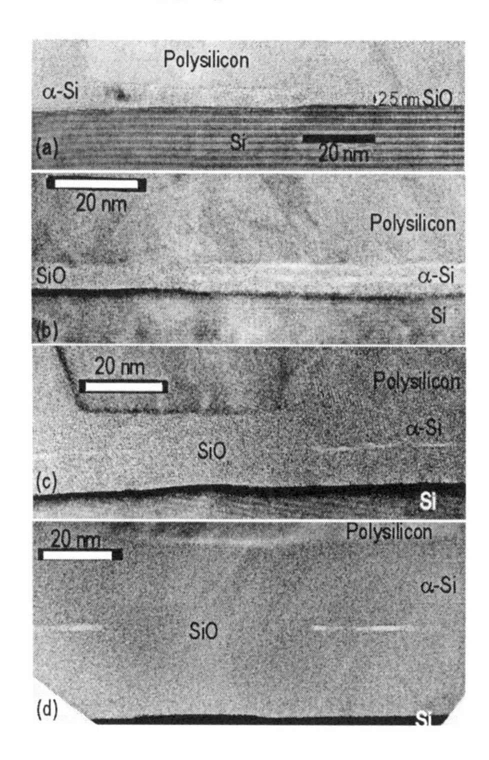

Figure 17.3 TEM micrographs of laterally etched cavities that have been refilled with α-Si. The cavity height (oxide thickness) is 2.5 nm in (a), 6 nm in (b), 20 nm in (c), and 50 nm in (d).

By placing a hetero-junction at the base–emitter junction, these constraints can be considerably relaxed allowing continued scaling of bipolar devices. The expression for current gain of a hetero-junction bipolar transistor (HBT) is modified in the following way,

$$\beta \approx \frac{D_B x_E N_E}{D_E W_B N_B} \exp\left(\frac{\Delta E_g}{kT}\right), \tag{17.4}$$

where ΔE_g is the bandgap discontinuity between the large-gap emitter and smaller-gap base material. As ΔE_g approaches several kT, the doping levels in the emitter and base can be completely reversed, allowing a heavily doped base layer to coexist with a lightly doped emitter region. This enhancement in the injection efficiency by using dissimilar materials in the emitter and base regions permits bipolar devices to be scaled vertically while improvements in lithographic technology allow them to be scaled laterally.

17.4.1 *Epitaxial growth and material properties of Si$_{1-x}$Ge$_x$*

Si$_{1-x}$Ge$_x$ alloys represent an example of a perfectly miscible system, and therefore all compositions can be epitaxially grown as a single-crystal diamond lattice. Under certain growth conditions on silicon, such alloys will grow as isolated "islands" rather than as a continuous two-dimensional plane. This three-dimensional growth is appropriately termed "island growth". Early molecular beam epitaxy (MBE) investigations into the deposition of Ge on Si repeatedly encountered problems with island growth. Two-dimensional planar growth was not achieved until Kasper demonstrated that the islands disappear when $Si_{1-x}Ge_x$ alloys are grown on Si with Ge mole fractions less than 20 % [8–10]. Bean et al. later discovered that all compositions of $Si_{1-x}Ge_x$ will grow island free, if the substrate temperature is kept within a window enclosed by polycrystalline growth on one side and island growth on the other [11]. They found that a substrate temperature of 550°C is within this window for all Ge concentrations. The onset of island formation is dependent on the germanium fraction in the alloy and occurs at increasingly higher growth temperatures as the germanium fraction is lowered. This ground-breaking discovery paved the way for future implementations of $Si_{1-x}Ge_x$ alloys in devices.

Once two-dimensional planar growth is attained, there are still two options for the manner in which the alloy can grow on the Si substrate: strained and unstrained. In the case of an unstrained $Si_{1-x}Ge_x$ alloy on a Si substrate, the larger $Si_{1-x}Ge_x$ equilibrium lattice constant is maintained and, due to the difference in registration with the Si substrate, misfit dislocations occupy the inter-facial region. Misfit dislocations add energy to the system in the form of dangling bonds, making the other mode of planar growth, strained layer growth, energetically favorable under certain conditions.

For the case of strained layer growth of $Si_{1-x}Ge_x$ on a Si substrate, the in-plane lattice constant of the alloy layer compresses to match that of the underlying Si substrate. Because of the relative thicknesses of the epitaxial layer and the substrate, the change of the lattice constant of the Si substrate is negligible. Strain also adds energy to the system since the bond angles are altered. Strain energy increases with layer volume. Consequently, at some layer thickness, the energy to maintain strain in the growing layer will overcome the energy to produce misfit dislocations at the interface. Many researchers explain strained layer growth using equilibrium analysis. From equilibrium calculations, one may find a critical thickness at the point where layer growth reverts from strained to unstrained.

Fig. 17.4 shows the observed critical thickness for a Si-capped $Si_{1-x}Ge_x$ alloy layer on Si as a function of Ge mole fraction. Grown thicknesses above the curve indicate meta-stable or unstable layers that are prone to misfit dislocation formation. Layer thicknesses at or below the curve are completely stable and will not relax through the process of misfit dislocation formation. Relaxation can occur, however, through the process of Ge diffusion, which transpires at sufficient thermal treatments (1000°C for several minutes). On the other hand, boron diffusion of the base is a greater concern for HBTs at these thermal budgets.

From Eq. (17.4), we noted that the bandgap discontinuity, ΔE_g, is a very important parameter to determine the injection efficiency enhancement of an HBT device structure over that of a Si homo-junction structure. Fig. 17.5 displays the bandgap discontinuity for both strained [12] and unstrained [13] $Si_{1-x}Ge_x$ as a function of Ge composition. In the unstrained material, the bandgap changes rather slowly with Ge composition until a value of about 85 % is reached, at which point the bandgap difference rises quickly. In the case of strained material, ΔE_g is large with even small fractions of Ge and continues to rise. When the Ge fraction reaches values of only

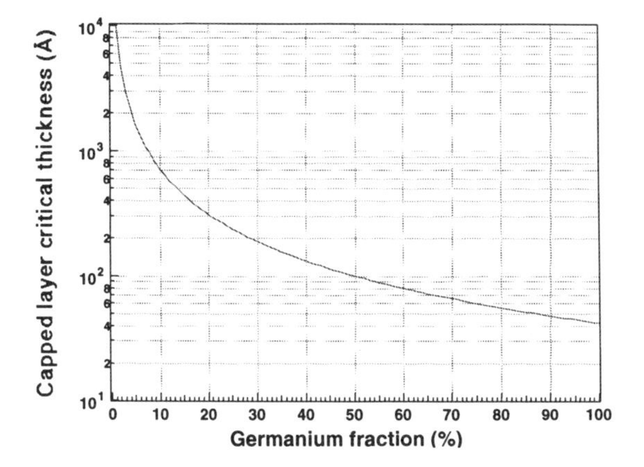

Figure 17.4 Critical thickness vs. Ge alloy composition.

60 % in the strained layer, the actual bandgap is smaller than that of pure Ge! Therefore, not only it is important for HBT applications to maintain complete strain in the alloy layer to avoid misfit dislocations, but also to obtain the maximum ΔE_g possible for a given Ge fraction.

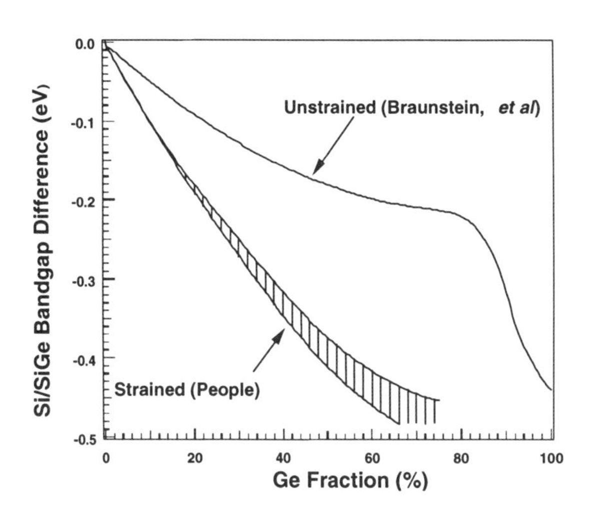

Figure 17.5 Bandgap difference for strained and unstrained SiGe as a function of Ge composition.

17.4.2 $Si_{1-x}Ge_x$ bipolar transistor structures

Since the first results that appeared in 1988, many researchers have begun examining the prospects of $Si_{1-x}Ge_x$ HBTs. Because other hetero-junction systems have been studied for years, many of the ideas conceived for HBTs in those materials are being applied to the $Si/Si_{1-x}Ge_x$ system. Compositional grading at the device junctions, compositional grading in the neutral base, and the use of a lightly doped emitter are among some examples of ideas retrofitted for $Si/Si_{1-x}Ge_x$ hetero-junctions. In addition, since this hetero-junction system is silicon based, researchers apply many concepts from silicon technology as well. Processes and ideas include poly-silicon emitters, ion implantation, rapid thermal and furnace annealing, and selective epitaxial growth. Most of the work to date can be divided into two groups. The first class of devices contains standard HBTs which utilize a single-crystal emitter and a base-emitter hetero-junction to achieve an increase in injection efficiency. The devices in the second class use a poly-silicon emitter to increase injection efficiency and grading of the bandgap in the neutral base region to introduce an electric field to speed the transport of minority carriers across the base.

Graded-base $Si_{1-x}Ge_x$ bipolar transistors. Patton et al. [14, 15] first introduced devices employing a poly-silicon emitter with a graded Ge profile. These devices use a Ge profile in the base that is graded from 0 % at the base-emitter junction up to a value of about 15 % at the base-collector junction. A secondary ion mass spectrometry (SIMS) plot of the Ge and boron profiles for this graded-base process is shown in Fig. 17.6. The graded Ge profile introduces a built-in electric field in the quasi-neutral base that reduces the transit time of injected electrons across the base thereby increasing f_T. The vertical doping profile, and thus the base sheet resistance, cannot be altered from the homo-junction case, since the band offset at the base-emitter junction is near 0 eV.

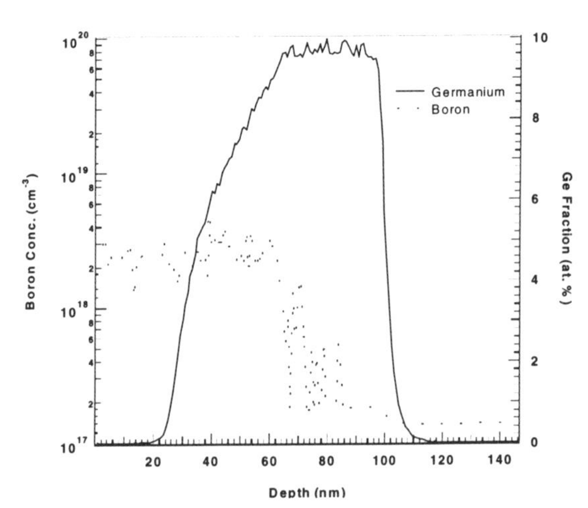

Figure 17.6 SIMS profile of graded SiGe base bipolar transistor.

Although grading the bandgap in the neutral base region lowers the base transit time, it does little to affect issues that address bipolar scaling. Since no hetero-junction is formed (aside from the poly-silicon/single crystal Si interface) at the base-emitter junction, base doping and current gain are still intimately coupled. Clearly, bandgap grading increases f_T over a homo-junction device, but it only weakly enhances injection efficiency. Therefore, further scaling of the base width cannot be accomplished with neutral base bandgap grading alone. For high-speed bipolar circuits, both low base resistance and high cutoff frequency are important. Graded-base devices provide a good initial step for using SiGe, since existing bipolar processes can largely be used requiring only the exchange of an implanted base with an epitaxially grown one. The true performance limits of SiGe bipolar devices will be attained with Si/SiGe HBTs.

$Si/Si_{1-x}Ge_x$ hetero-junction bipolar transistors. The strained-alloy bases with much higher Ge fractions (~30 %) [16, 17] which are used in Si/SiGe HBT structures allow much higher base doping (10^{19} to $10^{20}\,cm^{-3}$). $Si/Si_{1-x}Ge_x$ HBTs have demonstrated impressive figures of merit for both frequency response and noise performance [18]. However, these transistors were fabricated using blanket epitaxy coupled with a double-mesa process that makes integration very difficult. In our work at Lucent, we demonstrate a device structure usable for high-level integration which utilizes an aggressive vertical profile ($W_B = 18\,nm$, $N_B = 5 \cdot 10^{18}$ to $2.5 \cdot 10^{19}\,cm^{-3}$ and $x = 30\,\%$) in a planar transistor.

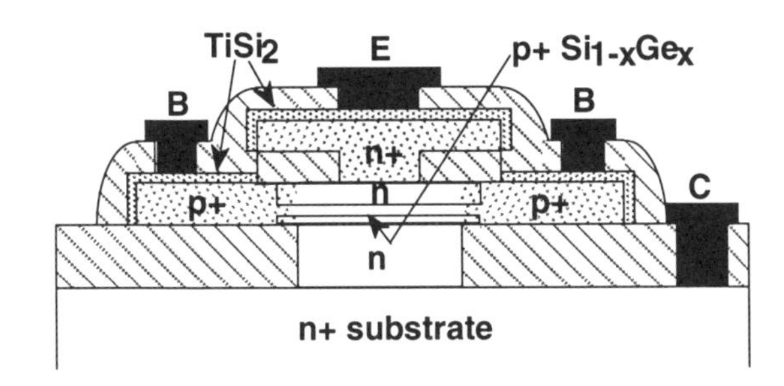

Figure 17.7 Si/SiGe HBT device structure.

Fig. 17.7 shows our device structure. We used rapid thermal epitaxy (RTE) to grow the layer structure, utilizing the limited-reaction-processing mode of growth [19]. A secondary ion mass spectrometry (SIMS) plot after the device fabrication of a structure with $N_B = 2.5 \cdot 10^{19}\,cm^{-3}$ is illustrated in Fig. 17.8. The very narrow base width is approximately 18 nm, which is below the equilibrium critical thickness for a capped $Si_{0.7}Ge_{0.3}$ layer ($R_{Bi} \sim 2.0k\Omega/\square$ for $N_B = 2.5 \times 10^{19}\,cm^{-3}$). The boron base doping is contained within the strained layer, indicating that transient enhanced diffusion effects are not important. The As profile shown in Fig. 17.8 is interrupted in the $Si_{0.7}Ge_{0.3}$ layer due to mass interference effects between Ge and As. These effects also lead to an exaggerated value of the As concentration close to the base-collector junction.

To evaluate the high-frequency performance of these transistors, we measured the S-parameters in common-collector mode from 2 to 18 GHz using a vector network analyzer and separated the effect of the parasitic pad capacitance using the Y-parameter subtraction technique. Devices with $N_B = 5 \times 10^{18}\,cm^{-3}$ yielded $f_T = 54\,GHz$, and with $N_B = 2.5 \times 10^{19}\,cm^{-3}$, $f_T = 44\,GHz$ for $V_{CE} = 1.5\,V$. Extraction of the intrinsic delay using τ_{ec} vs. I_E^{-1} for $N_B = 2.5 \times 10^{19}\,cm^{-3}$ yields an intercept of 2.9 ps and a

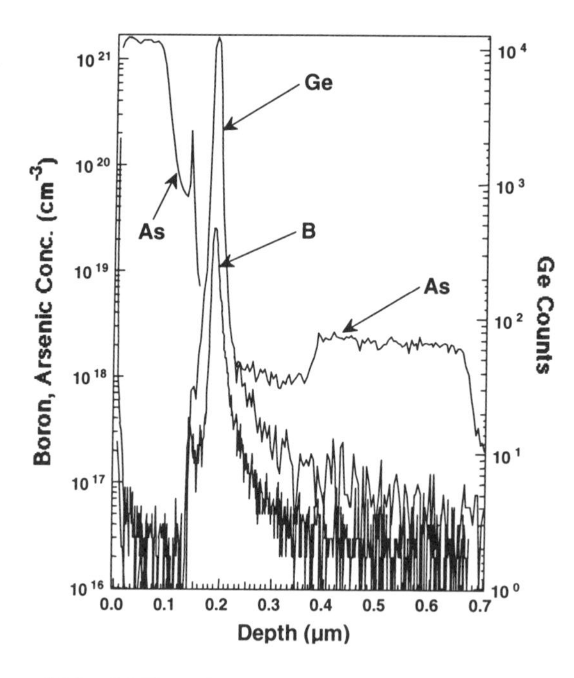

Figure 17.8 SIMS profile of HBT structure after all processing.

slope corresponding to $C_{BC}+C_{TE}=209\,\mathrm{fF}$. Table 17.2 shows the delay elements for this device. From this data, we estimate that the collector charging time accounts for about 72 % of the total delay, leaving much space for improvement in f_T and f_{max}, since the intrinsic delay is extremely small due to the thin 18 nm base width.

Table 17.2 Si/SiGe HBT delay components.

Component	*Delay*
T_b+t_d	1.0 ps (28.1 %)
$KT/qI_E(C_{BC}+C_{BE})$	0.67 ps (18.9 %)
$(R_C+R_E)C_{BC}$	1.88 ps (53 %)

17.5 SUMMARY

Low-cost Si BiCMOS will continue to be an important player in the wireless communication equipment business especially with the push toward lower power and higher integration. However, Si bipolar devices are reaching fundamental limits in terms of base doping and thickness which can be alleviated by the use of SiGe bipolar devices. Si/SiGe HBT technology has progressed steadily since the first reports. Indeed, initial products are now beginning to emerge, and with the demand for wireless goods expected to expand in the coming years, the need for this technology will surely grow. Graded-base devices are a first step to increase the cutoff frequency, but further device scaling will require an HBT structure with a high Ge fraction at the base-emitter junction. In this paper, we introduced a process which utilizes selective and non-selective rapid thermal epitaxy for growing a structure usable for high level integration and which uses a high Ge fraction of 30 % thereby permitting high base doping. High-frequency measurements of a $0.5 \times 10\,\mu m^2$ transistor yielded a peak f_T of 54 GHz. The performance is expected to be greatly enhanced by reducing the junction capacitances (C_{BC} and C_{TE}) through minor changes in processing conditions.

References

[1] Y.-F. Chyan, M. S. Carroll, T. G. Ivanov, A. S. Chen, W. J. Nagy, S. Chaudhry, R. W. Dail, V. D. Archer, K. K. Ng, S. Martin, M. Oh, M. R. Frei, I. C. Kizilyalli, R. Y. Huang, M. J. Thoma, C. A. King, W. T. Cochran, and K. H. Lee, in *Proc. of the Bipolar/BiCMOS Circuits and Technolgy Meeting*, pp. 128–129, (1998).

[2] Y. Kinoshita, K. Imai, H. Yoshida, H. Suzuki, T. Tatsumi, and T. Yamazaki, in *IEDM Tech. Digest*, pp. 441–444, (1994).

[3] M. Ugajin, J. Kodate, Y. Kobayashi, S. Konaka, and T. Sakai, in *IEDM Tech. Digest*, pp. 735–738, (1995).

[4] J. Liu, Y.-C. Tai, J. Lee, K.-C. Pong, Y. Zohar, and C.-M. Ho, in *Proceedings of IEEE Micro Electro Mechanical Systems*, Ft. Lauderdale, FL, 1993 (IEEE), pp. 71–76.

[5] D. J. Monk, D. S. Soane, and R. T. Howe, "Hydrofluoric Acid Etching of Silicon Dioxide Sacrificial Layers", *J. Electrochem. Soc.*, vol. 141, pp. 264–269, (1994).

[6] G. Timp, A. Agarwal, F. H. Bauman, T. Boone, M. Buonanno, R. Cirelli, V. Donnelly, M. Foad, D. Grant, M. Green, H. Gossman, S. Hillenius, J. Jackson, D. Jacobson, R. Kleiman, A. Kornblit, F. Klemens, J. T.-C. Lee, W. Mansfield, S. Moccio, A. Murrell, M. O'Malley, J. Rosamilia, J. Sepjeta, P. Silverman, T. Sorsch, W. W. Tai, D. Tennant, H. Vuong, and B. Weir, in *IEDM Tech. Digest*, pp. 930–931, (1997).

[7] E. Kim and G. M. Whitesides, "Imbibition and flow of wetting liquids in noncircular capillaries", *J. Phys. Chem. B*, vol. 101, pp. 855–863, (1997).

[8] E. Kasper, H. J. Herzog, and H. Kibbel, "A one-dimensional SiGe superlattice grown by UHV epitaxy", *Appl. Phys.*, vol. 8, pp. 199–205, (1975).

[9] E. Kasper and W. Pabst, "Profiling of SiGe superlattices by He backscattering", *Thin Solid Films*, vol. 37, pp. L5–L7, (1976).

[10] E. Kasper and H. J. Herzog, "Elastic and misfit dislocation density in $Si_{0.92}Ge_{0.08}$ films on Si substrates", *Thin Solid Films*, vol. 44, pp. 357–370, (1977).

[11] J. C. Bean, T. T. Sheng, L. C. Feldman, A. T. Fiory, and R. T. Lynch, "Pseudomorphic growth of Ge_xSi_{1-x} on silicon by molecular beam epitaxy", *Appl. Phys. Lett.*, vol. 44, pp. 102–104, (1983).

[12] R. People, "Indirect band gap of coherently strained $Si_{1-x}Ge_x$ bulk alloys on <100> silicon substrates", *Phys. Rev. B*, vol. 32, pp. 1405–1408, (1985).

[13] R. Braunstein, A. R. Moore, and F. Herman, "Intrinsic optical absorption in germanium-silicon alloys", *Phys. Rev.*, vol. 109, pp. 695–710, (1958).

[14] G. L. Patton, D. L. Harame, J. M. C. Stork, B. S. Meyerson, G. J. Scilla, and E. Ganin, in *Symp. on VLSI Technol. Dig. Tech. Papers*, pp. 95–96, (1989).

[15] G. L. Patton, J. H. Comfort, B. S. Meyerson, E. F. Crabbé, G. J. Scilla, E. D. Fresart, J. M. C. Stork, J. Y.-C. Sun, D. L. Harame, and J. N. Burghartz, "75-GHz f_T SiGe-base heterojunction bipolar transistors", *IEEE Electron Dev. Lett.*, vol. 11, pp. 171–173, (1990).

[16] T. I. Kamins, K. Nauka, J. B. Kruger, J. L. Hoyt, C. A. King, D. B. Noble, C. M. Gronet, and J. F. Gibbons, "Small-geometry, high-performance, Si-Si1-xGex heterojunction bipolar transistors", *IEEE Electron Dev. Lett.*, vol. 10, pp. 503–505, (1989).

[17] A. Schüppen, U. Erben, A. Gruhle, H. Kibbel, H. Schumacher, and U. König, in *IEDM Tech. Digest*, pp. 743–746, (1995).

[18] A. Schüppen, A. Gruhle, U. Erben, H. Kibbel, and U. König, in *IEDM Tech. Dig.*, pp. 377–380, (1994).

[19] J. F. Gibbons, C. M. Gronet, and K. E. Williams, "Limited reaction processing: Silicon epitaxy", *Appl. Phys. Lett.*, vol. 47, pp. 721–723, (1985).

18 GaAs-BASED RFIC TECHNOLOGY FOR CONSUMER RADIOS

Rob Christ

TriQuint Semiconductor 2300 NE
Brookwood Parkway Hillsboro
Oregon 97124, USA

Abstract: GaAs-based IC technologies are now firmly established in mobile phones and wireless data transmission applications. The success of GaAs MESFET RFICs in recent years comes from the need for improved performance for digital standards and from a significant reduction of the manufacturing costs for wafer fabrication. While there is little doubt that the GaAs IC industry is important today, there are differing opinions about the role of RFIC technologies in future wireless systems. Some believe that GaAs MESFET technology will take over the entire RF section of future radios, while others believe the present foothold will be chipped away as higher performance silicon technologies come on line. Yet others believe that alternate GaAs technologies, such as pHEMT and HBT, will gain greater acceptance for specific RF functional blocks. A minority believe that "radio on a chip" solutions in silicon will come to pass and some think it will include SiGe. In this paper, the author discusses the future of GaAs IC technologies in future commercial radio systems. The specific cost drivers for various process technologies are described and the key performance criteria that the mobile phone makers demand are reviewed. The combination of these factors shows trends in system architectures for future mobile wireless systems.

18.1 THE RF-INTEGRATION PARADIGM IS DIFFERENT

As enabling components of the Information Age, wireless technologies are quickly moving from mere curiosities and conveniences to indispensable pieces of modern life, much in the same way that digital technologies have made the computer the backbone of modern enterprise. The advent and maturing of silicon IC process technologies have, over the last three decades, made the stunning achievements in modern computing possible. In hindsight it may seem that the current dominance of CMOS IC technologies was inevitable. However, only since the massive worldwide investment in CMOS production capacity in the 1980's has it become clear that the staggering achievement represented in Moore's Law (in which former Intel Chairman Gordon Moore stated that the density of digital ICs would double at half the cost every three years) could be realized.

It is natural to extrapolate the success exemplified by Moore's Law to all subsequent IC technologies, but this ignores the fact that Moore's Law has not worked for high-performance analog and RF ICs. RF applications especially have stubbornly resisted integration. Some workers in the analog IC community, well represented by the participants in this publication, believe that this resistance will be broken by the inevitable improvements in the high-frequency performance of CMOS. This view taken to its extreme gives rise to the dream of "radio on a chip" integration. Following the digital integration paradigm, there are several assumptions built into the argument for ultimate RF integration in CMOS. Some of these assumptions are:

1. All Silicon is Cheap

2. Transistor Frequency Performance is the Limiting Factor

3. "Radio on a Chip" is Possible or Desirable

It is shown below that these assumptions do not stand up to close scrutiny, and that future wireless systems will contain a mixture of silicon and GaAs IC technologies.

18.2 WHERE IS RF INTEGRATION BEING USED?

Mobile telephones are currently the highest-volume application for high performance RF circuits. Until recently RF front-end circuits have been dominated by discrete devices. As the complexity of mobiles has increased and as TDMA and CDMA digital systems have come on line, we have seen a dramatic increase in the level of integration in the baseband and signal processing functions with the widespread use of DSP and integrated silicon for low frequency IF and analog functions. This has resulted in a commensurate decrease in the number of components needed for these functions. However, the RF portion of the radio has resisted integration until very recently.

The greatest success for RF integration has been in receiver front-end circuits for GSM and DCS-1800 systems. Through the use of very sophisticated circuit design techniques in high performance silicon BiCMOS, suppliers like Siemens and Philips have successfully penetrated the market with modestly priced ICs. Dual band ICs for GSM and DCS-1800 are selling for approximately 2–3 US$ in million-piece quantities. These circuits can be built mainly because the GSM standard is undemanding for IIP3 (input 3rd order intercept point) and Noise Figure, and because these ICs make use of a large number of external discrete passive components. Current silicon BiCMOS receiver ICs require more than 60 external components, for example.

Since GSM standards are very well established and mobile handsets are becoming more like a commodity, there are few opportunities left for the introduction of new IC process technologies. The level of silicon integration in GSM and DCS-1800 handsets is likely nearing its peak. It will be in the newer and less widespread standards, such as TDMA IS-136 in the US, narrow-band CDMA IS-95 and the new wide-band CDMA likely to be used for 3rd generation systems, where most of the challenging IC technology choices will still be made. Unfortunately for integrated silicon, whether CMOS or BiCMOS, the requirements for noise, linearity and power efficiency are much tougher than those for GSM. It is easy to have a GSM-centric viewpoint, especially outside the US, since GSM has become so dominant with so much popular appeal throughout the world. However, it is not a good idea to concentrate solely on GSM when looking at trends for future systems. Other systems like IS-136 and IS-95 offer important examples of the demands to be put on IC technologies for RF sections of future mobile phones. With this in mind, it is illustrative to compare what IC technologies are being used in today's digital mobile phones of different standards (see Table 18.1).

18.3 GaAs FOR MOBILE POWER APPLICATIONS

It is noteworthy that the GaAs content of an RF system increases generally as the linearity requirements of the system increases. This is true for both transmit and receive sides of the RF section. Based on the clear demand for maximum power efficiency, GaAs has dominated power amplifiers, especially in 3-volt systems. There has been some competition from LDMOS (laterally diffused MOS) in integrated modules, primarily from Japanese manufacturers, but these have been limited to 5-volt systems (or 3-volt systems with built-in voltage doubler) with lower linearity requirements.

With existing and emerging TDMA and CDMA systems, adjacent channel power (ACP) performance becomes much more important. This is where GaAs MESFET technology has a distinct advantage. For fully saturated power amplifiers, power added efficiency (PAE) values exceeding 60 % are routinely achieved, exceeding the efficiency of Bipolar and LDMOS power amplifiers. Narrow-band CDMA systems in particular have very stringent ACP requirements. For these systems, especially at 3 volts, no other process technology has achieved greater than 35 % PAE in commercially available circuits. This is a result of the very low "knee" voltage (voltage that denotes the transition from linear to saturated operation of the FET) of GaAs MESFETs. State-of-the-art power MESFETs have $V_{\mathrm{knee}} = 0.5\,\mathrm{V}$ or less, while silicon MOSFETs have $V_{\mathrm{knee}} > 1.5\,\mathrm{V}$ or more. Therefore, as wide-band CDMA systems come into being for 3rd generation systems, it is clear that GaAs MESFET technology will continue to have a bright future for mobile power amplifiers.

Two other GaAs technologies also show significant promise for power amplifiers. GaAs based hetero-junction bipolar transistors (HBT) are currently being used in high volume power amplifiers for GSM and IS-136 mobiles. The major benefits of HBT technology are low leakage, high PAE and reasonable linearity with a single power supply. The main disadvantages are the high cost and control issues associated with GaAs epitaxy growth. Much of this cost can be mitigated by the smaller die sizes achievable in HBT technology. It is not clear at this time whether MESFET or HBT

Table 18.1 RFIC Technologies Used in Today's Mobile Phones.

GSM 900	**DCS1800 & PCS1900**
Tx PA: 5 V = LDMOS Module 3 V = GaAs IC or Module	*Tx PA:* 5 V = LDMOS Module 3 V = GaAs IC or Module
Tx UPC & DA: BJT or BiCMOS IC	*Tx UPC & DA:* BJT or BiCMOS IC
Rx LNA + Mixer: BJT or BiCMOS IC	*Rx LNA + Mixer:* GaAs Discrete BiCMOS IC
TDMA 900 (IS-136)	**TDMA Dual Band**
Tx PA: 5 V = GaAs Module 3 V = GaAs MESFET IC	*Tx PA:* 2 GaAs MESFET IC 2 GaAs Modules Dual band GaAs IC
Tx UPC & DA: 5 V = BJT Discrete 3 V = GaAs MESFET IC	*Tx UPC & DA:* 5 V = BJT Discrete 3 V = GaAs MESFET IC
Rx LNA + Mixer: GaAs MESFET IC	*Rx LNA + Mixer:* Dual band GaAs IC
CDMA 900	**CDMA 1800/1900**
Tx PA: GaAs MESFET IC HBT IC & Modules	*Tx PA:* GaAs MESFET IC HBT IC & Modules
Tx UPC & DA: GaAs MESFET IC	*Tx UPC & DA:* GaAs MESFET IC
Rx LNA + Mixer: GaAs MESFET IC	*Rx LNA + Mixer:* GaAs MESFET IC

will, in the long run, be more economical for power amplifiers. The other GaAs-based technology that has promise for power amplifiers is pseudo-morphic high electron mobility transistors (pHEMT). pHEMTs make the most power-efficient power amplifiers, especially at very low voltages and high frequencies. However, since pHEMT power amplifiers cannot be made significantly smaller than MESFET power amplifiers, the high cost of epitaxy material limits the use of pHEMTs in L-band and S-band mobiles. pHEMTs will likely remain in the domain of high frequency systems, at X-band and above.

18.4 GaAs IN MOBILE RECEIVERS

Even most "radio on a chip" proponents admit that GaAs has a strong future in power amplifiers. However, they feel that small-signal RF parts, especially receivers, are the eventual exclusive domain of integrated silicon. Again the picture is not so clear. As seen above, standards that require higher linearity (as designated by ACP specifications) tend to use GaAs for receiver LNAs plus down-conversion mixers. CMOS solutions, as presented by other authors in this book, require ever more elaborate circuit techniques to meet even the very modest requirements of GSM and DCS-1800. Again, meeting the linearity requirements of more challenging standards with simultaneously low current draw favors the use of GaAs MESFET technology. The current consumption of receiver circuits is critical, since receivers are continuously on in mobile systems during standby. Today's extremely long standby times in IS-136 systems, for example, are largely due to the use of GaAs MESFET technology in receiver LNA plus mixer ICs. The four major specifications that are important for receivers are: 1) Noise figure, 2) Linearity (as in IIP3 or ACP), 3) Gain, and 4) Current draw. GaAs MESFETs allow the simultaneous optimization of all four of these specs in a way not possible in CMOS. For example, the following specifications are only simultaneously possible and commercially available in a GaAs MESFET based IC for 900 MHz IS-136:

- ACP/ALT $> -29\,\text{dBc}/-49\,\text{dBc}$
- Rx LNA/Mixer IIP3 $> -8.5\,\text{dBm}$
- NF $< 2.7\,\text{dB}$
- G $= 18.5\,\text{dB}$
- Idd $< 10\,\text{mA}$

(Refer to TriQuint's TQ9222 data-sheet.)

GaAs is not suitable for many of the low-frequency and digital functions of mobile handsets, but the technology has significant applicability for Tx up-converters, driver amplifiers, power amplifiers, and Rx LNAs and down-converter mixers. If it is so clear that technical advantages can be realized by using GaAs MESFET technology, why is there so much talk about stopping the spread of GaAs circuits and even replacing GaAs in mobile systems? This effort is fueled largely by the silicon digital integration paradigm discussed above.

18.5 TESTING THE ASSUMPTIONS

The mobile phone and wireless data hardware markets are brutally competitive. Any small cost advantage for performance that is "just good enough" is quickly exploited. With product life cycles of less than a year, there are continuous opportunities to insert new components into these systems. These pressures have led to dramatic improvements for integrated DSP and baseband functions. But, as mentioned earlier, the RF section has steadfastly resisted integration in silicon except in a few limited cases. The reasons go to the heart of the assumptions built into the "radio on a chip" paradigm. Let's examine these assumptions.

Assumption #1: All Silicon is Cheap. This assumption comes from the notion that all silicon processes can take advantage of the economies of scale available to digital circuits. For state-of-the-art processing, the lowest cost per mm^2 comes from CMOS used for memories and microprocessors. The costs for DRAM processing are fairly well known since these costs are monitored by the United States Department of Commerce as part of anti-dumping cases against Japanese and Korean DRAM makers. DRAM CMOS costs roughly 4 US cents per mm^2 for 200 mm wafers depending on utilization and yields.

This may be well and good, but digital CMOS processes are not useful for RFICs. RF silicon is produced in much lower volumes, typically using 6-inch or smaller wafers. Rather than being defect-density limited, RFIC yields are strongly driven by parametric specifications. As a result, even for the highest volume silicon RF processes the costs per mm^2 are substantially higher. Unfortunately, these costs are not as well reported, but judging from foundry prices, these costs are at least 50 % higher for high volume on 6-inch wafers, up to three or four times the cost for the highest-performance (Ft> 20 GHz) BiCMOS processes. These costs are similar to the cost of GaAs MESFET processing for 4-inch wafers in moderate volume.

There is little mystery about the cost drivers in the semiconductor industry, whether CMOS, silicon bipolar or GaAs. If we assume that the cost of packaging and testing is roughly the same whether the chip is silicon or GaAs, it is the cost of the processed chip where the real cost differences arise. The cost of a processed RFIC chip is a function of:

- Materials: wafers, chemicals, metals
- Number of Masking Layers: time in the fab, equipment utilization, labor
- Fab Capacity Utilization: spreading fixed cost
- Yield: wafer level, die level

The primary disadvantage of GaAs has traditionally been the cost per mm^2 of the starting wafer, which can be 10 times the cost of a silicon wafer or about 4 to 5 times the cost of silicon epitaxy wafer. However, this is offset by the second component of cost. GaAs MESFET processing is significantly shorter than competing silicon processes. For example, a GaAs process with 2 FET types, capacitors, thin film resistor and 3 layers of metal is 14 layers. By comparison, RF bipolar processes with the same functionality have 21 layers, and BiCMOS processes run 25 or more layers. The main advantage of silicon processing is that it is run in higher volumes with better fab capacity utilization. As GaAs moves up the volume curve and to 6 inch wafers, this advantage of silicon will evaporate. Finally, yield is often the biggest lever in any fab, given the variable nature of yields in any analog processing line. For the larger-volume GaAs suppliers, wafer line yields of 80–90 % are common today, equivalent to silicon RF processes.

Assumption #2: Transistor Frequency Performance is the Limiting Factor. There have been steady improvements in the frequency performance of CMOS and silicon bipolar transistors. As gate lengths decrease to 0.25 μm and below, the cutoff frequency (F_t) of n-channel MOS transistors exceed 10 GHz, which is in the range of suitability for building RF circuits with usable gain. Likewise, RF bipolar processes with $F_t > 20$ GHz are now commercially available. In the extreme, some SiGe HBT transistors have $F_t > 60$ GHz. While transistor frequency performance is important, it is important to look deeper. First, F_t is a misleading, some would say irrelevant, parameter for RF circuits as opposed to baseband linear circuits. Instead, $F_{\max}$ is a better measure of RF performance, since it applies to power gain. $F_{\max}$ is proportional to F_t by the factor of input resistance (Gate resistance for FETs, or Base resistance for bipolar). So while silicon technologies may have impressive F_t values, the high resistance of poly-silicon gates in MOS and thin bases in bipolar, limit the F_{max} values to barely more than F_t and sometimes even less. It is difficult to find MOS or bipolar transistors with F_{max} values that exceed 20 GHz. In contrast, the very low gate resistance of GaAs MESFETs (typically less than $1\,\Omega$) give F_{max} values 3 to 4 times F_t. For example, a $0.6\,\mu$m GaAs MESFET with $F_t = 20$ GHz has $F_{max} > 75$ GHz. The importance of this for RF circuits cannot be overstated. High F_{max} allows circuits to have enough gain overhead to produce very low noise figure with high linearity simultaneously.

But even high F_{max} is not enough to produce competitive RF circuits. The other important factor is the influence of parasitics. The conductive silicon substrate severely limits the ability to build high-Q passive structures and to achieve isolation on chip. As a result, all "radio on a chip" circuits still have a large number of external passive components. Also, these circuits must use fully differential topologies in order to deal with the cross-coupling and isolation issues. As a result, practical CMOS based RF circuits ironically consume more current than discrete implementations. Therefore, the main advantage of GaAs MESFET RFICs is not necessarily higher performance than silicon RFICs or discretes, rather it is achieving the performance standards for challenging standards with lower current.

Assumption #3: "radio on a chip" is Possible or Desirable. The term "radio on a chip" is somewhat misleading. Even the strongest proponents of this approach, represented in other chapters of this book, readily admit that many of the important parts of the radio cannot be effectively integrated in silicon. These include filters, high-Q passive components, duplexers/switches, and most importantly power amplifiers. If these are not considered part of the radio, then the definition of "radio on a chip" is arbitrarily limited to that which can be integrated in silicon. Since even the most integrated approaches have many dozens of external components, pushing all functions into a single IC becomes merely an academic exercise.

It is telling that the only serious discussion of "radio on a chip" topologies is being done by university professors. One possibility for this may come from the belief that it is only a matter of advancing research to the point where industry can implement the new topologies. While this may be the case in some circumstances, "radio on a chip" research in fact misses a key marketing reality. The more you integrate, the more flexibility is sacrificed. Creating a new IC product requires huge investments in engineering and time. Therefore, IC designs must be useful long enough to produce an adequate return on investment. With the very short life cycle of mobile phones, IC designs must be useful for several generations. Also, it is impossible to completely know in advance exactly what performance and features ICs will need to have before the mobile phone designs have been completely worked out. Therefore, flexible designs are crucial for making a commercially viable IC.

Digital circuits have overcome the need for flexibility in three ways:

1. When the function has become fully commoditized, like DRAM or Flash ROM. Therefore, the standards are known long in advance, *de facto*.

2. When the circuits are customizable or programmable. Programmable logic arrays and DSP can be used for several generations of systems, since their programs can be changed.

3. When the standards are set by the semiconductor maker. Intel Corporation enjoys the ability to make microprocessors that determine the direction of computer development. All others must follow.

None of these conditions are likely to be present in the RFIC market any time soon, if ever. When integration occurs, it will be driven by powerful economic and political forces. As of this writing, there are no compelling reasons for semiconductor makers to invest in a single-chip RF solution in CMOS, when a multiple-chip solution using IC technologies optimized for different parts of the radio provide the flexibility and performance that system makers desire. The goal for system makers is for optimum integration, rather than total integration. Even in the digital world, chip partitioning is the key for optimum cost and yield. For this reason, Pentium® II modules are actually composed of multiple chips each optimized for different functions. Likewise, with the widely divergent demands on IC technologies in RF systems, there is little incentive to push all functions into a single sub-optimized CMOS IC.

18.6 ADVANTAGES AND DISADVANTAGES OF DIFFERENT RFIC TECHNOLOGIES

Each RFIC technology has its advantages and disadvantages. Tab. 18.2 is a brief summary of the features of each technology. As this is a simplistic list of "pluses" and "minuses" rather than a comprehensive analysis, and because these technologies will change overtime, it is important to not read too much into the details of the list. Rather, the purpose of the list is to illustrate that each RFIC technology has its place in the range of choices available for RF systems. Tab. 18.3 shows the relative advantages and disadvantages of RFIC technologies for various RF blocks.

18.7 PREDICTING THE FUTURE: WHERE ARE CONSUMER RADIOS GOING?

With little doubt, the picture will remain complicated for mobile phones. As long as there are multiple standards, and with product life cycles so short, there will remain many opportunities for the various IC technologies to be used. Fig. 18.1 is a diagram of a typical mobile today. It schematically illustrates the fact that the RF section remains fragmented, dominated by discrete devices and single-function ICs.

Fig. 18.2 shows a likely scenario for future systems. There may be opportunities for further integration in digital and analog sections, but the greatest change will be in the RF section. Since there will be continuing needs for various technologies optimized for the different RF blocks, multi-chip RF modules pulling together these technologies is a more likely than "radio on a chip" integration in CMOS.

Likely future RF developments favor increased use of GaAs for two major reasons:

1. Increased demand for linearity, noise and power efficiency performance as described above.

2. Insulating substrate allows for integration of high quality passives ($Q > 17$ @ 2 GHz, for example) and excellent pin to pin isolation.

Table 18.2 RFIC Technologies Considered.

Silicon BJT and BiCMOS	
+	Established Infrastructure, Flexible Design
–	Low Q Passives Limits Integration
–	Parasitics Limits RF Integration
Silicon MOS	
+	Lowest Die Cost
–	Can't Get RF Performance With Low Current Draw
–	Parasitics Limits RF Integration
GaAs MESFET	
+	Established Infrastructure, Similar Cost to RF BiCMOS
+	RF Performance: Low NF, High PAE, High Linearity (ACP, IP3)
+	High Q Passives Enable RF Integration
–	Need Negative Supply (or Bias Generator) for Power
–	Need External Power Switch
GaAs HBT	
+	Single Supply and Easy to Design, Small Power Amplifiers
+	Very Good Linearity and PAE
–	Limited Material Availability, Expensive Epi
GaAs pHEMT	
+	Best PAE, Especially at Low Voltage
–	Difficult to Manufacture, Very Limited Availability of Starting Material
–	Cost Limits Integration
SiGe HBT	
–	No Real Products or Production Heritage
–	SiGe Epitaxy is hard to do in reality
–	Low BV_{cbo}, Bad for Power
–	Si bipolar nearly matches performance with far fewer process layers
–	Many more mask layers, > 30 layers

As a result for half duplex TDMA systems, ICs merging PA plus Switch plus LNA in GaAs are attractive without compromising RF performance. For CDMA systems, which have very large dynamic power control requirements favoring GaAs, ICs merging PA plus Driver Amp plus Up-converter in GaAs are also attractive. Since CDMA systems are full duplex, it is unlikely that it will be possible to merge Rx and Tx in a single IC of any technology, even GaAs. Therefore, receiver ICs will likely stay separate from transmit ICs in CDMA. Direct Conversion Rx in Si CMOS remains a strong possibility for less challenging standards like GSM. Regardless of process technology,

Table 18.3 RFIC Technologies by RF Block Function.

Function	*BiCMOS*	*CMOS*	*GaAs FET*	*GaAs HBT*	*SiGe HBT*
LNA	Good	Fair	V. Good	Good	Good
Mixers	Good	Good	V. Good	V. Good	?
PA	Fair	Good	V. Good	Good	Fair
RF Switch	Poor	Poor	V Good	Poor	Poor
VCO	V. Good	Good	Fair	Good	?
Pre-Scalers	V. Good	Poor	Fair	Good	?
Dig. Functions	V. Good	Good	Fair	Fair	?

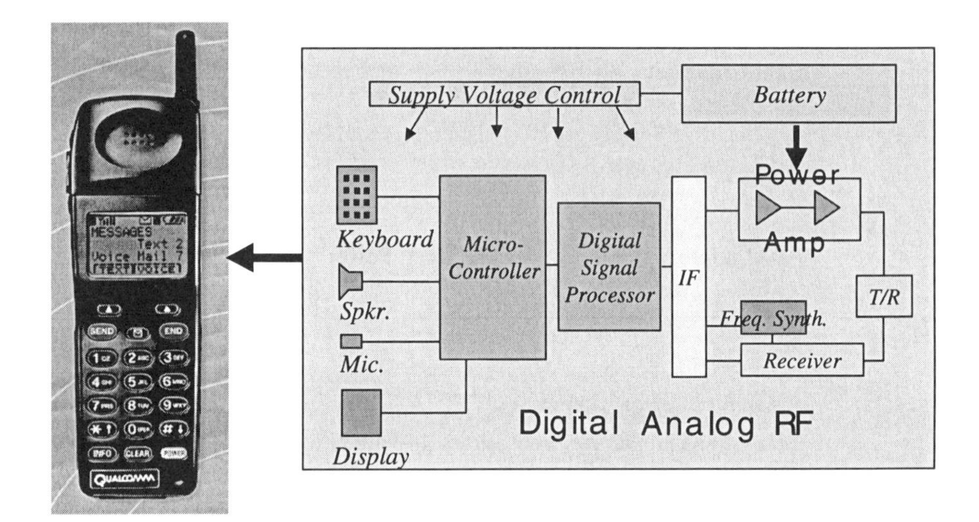

Figure 18.1 Today's mobile phone system.

it appears that RF circuits will modularize into commodity chip sets by standard. Only then will a consensus emerge. And only then will it be clear what the optimum level of RF integration will be for each RF block.

18.8 LOW-TECH GaAs MESFETS: COST-EFFECTIVE RF INTEGRATION

The main reason why GaAs has successfully penetrated consumer radios is that ion-implanted MESFETs have become a main-stream IC technology. Using production IC process tools and processing techniques including ion implantation, along with low cost surface mount packaging, GaAs MESFET manufacturing can take advantage of the economies of scale of silicon IC manufacturing. When this is parleyed with the inherent RF advantages of an insulating substrate, low-loss metalization and low-dis-

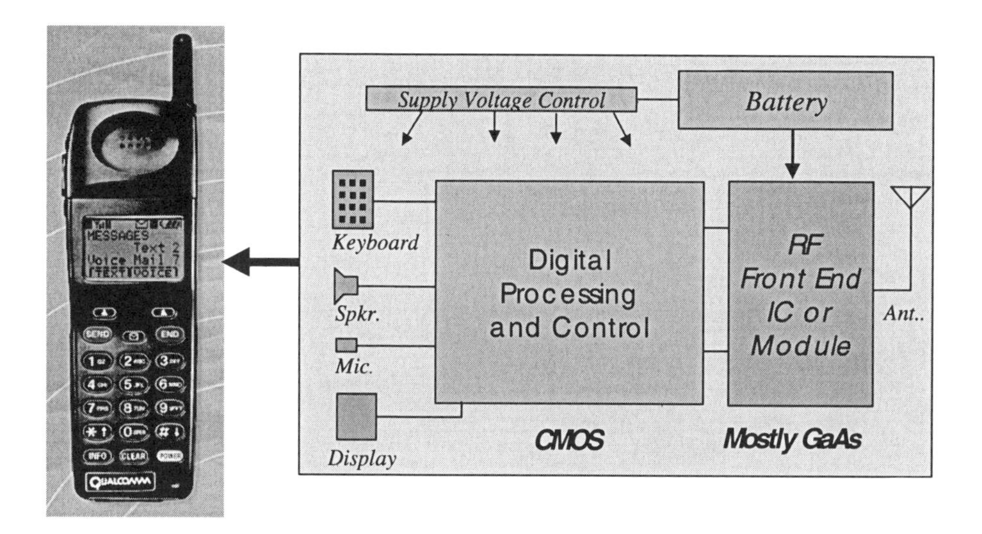

Figure 18.2 Mobile phone system Vision.

tortion/low-noise FETs, GaAs MESFET technology provides a natural foundation for optimum RF integration.

The state of the art in production ion implanted GaAs RFIC technology for mobile phones is well represented by TriQuint Semiconductor's TQTRx process. As its name implies, the process is useful for receive and transmit applications. Currently, tens of millions of LNA, Mixer and Power Amplifier ICs built in TQTRx are being shipped per year. As shown in Tab. 18.4, TQTRx has Enhancement (normally off), Depletion (normally on) and Power FETs integrated with thin film resistors, MIM capacitors and three layers of gold metalization. Fig. 18.3 shows a cross sectional diagram of TQTRx.

18.9 CONCLUSION

The Role of GaAs in Consumer Radios: It is rare in any high-technology business to see clear directions for which component technologies will dominate future systems. When a very promising new application arises, technology proponents and component suppliers scramble to take advantage of the opportunity. Wireless systems including mobile phones, pagers and wireless data systems are certainly "killer applications" for the late 1990s.

In the inevitable response to the opportunities in wireless, the proponents of various RFIC technologies have lined up to take part in the excitement. Some proponents of highly integrated CMOS "radio on a chip" ICs are quick to say that the dominance of CMOS in digital systems will naturally extrapolate to RF. Even if it were technically possible to produce these circuits, to say that they would be commercially viable ignores some fundamental market realities. By testing the assumptions built

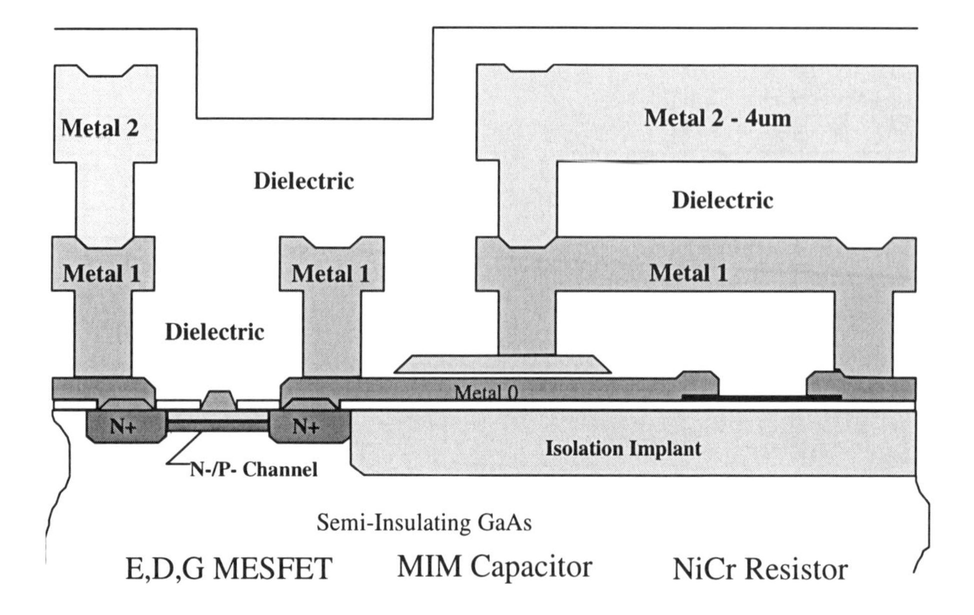

Figure 18.3 TQTRx process cross section.

Table 18.4 Key Parameters of TriQuint's TQTRx Process.

Parameter	*Units*	*TQTRx*
MESFET Types		E, D, Power
E-FET V_{th}	V (E)	+0.15
D-FET V_p	V (D, Power)	−0.6, −2.2
Gate Length	μm	0.6
Power Output	mW/mm	> 300
F_t, F_{max}	GHz	20, > 70
F_{min}	dB	< 1
Avalanche Breakdown	V	> 15
Interconnect	Total Layers	3, Total 6.5μm thick Gold
Inductor Quality	Q Factor	>17 at 1.9 GHz
MIM Caps	pF/mm^2	1200
Mask Layers		16

into the reasons for integration, it is clear that multiple IC technologies will be used in consumer radios for the foreseeable future. This will be especially true as more challenging standards come on line.

Because of the inherent advantages of GaAs IC processes for RF circuits, GaAs has a very bright future in the RF blocks in consumer radios. Instead of searching for "one size fits all" solutions, RF systems designers search for components that provide optimum solutions, in both cost and performance, for each part of the system. As a mature IC technology, GaAs MESFET provides many opportunities for low-cost system insertions with the necessary RF performance and higher power efficiency than silicon. While virtually everyone admits that GaAs has advantages for power amplifiers, closer examination reveals that GaAs also offers significant power advantages throughout the RF section of mobile phones. These advantages will become enabling components for 3rd generation systems based on wide-band CDMA. State-of-the-art, high-volume GaAs MESFET RFIC processes like TriQuint's TQTRx will play an important role to making these systems a reality.

19 MONOLITHIC INTEGRATED TRANSCEIVER CIRCUITS FOR GHZ FREQUENCIES

Urs Lott and Werner Bächtold

Lab. for Electromagnetic Fields and Microwave Electronics
Swiss Federal Institute of Technology (ETH) Zurich
ETH-Zentrum, CH – 8092 Zurich, Switzerland

Abstract: Different RF IC technologies are compared regarding their application, frequency range, and suitability for RF and baseband integration. The fully monolithic "phone on a chip" may be technically challenging to realize, and its commercial success may be challenged by progress in multi-chip integration. In the second part, examples of state-of-the-art RF monolithic circuits for applications in the frequency range of 2 to 17 GHz are presented.

19.1 SOME MYTHS ABOUT RF INTEGRATED CIRCUITS

19.1.1 Is using a single technology an advantage?

There is a lot of hype about the single technology which will enable a complete RF and baseband section of a radio transceiver to be integrated on a single chip. The presumed advantages of such a complete integration are stated as follows:

1. Complete integration will deliver a performance superior to a multi-chip radio.

2. Completely integrated radios will have a smaller size.

3. Complete integration will yield cheaper radio transceivers.

4. Completely integrated radio chips will be easier to apply in large quantities.

5. Complete integration will have a cost advantage over multi-chip solutions.

Before testing the validity of each of these statements, let us briefly analyse the technologies available for RF integration and their future development over the next years. Then, a summary of current baseband IC technologies is needed for investigating a possible fully integrated transceiver chip.

19.1.2 Present technologies for RF integrated circuits

Silicon bipolar. The Silicon Bipolar transistor (Si-Bip) in today's technologies typically has a transit frequency f_T of approximately 27 GHz and a maximum frequency of oscillation f_{max} on the order of 30 GHz. This technology is well suited for RF applications up to 2.5 GHz, for some functions up to 10 GHz. It is readily available from different foundries as a full-custom or semi-custom (array-type) technology. Future evolution will be towards hetero-junction bipolar devices, e.g. using SiGe as a base material. This will increase f_T to over 50 GHz. Standard bipolar technology will gradually be replaced by CMOS.

Silicon CMOS. Today, CMOS technology is in a process of rapid reduction of the minimum gate length of the MOS transistor. This reduction is mainly driven by digital CMOS applications, e.g. high-speed microprocessors, and the ever increasing storage density of dynamic random access memory (DRAM) chips. In line with the shrinking gate length, the breakdown voltage and thus the operating voltage is reduced to below 2.5 V.

With the 0.25 μm gate length transistors used in state-of-the-art microprocessors it is possible to build RF circuits with good performance up to 1...2 GHz on standard substrates. The intrinsic < 0.2 μm transistor will perform well up to 10 GHz, as has been shown by Silicon-on-Sapphire (SOS) CMOS circuits [1]. However, the substrates used in standard CMOS processes are highly lossy for RF. This reduces the performance of passive elements or even the interconnections in RF circuits considerably.

Future 0.18 μm and 0.13 μm CMOS technologies will yield devices with a performance comparable to today's GaAs MESFET device.

Silicon BiCMOS. BiCMOS technologies try to blend the high integration level of CMOS with the current drive capabilities of the bipolar device. This generally results in a quite complex manufacturing process using more than 20 mask levels. BiCMOS is therefore more expensive than other technologies, particularly for low volume application-specific chips. In addition, some compromises have to be made in the design of the CMOS and bipolar transistors to make them compatible in a single process.

With the improved performance of plain CMOS at reduced gate lengths it is likely that the bipolar device of the BiCMOS process is no longer needed in many applications.

Gallium arsenide MESFET. The RF performance of Metal-Semiconductor FETs (MESFETs) can profit from the higher electron mobility of GaAs compared to silicon. A typical MESFET with 0.6 µm gate length has an f_T of approximately 20 GHz and an f_{max} of 40...50 GHz. The higher f_{max} makes most RF functions feasible up to 5 GHz. An advantage of GaAs over silicon for RF integrated circuits is the high resistivity of the undoped substrate. Passive elements like inductors can be realized on chip with adequate performance. For RF circuits using reactive (L-C) matching, the limiting device figure of merit is f_{max} rather than f_T. Technologies like GaAs MESFET with good f_{max} and good passive elements have therefore a performance advantage not directly visible from f_T alone.

GaAs and silicon germanium HBT. In a Hetero-junction Bipolar Transistor (HBT), the base is made of a different semiconductor material than the emitter and collector regions, e.g. a GaAlAs emitter with a GaAs or InGaAs base for GaAs based transistors or a SiGe base for silicon based devices. Due to the lower bandgap base material the base can be doped higher without losing the current gain. The reduced base resistance boosts f_T and f_{max} towards 40 GHz and higher. GaAs based HBT ICs are already in volume production [2], SiGe HBT processes are just becoming commercially available [3].

High-electron-mobility transistor (HEMT). The HEMT is a field effect transistor in which the doping layer is separated from the conducting channel by using a special layer design with different bandgap materials, such as AlGaAs in the GaAs material system. Thus the mobility of the electrons is higher in the channel compared to a standard MESFET. GaAs HEMTs can achieve an f_T of above 60 GHz at 0.25 µm gate length, while HEMTs on Indium Phosphide reach an f_T above 100 GHz at similar gate lengths. Together with the high f_{max} (> 200 GHz) this allows to build circuits with operating frequencies of 100 GHz and more.

Comparison of RF device technologies. Figure 19.1 compares the application frequency ranges of different semiconductor technologies available for advanced integrated RF subsystems. The broken bars in the chart indicate the foreseen future expansion of silicon based technologies towards higher frequencies.

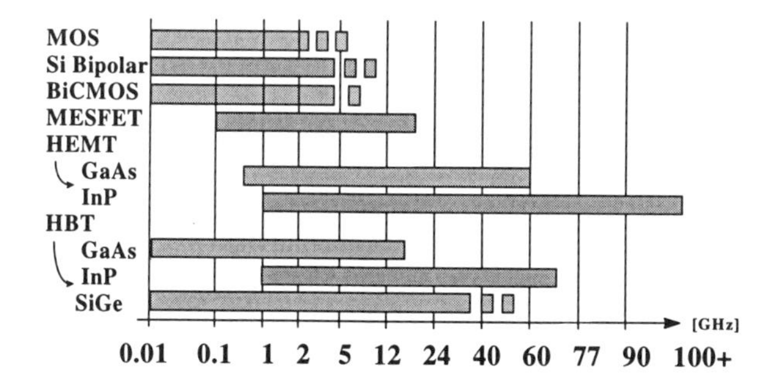

Figure 19.1 Application frequency range for different RF IC technologies.

Trying to avoid the common mistake of comparing a technology 3 years in the future with a competing one available today, one should be aware that today's > 10 GHz technologies (MESFET, HBT, HEMT) will also improve over time. However, most high-volume commercial applications for RF ICs within the next 5 to 10 years will be in the frequency range up to 10 GHz.

19.1.3 Technology choices for baseband circuits

In the baseband IC sector, CMOS is expected to keep its leading role. With the shrinking gate length of the MOS transistor, higher performance CMOS digital circuits can be designed. Thus the application areas for BiCMOS circuits are likely to diminish due to the higher manufacturing complexity of BiCMOS processes.

When considering RF and baseband integration, the highest potential therefore exists in the integration of pure CMOS RF and baseband circuits. The operating voltage of CMOS circuits is being reduced with shrinking gate length from 3.3 V down to 2.5 V and 1.8 V, eventually going down to 1 V or less.

19.1.4 Pros and cons of complete RF and baseband integration

Coming back to the often cited performance advantages of full RF and baseband integration:

1. higher performance,
2. smaller size,
3. lower cost,
4. easier application,

and considering the diversity and high performance of the above-mentioned RF IC technologies, it may be difficult to achieve the first advantage very soon. Concerning the other goals, there are also a few major obstacles, in particular related to the diverse properties of RF and baseband circuits:

1. The development tools for RF and digital baseband circuits are still very different, with RF design tools usually not offering as many top-down design possibilities as digital design tools.
2. As a result, development times for RF circuits are often longer.
3. Flexibility of a fully integrated solution can be much reduced over a multi-chip system, which is of special concern in applications that do not require millions of identical circuits.
4. Testing combined RF and baseband circuits can require a very complex and expensive equipment setup, it may not be possible to have test point access at all required instances, and built-in-self-test for RF is still in its infancy.
5. Processes used for RF and baseband, even when both are CMOS, may still be incompatible, e.g. when combining a low voltage digital part with a power amplifier in the 1 W class.

6. Integration of RF and baseband requires a technology which can do both with good performance, high yield and low cost.

7. Since competing technologies like Multi-Chip Modules (MCM) are progressing as well, the cost of a fully integrated radio may not be competitive for some time.

Even if many people are working on integrated RF and baseband transceivers, the fully integrated "radio on a chip" will need time to develop. The question is not if the technology allows to build it—it will soon—but rather if it is commercially competitive with a set of a few optimised single chips. The decision for or against full RF-baseband integration will also be influenced by the manufacturing volume and the flexibility needed in a particular product design. A device such as the proposed "Bluetooth" [4] could soon be integrated in a single chip, whereas a true single chip GSM phone is still a dream of the future—and may continue to be a dream.

In the following section, a few state of the art RF chips in the GHz frequency range will be reviewed.

19.2 EXAMPLES OF GHz TRANSCEIVER CIRCUITS

19.2.1 Low noise amplifiers in the 2 GHz range

LNA on Si bipolar array. As an example of a design on a silicon bipolar array, a two-stage low noise amplifier (LNA) is shown in Fig. 19.2. A high-gain common emitter stage (E1-RL) is followed by an emitter follower output stage (E2). DC feedback around the first stage transistor is used for bias stabilisation. External inductors are used for RF noise and gain matching.

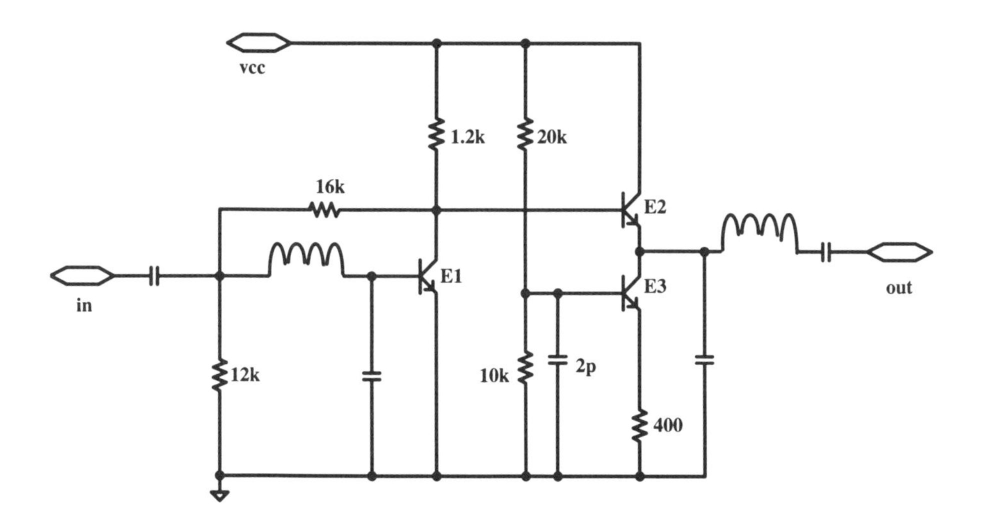

Figure 19.2 Schematic diagram of 1.9 GHz LNA on silicon bipolar array.

On the test circuit (Fig. 19.3), the matching inductors are realized as printed inductors on the substrate. They have a higher Q-factor than inductors integrated on the silicon chip. A noise figure of 2.3 dB could be achieved at 1.9 GHz [5].

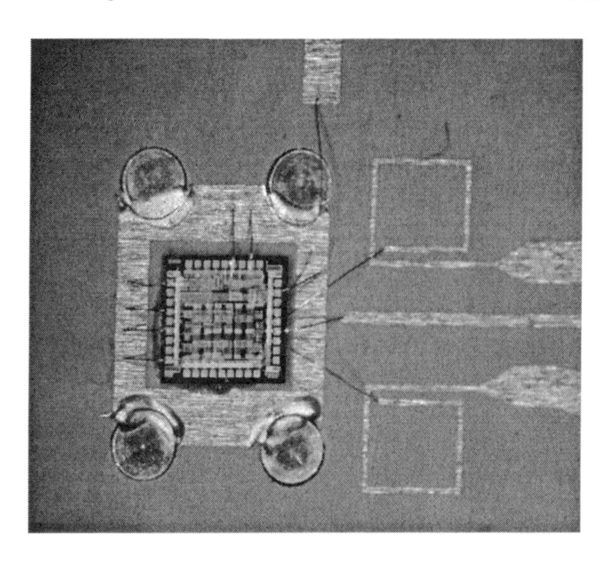

Figure 19.3 The 1.9 GHz LNA mounted on a test board with matching inductors.

This amplifier has a power consumption of only 5.2 mW at 3 V supply voltage. Together with the gain of 15 dB, this results in a high figure of merit for gain per DC power of 2.9 dB/mW. As can be seen from the comparison chart in Fig. 19.4, this amplifier is competitive with LNAs fabricated with other technologies. The chart also shows that it is difficult to achieve the same low noise figure on silicon that is possible with amplifiers on a high resistivity substrate such as GaAs.

LNA in GaAs MESFET technology with sub-1 V supply voltage. For portable communications applications, the power consumption of the receiver can be minimised by reducing the operating voltage to below 1.5 V. This may not be possible for the transmitter, where high efficiency is very difficult to achieve at supply voltages less than 2 V.

The schematic diagram of a low-noise amplifier operating with less than 1 V supply voltage is shown in Fig. 19.5 [6]. Key features are a RF load at the drain of the FET with almost zero DC voltage drop, and a bias stabilisation circuit working down to less than 1 V, as the simulation in Fig. 19.6 shows. Furthermore, this LNA was designed for a 200 Ω load impedance of the following mixer stage. In Fig. 19.7 the gain is plotted, which was measured in a 50 Ω system and recalculated for 200 Ω load impedance. With 0.9 V supply voltage, the gain is still higher than 10 dB. The gain is varying less than 1.5 dB for supply voltages between 0.7 and 1.2 V.

The measured noise figure is below 3 dB between 1750 and 2000 MHz (Fig. 19.8). The chip area is mainly used by the inductors and the bond pads, as the micro-photograph in Fig. 19.9 shows. Other performance data include a -1 dB input gain compres-

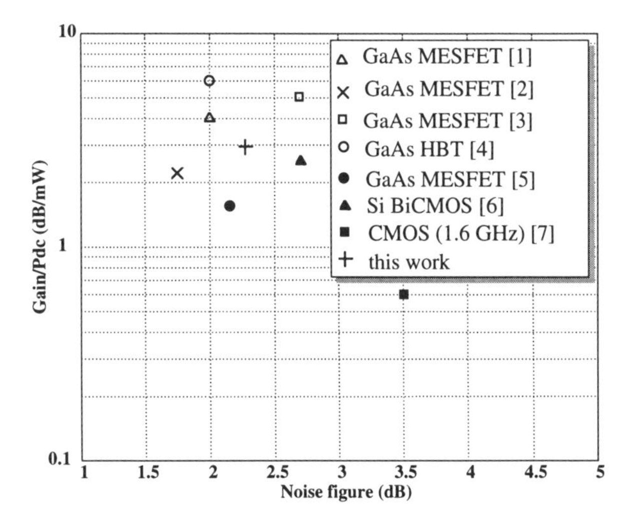

Figure 19.4 Comparison of 1.9 GHz LNAs (from [5]).

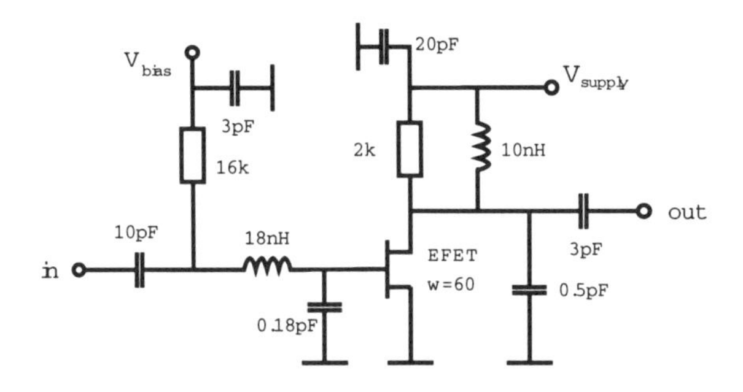

Figure 19.5 Schematic diagram of sub-1 V LNA.

sion point of -14.5 dBm and an input third-order intercept point of -1.5 dBm at 0.9 V supply voltage.

19.2.2 5 GHz LNA with switch for antenna diversity

There are a number of proposals which require diversity reception in wireless mobile terminals. Most of them use separate low noise amplifiers and a diversity switch. Recently, a MCM integration of a diversity front-end was demonstrated. Power consumption and chip size of the receiver front-end can be further minimised with a monolithic one-chip integration of the LNA and the diversity switch function. The block diagram of such a "diversity LNA" is shown in Fig. 19.10.

The diversity LNA achieves a minimum noise figure of 3.5 dB at 4.5 GHz with a maximum gain of 14 dB using the enhancement device of a standard E/D-MESFET foundry process. For comparison, a single LNA was designed on the same process

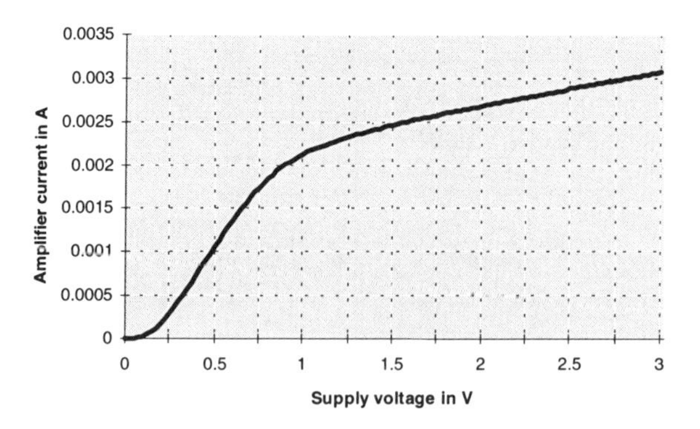

Figure 19.6 Supply current variation of the sub-1 V LNA versus supply voltage.

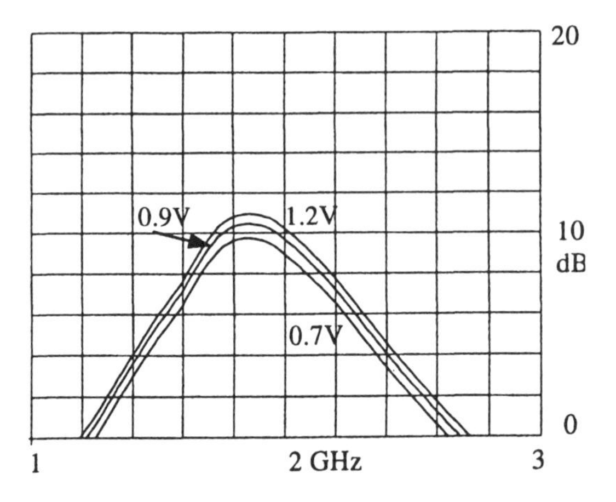

Figure 19.7 Measured gain versus frequency of the sub-1 V LNA versus frequency.

showing very similar gain and noise performance [7]. Both chips use full on-chip matching circuits. By comparison of the chip sizes in Fig. 19.11 and 19.12 it becomes clear that the designed diversity LNA requires at least 30 % smaller chip area than a multi-chip solution without performance penalty.

The ratio between the on and off state of a diversity channel is more than 20 dB (Fig. 19.13). The noise figure of each channel is plotted in Fig. 19.14. These measurements were made on chips mounted on a test board. Due to a longer bond wire, the frequency response of one channel is shifted to a slightly lower frequency.

The diversity LNA uses a single 3.3 V supply voltage for compatibility with low-power CMOS circuits. Power consumption is 13 mW (4 mA from a 3.3 V supply) with one diversity channel switched on. This is a reduction of roughly 50% compared to a conventional diversity receiver using individual LNAs.

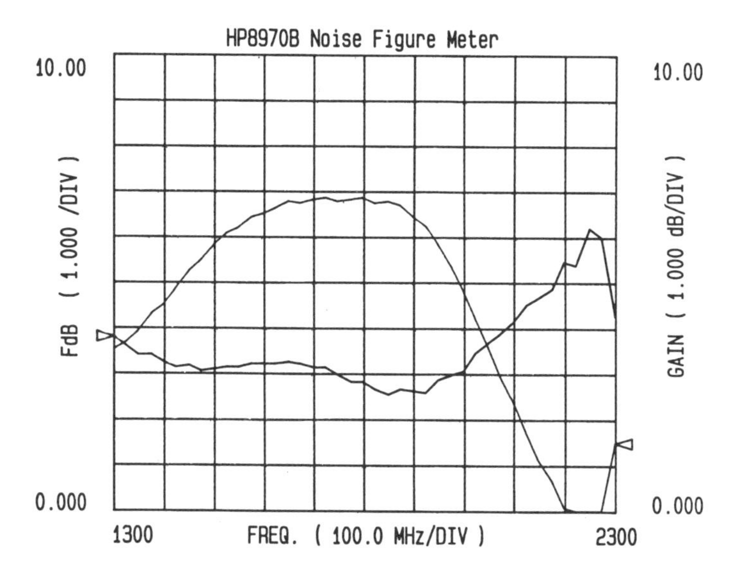

Figure 19.8 Measured noise figure of the sub-1 V LNA at 0.9 V/2 mA bias.

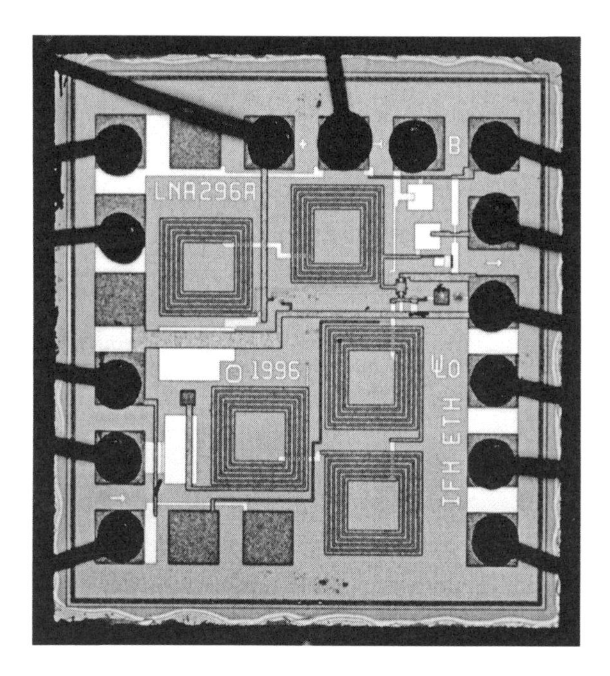

Figure 19.9 Micro-photograph of the sub-1 V LNA (chip size $1 \times 1\,\mathrm{mm}^2$).

19.2.3 17 GHz PHEMT power amplifier

High-speed wireless local area networks (WLAN) for indoor use are becoming more and more important. WLAN systems with a data rate up to 2 Mb/s are already stan-

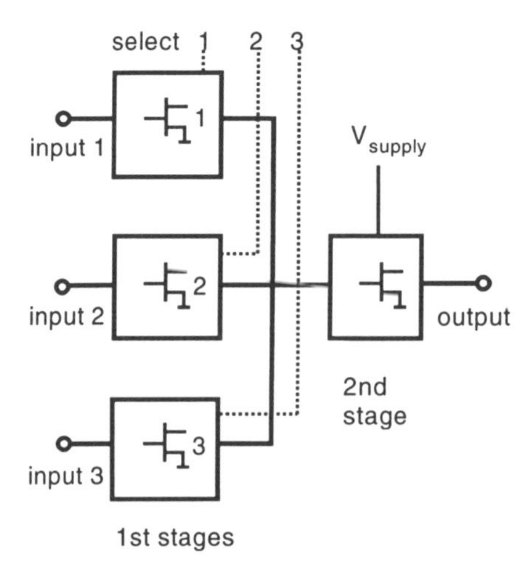

Figure 19.10 Block diagram of the 5 GHz diversity LNA.

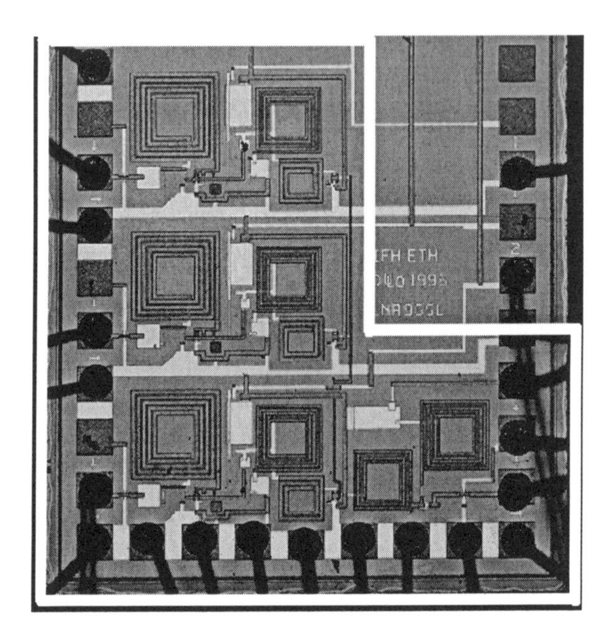

Figure 19.11 Chip photo of the 5 GHz diversity LNA (size 1.5 x 1.5 mm^2).

dardised in IEEE 802.11. To provide higher transmission capacity, new frequency bands have to be allocated. The emerging European HIPERLAN standard addresses frequency spectra in the 5 GHz and 17 GHz range. In the European ACTS project "Wireless ATM Network Demonstrator", a system with data rates of 20 Mb/s in the 5 GHz range has been developed [8] and investigations of the requirements for data

Figure 19.12 Chip photo of single LNA for comparison (size 1.5 x 0.7 mm^2).

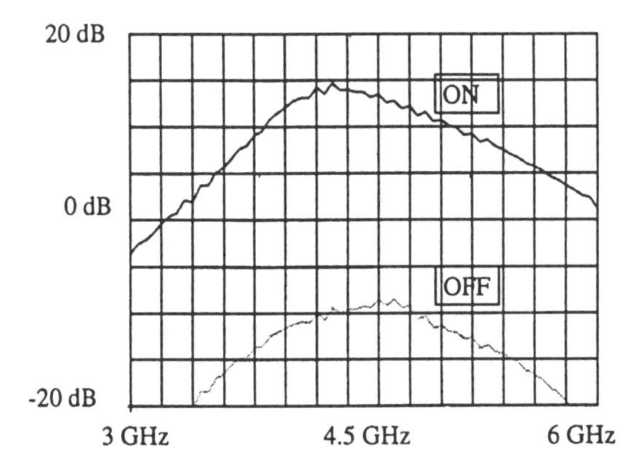

Figure 19.13 Measured gain versus frequency of the diversity LNA channel 1 switched ON and OFF.

rates up to 155 Mb/s in the 17 GHz band have been made, taking into account system architecture and technological constraints.

As one of the key components of a future 17 GHz wireless LAN, a monolithic integrated power amplifier has been designed. A two-stage on-chip matched power amplifier with 200 mW output power was realized on a commercial pseudo-morphic HEMT process (Philips D02AH). Fig. 19.15 shows a block diagram of the amplifier.

The transistor of the driver stage has a gate width of 300 μm. The input matching, the gate and drain biasing network as well as the interstage matching network are integrated on the chip. In the power stage, the 1200 μm transistor and its gate biasing and the output matching network are integrated. The drain bias network is external due to the high DC current required by the transistor. The low-pass-type output matching network, shown in the basic schematic diagram (Fig. 19.16), allows for external biasing. No harmonic tuning was done at the output. The output stage is designed to be conditionally stable. Negative feedback stabilises the driver stage.

The whole circuit was simulated with a harmonic balance simulator. The HEMTs are simulated with a Statz MESFET model modified with an RC network between

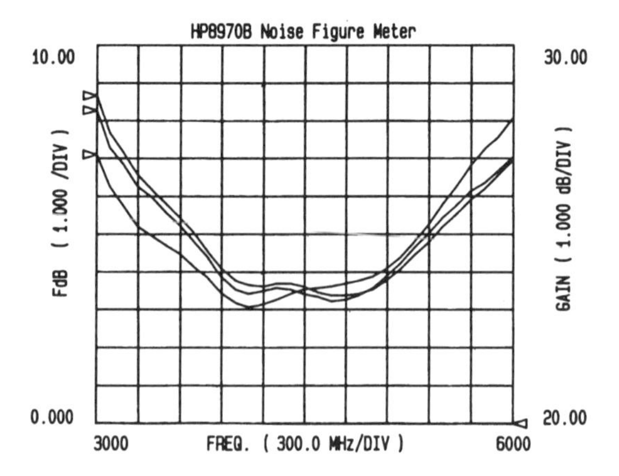

Figure 19.14 Measured noise figure of the three channels of the diversity LNA.

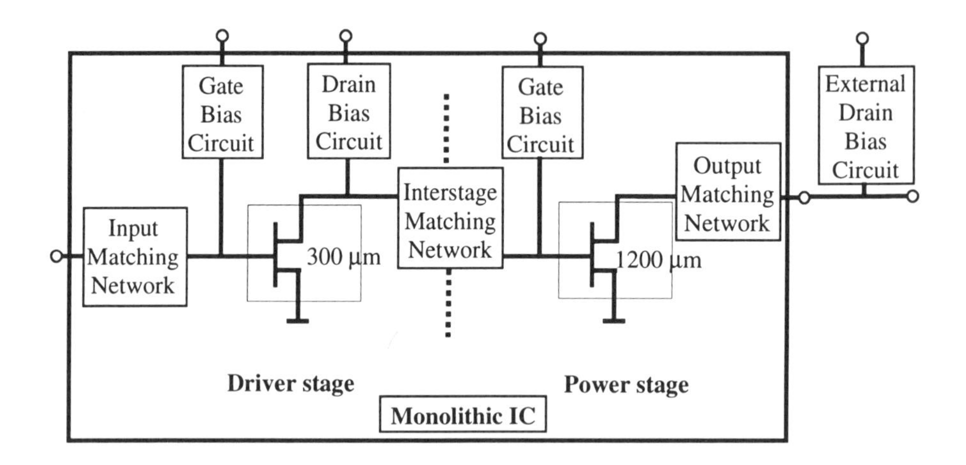

Figure 19.15 Block diagram of the 17 GHz PHEMT power amplifier IC.

drain and source. Thus, small inaccuracies in the large signal simulation results are expected.

As the chip photograph in Fig. 19.17 shows, via holes to the back metal provide an optimum source grounding, but limit a maximum layout density by the minimum distance required between two adjacent via holes of 200 μm. Nevertheless, the layout is very compact with a chip size of $1 \times 1.5\,\text{mm}^2$.

The measured fundamental power, the power added efficiency and the drain efficiency at 17.2 GHz agree with the harmonic balance simulation as shown in Fig. 19.18. The output power at -1 dB gain compression is 20 dBm, the saturated output power is 23 dBm at 3.3 V supply voltage. The measured power added efficiency has a maxi-

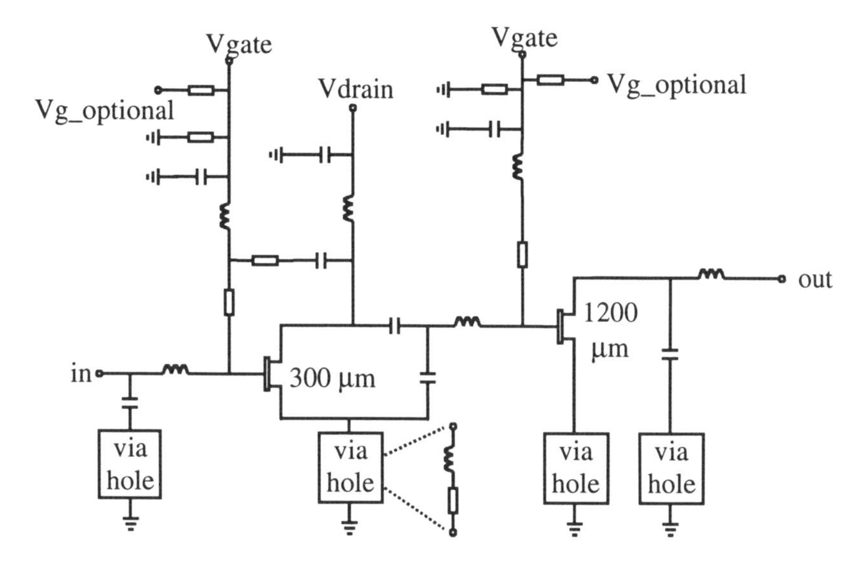

Figure 19.16 Basic schematic diagram of the 17 GHz power amplifier IC.

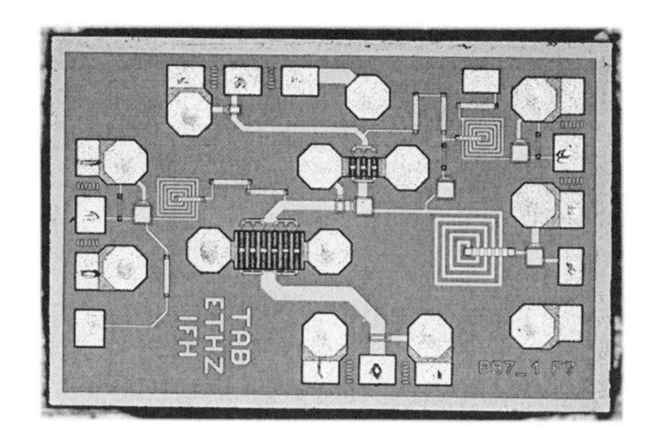

Figure 19.17 Chip photograph of the 17 GHz power amplifier (size 1 x 1.5 mm^2).

mum of 20 % and the drain efficiency is 31.5 %. More measurement data can be found in [9].

19.3 CONCLUSIONS

In the foreseeable future, special RF IC technologies will still be required for applications in the GHz range, particularly when high transmit power or a low noise figure is important. For low-cost, lower performance applications such as the proposed Bluetooth, full monolithic RF and baseband integration in CMOS must be the goal.

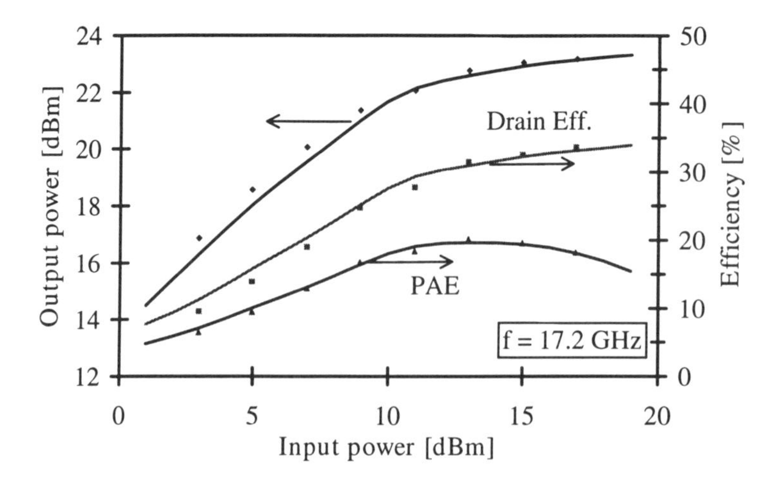

Figure 19.18 Output power, drain efficiency and power added efficiency (PAE) of the 17 GHz monolithic power amplifier.

Acknowledgments

The authors would like to thank their colleagues Thomas Bös and Jakub Kucera for being allowed to use some of their circuits as examples of today's RF integrated circuits in the GHz range.

References

[1] K-H. Kim *et al.*, "4 GHz and 13 GHz tuned amplifiers implemented in a 0.1 μm CMOS technology on SOI and SOS substrates", *1998 IEEE Solid-State Circuits Conference Digest*, pp. 134–135.

[2] K. W. Kobayashi *et al.*, "Sub-1.3 dB noise figure direct-coupled MMIC LNAs using a high-current gain 1 μm GaAs HBT technology", *IEEE 1997 GaAs IC Symposium Digest*, pp. 240–243.

[3] M. Soyuer *et al.*, "A cost-effective approach to a short-range, high-speed radio design in the U-NII 5.x GHz band", *IEEE 1998 Radio and Wireless Conference Proceedings*, pp. 133–136.

[4] see `http://www.bluetooth.com`

[5] J. J. Kucera, U. Lott "Low-power silicon BJT LNA for 1.9 GHz", *IEEE Microwave and Guided Wave Letters*, vol. 8, no. 3, March 1998, pp. 136–137.

[6] U. Lott, M. Schmatz, "2 GHz monolithic low noise amplifier using lower than 1 V supply voltage", *Proceedings of the 1997 IEEE Wireless Communications Conference*, pp. 138–140.

[7] U. Lott, "Triple input single chip LNA", *IEEE 1998 Radio and Wireless Conference Proceedings*, pp. 335–337.

[8] Information on the ACTS project WAND can be found on the internet at `http://www.tik.ee.ethz.ch/~wand`

[9] T. A. Bös, U. Lott, W. Bächtold, "A two stage, monolithic integrated 200 mW HEMT amplifier for wireless ATM", *IEEE 1998 Radio and Wireless Conference Proceedings*, pp. 125–128.

V DSP for Wireless Communications

20 DSP FOR WIRELESS COMMUNICATIONS

Urs Fawer* and Gertjan Kaat

Philips Semiconductors AG
Binzstrasse 44
CH–8045 Zürich, Switzerland

20.1 INTRODUCTION TO THE FOLLOWING PAPERS

Recently, wireless communication has turned into a very fast-growing market. GSM[2] gives a good example with a total number of 135 million users world-wide by the end of 1998, of which 65 million new customers subscribed in the last year. This impressive growth is based on the underlying technologies. The latter are the topic of this part: the study of complex communication functionality and related algorithms as well as their realization in hardware and software.

Digital Signal Processors (DSPs) are a key component for implementing communication algorithms. Very often, we encounter embedded DSP cores where the DSP is integrated on a silicon chip together with a controller core, memories, hardware accelerators, and dedicated periphery blocks. For many algorithms, the DSP represents the essential block in the implementation, e.g. voice codecs, channel equalizers, error correction, encryption, interference suppression, multi-channel and multi-beam reception as well as additional features like voice recognition. DSP performance is often mea-

*Urs Fawer is now with diAx, Birgistr. 4a, CH–8304 Wallisellen, Switzerland.

[2] Global System for Mobile Communications; second-generation digital cellular mobile system

sured in terms of million instructions per second (MIPS). Since the influence of the considered architectures is not reflected in this MIPS figure, a benchmarking exercise based on well-defined algorithms is recommended to compare DSP cores.

Wireless communication systems impose demanding requirements on the DSP, especially for mobile terminals with inherent constraints on size, weight, and battery capacity. In consumer applications, overall system cost and low power consumption represent the main key driving factors.

20.2 TRENDS

The trend towards digitization will continue: functionality which was previously realized in the analog domain is more and more implemented by digital signal processing. This tendency follows the CMOS process road-map with continuously decreasing feature size along the time axis. It also results in a continuous cost reduction in integrating digital circuitry because the number of building blocks integrated on one piece of silicon can be further increased. This leads more and more to a fully integrated solution which is often referred to as a "system on silicon."

The demand for wireless transmission of data, and with it the need for more bandwidth, is growing. Yet, bandwidth is a limited resource. Therefore, the increasing capacity demands for wireless voice and data transmission will result in a higher complexity and efficiency of future systems.

These trends clearly indicate that the requirements increase with respect to MIPS figures, low power consumption, high-level design and programming approach. Improved support for implementing floating-point algorithms is another important issue. It addresses the rising numerical complexity and the short time-to-market requirements. It will be interesting to see how all these trends will be realized. We can certainly expect a lot of innovation from future DSP architectures and applications.

20.3 PRESENTATION OVERVIEW

The first contribution of H. Keding et al. deals with the implementation of floating-point algorithms in a fixed-point architecture.

The next two papers investigate different hardware solutions for implementing digital signal processing algorithms. E. Lambers et al. present the "R.E.A.L. DSP," a programmable architecture with a flexible application-specific instruction set, which can be tailored towards the algorithmic needs. Advanced algorithms for implementation in receivers are described in B. Haller's contribution. They lead to a dedicated, yet configurable hardware approach. The fourth paper, authored by J. Sanchez, summarizes the requirements and trends of cellular mobile communication terminals.

In the fifth contribution, G. Miet describes future implementations of speech codecs and their impact on the system complexity.

The last paper of this part, authored by R. L. J. Roovers, investigates wide-band analog-to-digital converters, which are fundamental building blocks of multi-channel receivers.

21 EFFICIENT DESIGN FLOW FOR FIXED-POINT SYSTEMS

Holger Keding, Martin Coors, and Heinrich Meyr

Institute for Integrated Systems in Signal Processing (ISS)
Aachen University of Technology
Templergraben 55, 52056-Aachen, Germany

Abstract: The complexity of DSP systems grows at an ever-increasing rate while the implementation of these designs must meet criteria like minimum cost and a short time to market. Hence there is a growing need for efficient design automation and a seamless design flow that allows the execution of the design steps at the highest suitable level of abstraction.

Especially for the design of fixed-point algorithms and systems, tool support is very rare, though for most digital systems the design has to result in a fixed-point implementation. This is due to the fact that these systems are sensitive to power consumption, chip size and price per device. Fixed-point realizations outperform floating-point realizations by far with regard to these criteria. On the other hand, algorithm design starts from a floating-point description in order to initially abstract from all fixed-point effects. The resulting gap between the floating-point prototype and the fixed-point implementation represents one of the major bottlenecks in today's digital designs.

This paper will give a survey of FRIDGE[1], a tool suite that permits a seamless design flow, starting from an ANSI-C floating-point algorithm which is then converted to

[1] **F**ixed-point p**R**ogram**I**ng **D**esi**G**n **E**nvironment

a fixed-point description in *fixed-C*[2]. This bit-true description of the algorithm can finally be mapped to different implementation targets, programmable DSPs or dedicated hardware structures.

21.1 INTRODUCTION

Digital system design is characterised by ever-increasing complexity and ever-tighter schedules, resulting in minimum costs and short time to market. This requires a seamless design flow that allows the execution of the design steps at the highest suitable level of abstraction.

For most digital systems, the design has to result in a fixed-point implementation, either in hardware or software. This is due to the fact that digital systems have to be optimised with regard to power consumption, chip size, throughput and price per-device. Fixed-point realizations outperform floating-point realizations by far with regard to these criteria.

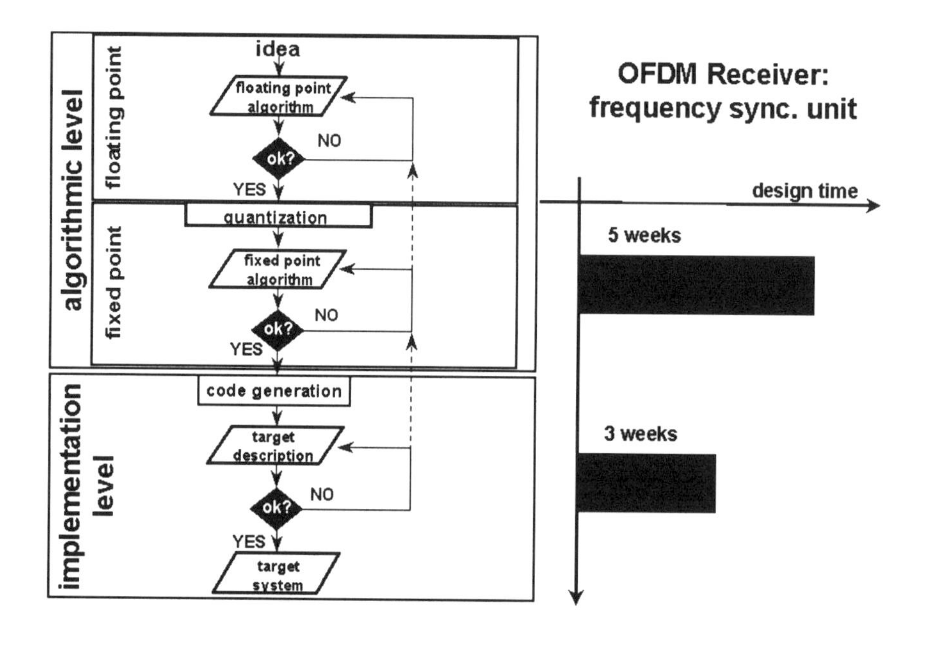

Figure 21.1 Fixed-point design process.

A typical fixed-point design flow is shown in Fig. 21.1. Algorithm design starts from a floating-point description that is analysed by means of simulation. This abstraction from all implementation effects allows an evaluation of the algorithm space with-

[2]fixed-C is a Fixed-Point Language based on ANSI-C, but with additional generic fixed-point data types.

out taking into account the quantisation effects on the algorithmic behaviour. Thus, modeling efficiency is higher at the floating-point level, and the use of floating-point models offers a maximum degree of re-usability.

The transformation to the fixed-point level is quite tedious and error-prone, since a fixed word length and a fixed exponent needs to be assigned to every operand manually. To provide a more quantitative picture of this time-consuming process, in Fig. 21.1 we exemplarily displayed the design times for the implementation steps for an OFDM receiver frequency synchronisation unit [1] developed at ISS. In much the same way as for this example, often more than 50 % of the design time is spent on the algorithmic transformation to the fixed-point level for complex designs once the floating-point model has been specified. The major reasons for this bottleneck are:

- There is no unique transformation from floating-point to fixed-point.

    1. Different HW and SW targets put different constraints on the fixed-point specification.
    2. Optimisation for different design criteria, like throughput, chip size, memory size, or accuracy, are in general mutually exclusive goals and result in a complex design space as sketched in Fig. 21.2.
    3. The quantisation is generally highly dependent on the application, i.e. on the applied test vectors.

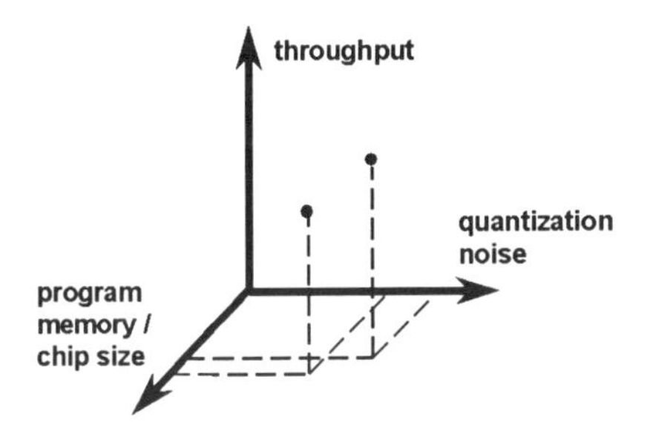

Figure 21.2 Fixed-point design space.

- The manual transformation of the floating-point algorithm to the fixed-point algorithm is known to be time-consuming and error-prone. Even for a single transformation, modeling efficiency is very low. Therefore the manual floating-to-fixed conversion appears to be no longer acceptable.

- The fixed-point simulation efficiency is low. This is due to the fact that the bit-true behaviour of the target system has to be emulated on the host machine.

These inefficiencies have been the motivation for FRIDGE [2–4], an interactive design environment for the specification, simulation and implementation of fixed-point systems.

In this paper we describe the principles and elements of FRIDGE and outline the seamless design flow which becomes possible using this design environment. We also give a global picture of the FRIDGE design flow by highlighting its constituents, briefly describing the fixed-point language fixed-C and the way to annotate operands with fixed-point attributes. Later we show the core algorithm of FRIDGE, the wordlength interpolation and the back ends, i.e. mapping the internal fixed-point representation of FRIDGE to a synthesisable/compilable description.

21.2 THE FRIDGE DESIGN FLOW

As pointed out above, a manual transformation into a fixed-point representation is hardly feasible. Since an efficient evaluation of the complex design space requires multiple transformations, this is even more of a design bottleneck.

In the FRIDGE design environment, system design starts from a floating-point algorithm in ANSI-C. The designer then may annotate *single* operands with fixed-point attributes, as illustrated in Fig. 21.3.

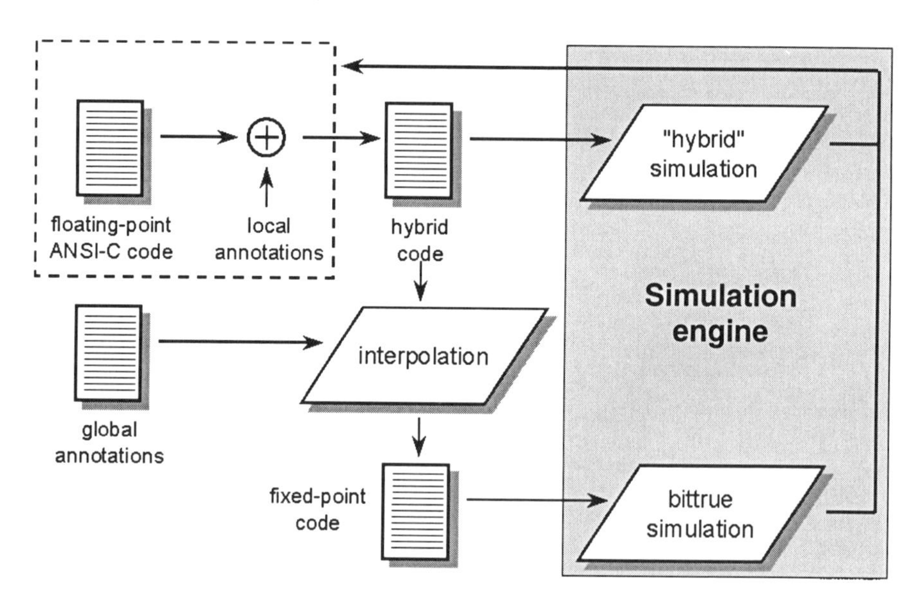

Figure 21.3 Quantisation methodology with FRIDGE.

Inserting these *local annotations* results in a *hybrid* description of the algorithm, i.e. some of the operands are specified bit-true, while the rest are still floating-point. A comparative simulation of the floating-point and the hybrid code within the same simulation environment shows whether the local annotations are appropriate, or if some annotations have to be modified.

Once the hybrid program matches the design criteria, the remaining floating-point operands are transferred to fixed-point operands by *interpolation*, where interpolation denotes the process of computing the fixed-point parameters of the non-annotated operands from the information that is inherent to the annotated operands and the operations performed on them. Additionally, the interpolator has to observe a set of *global annotations*, i.e. default restrictions for the calculation of fixed-point parameters. This can be e.g. a default maximum wordlength that corresponds to the register length of the target processor.

The interpolation supplies a fully annotated program, where each operand and operation is specified in a bit-true way. Co-simulating this algorithm with the original floating-point code will give an accuracy evaluation—and for any alterations only some of the local and global annotations have to be modified, while the rest is determined and kept consistent by the interpolator.

What is described above are the *algorithmic level* transformations as illustrated in Fig. 21.1 which change the behaviour or accuracy of an algorithm. The resulting completely bit-true algorithm in *fixed-C* is not yet suitable for implementation, thus it has to be mapped to a target, e.g. a processor's architecture or an ASIC. This is an *implementation level* transformation, where the bit-true behaviour normally stays unchanged. Within the FRIDGE environment different *back ends* map the internal bit-true specification to different formats/targets, according to the purpose or goal of the quantisation process:

- ANSI-C or fast simulation code
- Processor specific code
- HDL code, e.g. behavioural VHDL

This framework allows a seamless design flow from algorithm development to implementation and therefore tears down the well-known virtual brick wall that is often to be found between system level design and implementation.

21.3 FIXED-C AND LOCAL ANNOTATIONS

Since ANSI-C offers no efficient support for fixed-point data types [5,6], we developed the fixed-point language *fixed-C*, which is a superset of the ANSI-C language. It comprises different generic fixed-point data types, casting operators and interpolator directives. These properties enable fixed-C to be used for different purposes in the FRIDGE design flow:

- Since ANSI-C is a subset of fixed-C, the additional fixed-point constructs can be used as bit-true annotations to dedicated operands of the original floating-point ANSI-C file, resulting in a *hybrid* specification. This partially bit-true code can be used for simulation or as input to the interpolator.

- The bit-true output of the interpolator is represented in fixed-C as well. This allows maximum transparency of the results to the designer, since the changes to the code are reduced to a minimum and the effects of the designer's directives, such as local annotations in the hybrid code, become directly visible.

The language constructs to describe the bit-true behaviour are contained in a C++ class library which can be used in any design and simulation environment that comes with a C or C++ compiler, such as COSSAP [7], SPW [8] or Matlab/Simulink [9].

For a bit-true and implementation-independent specification of an operand, a three-tuple is necessary, the *wordlength* **wl**, the *integer wordlength* **iwl**, and the *sign* **s**, as illustrated in Fig. 21.4.

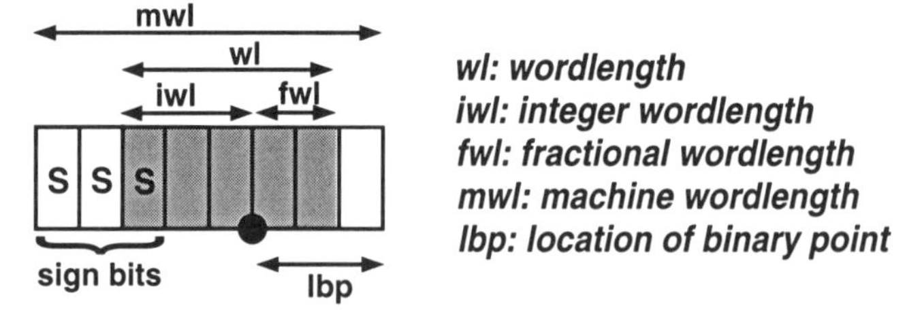

Figure 21.4 Fixed-point attributes of a bit-true description.

21.3.1 *The data type* `Fixed`

declaration: `Fixed<wl,iwl,sign>      d,*e,g[8];`

The data type `Fixed` receives its format when it is declared, thus a `Fixed` variable behaves according to these fixed-point parameters throughout its lifetime. This concept is called *declaration-time instantiation* (DTI). Similar concepts exist in other fixed-point languages as well [10–12]. Pointers and arrays, as frequently used in ANSI-C, are supported as well.

For every assignment to a `Fixed` variable, a data type check is performed:
Example:

```
Fixed<6,3,s>  a,b;
Fixed<12,12,u> c;
a = b; /* correct, both types match */
c = b; /* type mismatch */
```

`Fixed` is the data type of choice e.g. for interfaces to other functionalities or for look-up tables, since it behaves like a memory location of a specific length.

21.3.2 *The data type* `fixed`

In addition to the DTI data type concept, fixed-C provides the data type `fixed` which is specially tailored for the float-to-fixed transformation process.

declaration: `fixed      a,*b,c[8];`

A variable is declared to be of data type `fixed`, but no instantiation is performed and no fixed-point attributes are specified.

casting: `a=fixed(wl,iwl,sign,cast,*b);`

The variable a receives data of type `<wl,iwl,sign>`, the value of `*b` is cast according to the casting mode `cast`. The casting mode specifies how to handle overflow[3] as well as quantisation[4]. The different casting modes are shown in Tab. 21.1.

[3]Overflow handling specifies the behaviour in case of a wordlength reduction at the MSB side.
[4]Quantisation handling specifies the behaviour in case of a wordlength reduction at the LSB side.

Table 21.1 Different modes for cast to fixed.

casting modes	*overflow handling saturation*	*overflow handling wrap around*
quantisation by *rounding*	sr	wr
quantisation by *truncation*	st	wt

Every assignment to a variable *overwrites* all prior instantiations, i.e. one fixed variable may have different context-specific bit-true attributes in the same scope. This concept of **Assignment-Time Instantiation** (ATI) is motivated by the specific design flow: transformation starts from a floating-point program, where the designer abstracts from the fixed-point problems and does not think of a variable as a finite length register. Fig. 21.5 highlights the properties of ATI in a quantisation process.

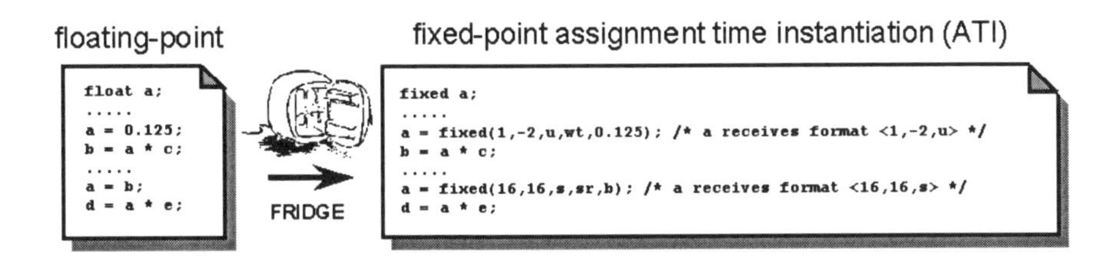

Figure 21.5 Example of assignment-time instantiation.

For the first assignment to a, a fixed-point format <1,-2,u> is sufficient to represent the constant 0.125, while for the second assignment a format <16,16,s> would be necessary. With ATI, this information is accessible without code changes, while for DTI, two options exist:

- merging over the fixed-point parameters for all assignments to the variable a to receive a super-format that matches the fixed-point requirements wherever a is used.[5] If this super-format were the input to an implementation tool, it would force the use of large registers, even when this is not necessary.

- to use the context-specific information, a variable must be split up manually into several different variables, which is a time-consuming and error-prone task as well.

In contrast, the concept of local annotations and ATI is an effective way to assign context-specific information without changing structures or variables when exploring the fixed-point design space.

[5]For the example in Fig. 21.5: 2 fractional bits from the first assignment to a, 16 integer bits from the second, so the merged format is 18 bits long.

21.3.3 Interpolator directives

The interpolator can extract information about the fixed-point parameters not only from direct information about wordlength, integer wordlength and sign,[6] but also from indirect or user-supplied information. For example one could specify only the range of an operand:

```
a=fixed(minim(-2.35),maxim(4.74), *b );
```

This information is *not* sufficient for a bit-true simulation, since there is no information extractable e.g. about the quantisation steps. Annotations like this are called *incomplete local annotation* or *interpolator directives*. They are ignored by the simulator, but can provide additional information to the interpolation process.

21.4 INTERPOLATION

The interpolator is the core of the FRIDGE design environment. As shown in Fig. 21.3, it determines the fixed-point formats for all operands of an algorithm, taking as input a user-annotated *hybrid description* of the algorithm and a set of global default rules, the *global annotation* file. Hence *interpolation* describes the computation of the fixed-point parameters of the non-annotated operands from the information that is inherent to the annotated operands.

21.4.1 Maximum precision interpolation

The interpolation concept is based on three key ideas:

Attribute propagation: The means of using the information inherent to the hybrid code to determine the fixed-point attributes of all operands.

Global annotations: The description of default rules and restrictions for attribute propagation.

Designer support: The interpolator supplies various feedback or reports to assist the designer to debug or improve the interpolation result.

For a better understanding, the first two points are now explained in more detail:

Attribute propagation. Given the information of the fixed-point attributes of some operands, the type and the fixed-point format of other operands can be extracted from that. For example, if both the range and the relevant fractional wordlength are specified for the inputs to an operation, the same attributes can be determined for the result.[7] A simple example for an addition is shown in Fig. 21.6.

Given the user-annotated ranges of variables `a` and `b`, FRIDGE automatically determines the range and the accuracy, i.e. the required number of fractional bits for the variable `c`. Note that this is a worst-case or maximum-precision interpolation (MPI),

---

[6] As provided by casting operators described earlier.

[7] An exception is the division, where the accuracy of the operation must be specified as well.

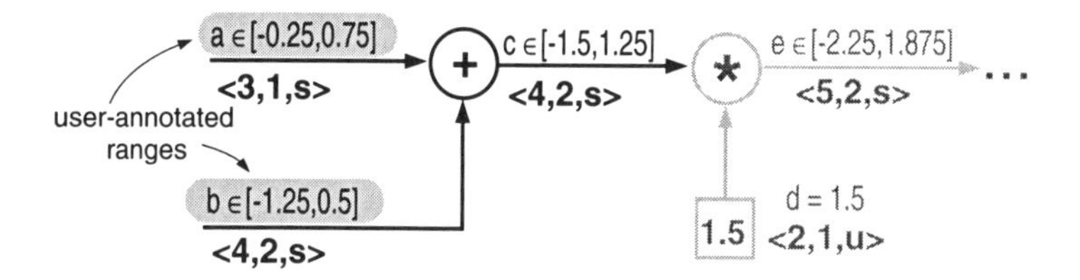

Figure 21.6 Example for interpolation of ranges/wordlengths.

so there is no way an overflow can occur. LSBs are only discarded if the wordlength exceeds the maximum wordlength specified in the *global annotation* file.

The fixed-point parameters *iwl* and *sign* of an operand can be determined once its range $[min, max]$ is known[8] [13]:

$$1. \quad min < 0 \quad \begin{cases} sign &= s \\ iwl &\geq \max\{\mathrm{ld}|min|, \mathrm{ld}(|max| + 2^{-fwl})\} + 1 \end{cases}$$

$$2. \quad min \geq 0 \quad \begin{cases} sign &= u \\ iwl &\geq \mathrm{ld}(max + 2^{-fwl}) \end{cases} .$$

Without the additional knowledge of *fwl* one has to spend one additional bit to guarantee that the maximum (minimum) can be represented.

wl is determined using the information about the fractional wordlength $fwl = wl - iwl$. It is only necessary to represent those fractional bits which actually carry any information.

Global annotations. While local annotations express fixed-point information for single operands, global annotations describe default restrictions to the complete design. For different targets, different global restrictions apply. For SW, the functional units to perform specific operations are already defined by the processor's architecture. Consider a 16×16 bit multiplier writing to a 32-bit register. A global annotation can inform the interpolator that the wordlength of a multiplication operand may not exceed 16 bits, while the result may have a wordlength of up to 32 bits.

21.4.2 Utilisation of statistical knowledge for interpolation

With the MPI, the wordlengths of the operands increase considerably during interpolation, while in many cases the obtained large wordlengths do *not* correspond to an adequate precision. This is due to the fact that each fixed-point input x to the algorithm represents a *quantised* value x_0, i.e. it consists of the exact value x_0 and the quantisation error Δx,

$$x = x_0 + \Delta x. \tag{21.1}$$

[8] We assume a two's complement representation here.

Taking this fact into account, one can compute a *noise floor* that can be used to judge the information content of each bit. The statistical properties of the noise are propagated by the interpolative approach described above. Assuming the error PDF is known for *some* operands, e.g. for the input, the resulting noise probability density function (PDF) at every point of the algorithm is calculated. In principle, the PDF of a linear arithmetic operation can be calculated if the input PDFs are known [14].[9]

To cut down this model to an implementable complexity we only take a few parameters to describe the quantisation error of a variable x, expressed by its mean value $m_{\Delta x}$ and variance $\sigma^2_{\Delta x}$. The propagation of these parameters is governed by a set of rules and the mean value m_{x_0} and variance $\sigma^2_{x_0}$ of each signal [15].

The effect of LSB reduction. The criterion for a wordlength reduction is its impact on the noise PDF. In other words: What is the deterioration of the ratio of signal power and quantisation noise (SNR) for a single operation (ΔSNR_{op}) if we skip the least significant bit (LSB)? If one additionally defines a global threshold ΔSNR_{gl} that must not be exceeded when reducing the wordlength, one can decide whether the LSB has to be preserved or not.

The PDF of the truncation error consists of two unit pulses at 0 and -2^{-fwl}, weighted with the set probabilities $P(0)$ and $P(1)$:

$$p(\Delta\mathbf{x}) = P(0)\delta(\Delta\mathbf{x}) + P(1)\delta(\Delta\mathbf{x} + 2^{-fwl}). \tag{21.2}$$

Thus the noise power and the loss of the signal-to-noise ratio for this operation (ΔSNR_{op}) can be calculated:

$$\begin{aligned}
m_{new} &\approx m_{old} - 2^{-fwl-1}, && (21.3)\\
\sigma^2_{new} &\approx \sigma^2_{old} + 2^{-2fwl-2}, && (21.4)\\
\Delta SNR_{op} &= 10\lg\left(\frac{\sigma^2_{new} + m^2_{new}}{\sigma^2_{old} + m^2_{old}}\right). && (21.5)
\end{aligned}$$

ΔSNR_{op} therefore is a quantitative measure for the information content of the LSB. According to a user-specified ΔSNR_{gl} threshold, the wordlength of an operand is either reduced or not, i.e. only if the calculated ΔSNR_{op} is lower than ΔSNR_{gl}, the LSB gets truncated. This is repeated recursively unless the sum of all calculated values ΔSNR_{op} exceeds the ΔSNR_{gl} threshold.

This optional interpolation method allows *quantitative* assessment of the impact of quantisation on the precision of operations and algorithms and permits the wordlengths of the algorithm to be kept comparatively short [15]. With the SNR criterion, the result is an optimised ratio of algorithm performance and the wordlengths used in the algorithm.

21.5 BACK ENDS

As stated above, the fixed-C output of the interpolator is very well-suited for algorithm design or for a floating-to-fixed transformation. Nevertheless, it is not directly suited

[9] For example, the PDF of a sum of two statistically independent variables is obtained by convolving the PDFs of these variables.

for implementation for HW Synthesis tools or for programmable architectures. Apart from the fact that there are as of yet only very few implementation tools that support the fixed-C language [16], there are other reasons that the fixed-C output is not directly suited for implementation:

- The fixed-C code contains all context-specific fixed-point information and retains the structure of the original floating-point code. For implementation purposes, most information is not directly applicable, but is has to be refined. Often the structure of the code has to be changed according to this information, according to the implementation target and to the design goals.
- The ATI concept is not supported by most imperative implementation languages, where each variable corresponds to some storage location.
- The weight of bit pattern, i.e. the *location of the binary point* (lbp) (see Fig. 21.4) in a finite register is handled automatically by the fixed-point data types of the fixed-C language. This is no longer the case for more implementation-related representations, like ANSI-C or VHDL.

Hence, a back end which maps the implementation-independent fixed-C representation to a specific target is necessary. It has to deal with embedding the fixed-point operands into registers, mapping them to signals or variables, optimising the code structures for the implementation architecture, etc.

Currently there are different research and development activities going on at the Institute for Integrated Signal Processing Systems. The main implementation targets are

- ANSI-C and fast-simulation back end,
- Processor-specific back end,
- Hardware-description-language back end.

To describe all of those back ends would be clearly out of the scope of this paper. Hence we focus on the ANSI-C and fast simulation back end as examples to highlight some problems and solutions found in this area.

21.5.1 ANSI-C and fast-simulation back end

The ANSI-C output of this back end serves two purposes: the implementation on programmable processors that come with a C compiler, and reduction of simulation time on a host machine. Fixed-point simulation based on a C++ class library as described earlier or in [8, 12, 17] increases the runtime by one or even two orders of magnitude compared to the corresponding floating-point simulation. This is due to the fact that the fixed-point specification has to be emulated on the host machine that supports different data types from the target.

The key idea for a fast fixed-point simulation is to take advantage of all *compile-time*[10] information to minimise the processing effort that has to be spent at *simulation-time*. Two areas for compile-time optimisation have been integrated into FRIDGE:

[10]In this case, a floating-point algorithm is compiled into a fixed-point representation, i.e. compile-time means interpolation-time here.

casting optimisation and *data type emulation.* FRIDGE performs a casting mode optimisation that has also been independently proposed by DeCoster [18]: whether an overflow check is necessary depends on the range that the result of an operation can take. If the possible range can be represented with the available wordlength, no overflow processing is necessary. The same holds for quantisation handling.

In addition, FRIDGE analyses the necessary *wl* for each operand at compile-time. This information is used to embed it into an appropriate ANSI-C integer data type. As can be seen from Fig. 21.4, there is a degree of freedom embedding the wl into a longer *machine wordlength* (mwl)[11], i.e. the *significant bit pattern* may be shifted in the *mwl.*

Besides these degrees of freedom, there is also a set of restrictions describing how the operands have to be embedded into the registers. For example, as demonstrated with the expression tree in Fig. 21.7, one restriction for an addition is that the *location of the binary point* (lbp) has to be the same for both input operands.

Using this concept it is possible to describe the transformation space by a set of inequalities and conditions. With the goal of minimising the cost of the casting and shift operations, this states an optimisation problem that is essential for integer-based C code generation [3], [2].

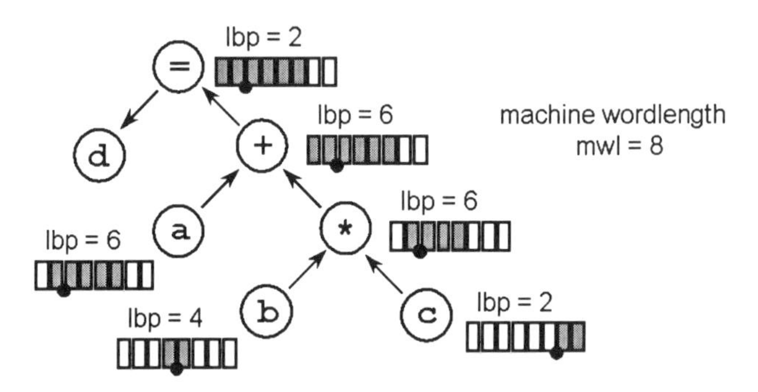

Figure 21.7 Location of the binary point (lbp) in an expression tree.

21.6 CONCLUSION

Traditional approaches make it necessary to annotate fixed-point specifications to **all** operands manually, which is an error-prone and time-consuming task. This is hardly acceptable even for a single transformation but becomes an unacceptable situation for an efficient evaluation of a complex design space. With the introduction of the *fixed-C* language and *interpolative approach*, an efficient transformation from a floating-point into a fixed-point algorithm becomes possible. With the support of FRIDGE, the de-

[11]mwl corresponds to the wordlength of the built-in data types on the host machine, e.g. `short`, `int` or `long`.

signer may concentrate on the critical issues for this transformation, thereby being able to explore the design space more efficiently.

The verification of the fixed-point algorithm has to be performed by means of simulation. Existing fixed-point simulation concepts increase simulation-time by one or even two orders of magnitude as compared to the corresponding floating-point simulation. The FRIDGE fast simulation back end applies advanced compile-time analysis concepts, analyses necessary casting operations and selects the appropriate built-in data type on the host machine, thus simulation time can be reduced considerably.

These features, in combination with target-specific back ends for implementation, make FRIDGE a powerful design environment for the specification, evaluation and implementation of fixed-point algorithms. This environment can easily be integrated into existing C/C++ based design environments.

References

[1] T. Grötker, E. Multhaup, and O. Mauss, "Evaluation of HW/SW Tradeoffs Using Behavioral Synthesis," in *Proceedings of the Int. Conf. on Signal Processing Application and Technology (ICSPAT)*, Boston, Oct. 1996.

[2] H. Keding, M. Willems, M. Coors, and H. Meyr, "FRIDGE: A Fixed-Point Design and Simulation Environment," in *Proceedings of the European Conference on Design Automation and Test (DATE)*, Paris, Feb. 1998, pp. 429–435.

[3] M. Willems, V. Bürsgens, and H. Meyr, "FRIDGE Enables Floating-Point Programming of Fixed-Point DSPs," in *Proceedings of the Int. Conf. on Signal Processing Application and Technology (ICSPAT)*, San Diego, Sep. 1997, pp. 1000–1005.

[4] M. Willems, V. Bürsgens, H. Keding, T. Grötker, and H. Meyr, "System Level Fixed-Point Design Based on an Interpolative Approach," in *Proceedings of the Design Automation Conference (DAC)*, Anaheim, Jun. 1997, pp. 293–298.

[5] W. Sung and K. Kum, "Word-Length Determination and Scaling Software for a Signal Flow Block Diagram," in *Proceedings of the IEEE International Conference on Acoustics, Speech and Signal Processing (ICASSP)*, Apr. 1994, pp.457–460 (II).

[6] B. W. Kernighan and D. M. Ritchie, *Programmieren in C*, Carl Hanser Verlag, 1990.

[7] Synopsys, Inc., 700 E. Middlefield Rd., Mountain View, CA 94043, USA, *COSSAP User's Manual.*

[8] Cadence Design Systems, 919 E. Hillsdale Blvd., Foster City, CA 94404, USA, *SPW User's Manual.*

[9] Mathworks Inc., *Simulink Reference Manual*, Mar. 1996.

[10] Mentor Graphics, 1001 Ridder Park Drive, San Jose, CA 95131, USA, *DSP Station User's Manual.*

[11] S. Kim, K. Kum, and W. Sung, "Fixed-Point Optimization Utility for C and C++ Based Digital Signal Processing Programs," in *Workshop on VLSI and Signal Processing '95*, Osaka, Nov. 1995, pp. 197–206.

[12] Frontier Design Inc., 9000 Crow Canyon Rd., Danville, CA 94506, USA, *A|RT Library User's and Reference Documentation*, 1998.

[13] Markus Willems, *Eine Methodik fuer den effizienten Entwurf von Festkommasystemen*, Dissertation an der RWTH-Aachen, 1998.

[14] A. Papoulis, *Probability, Random Variables and Stochastic Processes*, McGraw-Hill Book Company, 1991.

[15] H. Keding, F. Hürtgen, M. Willems, and M. Coors, "Transformation of Floating-Point into Fixed-Point Algorithms by Interpolation Applying a Statistical Approach," in *Proceedings of the Int. Conf. on Signal Processing Application and Technology (ICSPAT)*, Toronto, Sep. 1998.

[16] M. Jersak and M. Willems, "Fixed-Point Extended C Compiler Allows More Efficient High-Level Programming of Fixed-Point DSPs," in *Proceedings of the Int. Conf. on Signal Processing Application and Technology (ICSPAT)*, Toronto, Sep. 1998.

[17] W. Sung and K.I. Kum, "Simulation-Based Word-Length Optimization Method for Fixed-Point Digital Signal Processing Systems," *IEEE Transactions on Signal Processing*, vol. 43, no. 12, pp. 3087–3090, Dec. 1995.

[18] L. DeCoster, M. Engels, R. Lauwereins, and J.A. Peperstraete, "Global Approach for Compiled Bit-True Simulation of DSP-Applications," in *Proceedings of Euro-Par'96*, Lyon, Aug. 1996, vol. 2, pp. 236–239.

22 R.E.A.L. DSP: RECONFIGURABLE EMBEDDED DSP ARCHITECTURE FOR LOW-POWER / LOW-COST TELECOM BASEBAND PROCESSING

E. Lambers, C. Moerman, P. Kievits,
J. Walkier, and R. Woudsma

Embedded Systems Technology Centre (ESTC) ASIC Service Group
Philips Semiconductors
Prof. Holstlaan 4, 5656 AA Eindhoven, The Netherlands

Abstract: The **R.E.A.L. DSP** architecture for embedded applications features a high level of parallelism in the data path and the possibility to add specific peripheral modules, or to integrate dedicated application-specific execution units. Program code is compact by using a highly compressed instruction set, incorporating short (16-bit) instructions, orthogonal 32-bit instructions, and the possibility to define VLIW-type application-specific instructions. By its unique and leading architecture the **R.E.A.L. DSP** core is especially targeted for telecom terminals, in particular cellular phones. The core has been implemented in several ASIC technologies for different embedded applications, showing excellent performance and power figures.

22.1 INTRODUCTION

As opposed to general-purpose, stand-alone DSPs introduced in the eighties, deeply embedded DSP technology is characterised by a combination of high-functional performance, efficient chip area usage, high power efficiency, and a high degree of configurability. These characteristics are quite obvious when observing the key design constraints for high-volume, consumer-type applications: low-cost, short time-to-market, and low-power operation.

Simply boosting up DSP speed would result in fast execution of typical DSP functions like filter or transform algorithms, but this could implicitly also involve severe area and supply current penalties. Consequently, it appears to be much more appropriate to start with a thorough analysis of the target application domain(s), and tune the DSP architecture such that an optimal balance between specific application requirements and (re)programmability is achieved.

As Mobile Cellular Telephony is mainly driven by low power consumption and a high degree of integration, the above is especially true for this market.

The **R.E.A.L. DSP** core is optimised for cellular base-band processing in telecom terminals using instruction set colouring by means of application specific instructions (ASIs), making a large amount of instruction-level parallelism accessible to the programmer, and application-specific execution units (AXUs), which can accelerate carefully selected computationally intensive parts in DSP algorithms. Consequently, R.E.A.L. DSP firmware is very compact, resulting in smaller on-chip memories and significant power saving, since processing can be done in a relatively low number of clock cycles. For instance, it takes only two cycles to calculate a Viterbi butterfly section.

22.2 TOWARDS A NEW DSP ARCHITECTURE

Philips has a long experience of developing DSP cores. The EPICS DSP architecture has been developed for Cord-less Telephony and Digital Audio [1, 2]. This architecture allows a high degree of customisation and a wide variety of data-word lengths (12-, 16-, 18-, 20- and 24-bit variants have been realised). The KISS DSP core, featuring a 16-bit wide data-path, has been developed for cellular base-band processing in Telecom Terminals [3].

A number of considerations have led to the development of a next-generation embedded DSP technology:

1. The EPICS instruction set is highly concurrent and orthogonal. This results in high parallel processing power, and in easy and transparent programming at assembly level. For highly complex applications, including big parts of irregular control code, this approach is less feasible, since program code size is becoming a critical design and cost factor for embedded high-volume applications.

2. Increasing parallel processing power at the architecture level appears to be indispensable for boosting processor performance, either by adding more functional units operating in parallel, or by increasing the number of instruction pipelines.

3. There is a growing class of applications where different functions with diverging architecture requirements have to share the same DSP platform. Consequently, the

basic DSP core needs to be more generic than specific, while not destroying the advantages of efficient area and power usage.

4. For a short time to market, a more fixed and generic DSP architecture composition and associated instruction set must be applied, while still offering the option to add specific extensions when proven profitable for a given application.

These considerations have been the basis for the development of the next-generation DSP architecture. This new DSP is known as **R.E.A.L. DSP** (**R**econfigurable **E**mbedded DSP **A**rchitecture at **L**ow Power and Low Cost).

22.3 THE R.E.A.L. DSP ARCHITECTURE

The R.E.A.L. DSP shown in Fig. 22.1 is a dual Harvard architecture with 16-bit-wide data and address busses. Fig. 22.2 shows the heart of the DSP, the data computation unit (DCU), which contains two 16 × 16-bit signed multipliers and two 40-bit ALUs (32 data bits, 8 overflow bits). Depending on what type of instruction is executed, the ALUs can also be used as four separate 16-bit ALUs. Four input registers (x0|1, y0|1) are available to store values fetched from X and Y data memories. ALU results are stored in four 40-bit accumulators a0 through a3, which can also be used as eight 16-bit registers r0 up to r7. To provide good performance, the results of the multipliers are stored in pipeline registers p0 and p1. These are explicitly visible to the programmer. Shift and saturation units are available at a number of locations in the data-path. These can be configured with a control register.

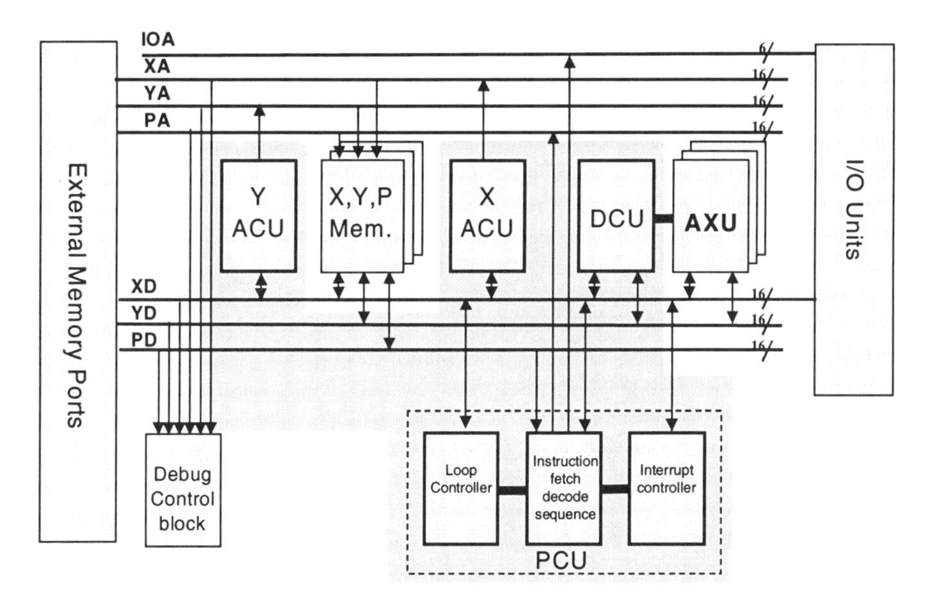

Figure 22.1 The R.E.A.L. DSP architecture template.

A stripped down version of the R.E.A.L. DSP with only one multiplier and single 40-bit ALU (configurable as two 16-bit ALUs) has been derived from this architecture for low-cost applications like, for instance, cord-less telephony and a digital telephone answering machine.

A distinctive feature of the R.E.A.L. DSP is the possibility to add application-specific execution units (AXU) to the data-path. A part of the instruction set has been reserved to control such blocks. Among the AXUs already available are a 40-bit barrel shifter, a normalisation unit, and a division support unit.

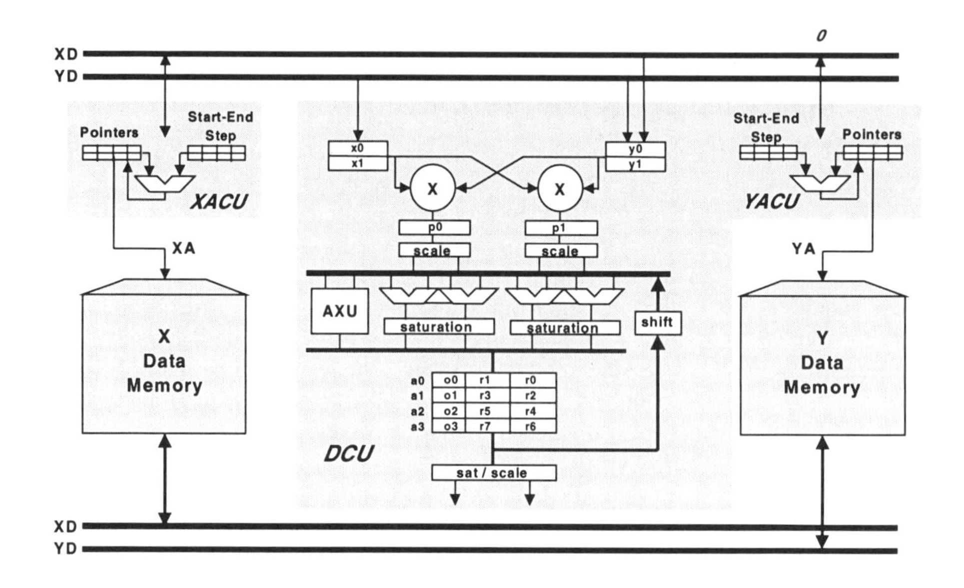

Figure 22.2 The R.E.A.L. DSP data-path.

The X and Y data memories can be as large as 64K words by 16 bits and are accessed by two address calculation units (ACU). Each of these contains eight pointers, grouped in four banks of two registers. This makes it possible to provide efficient context switching in case of interrupts. Modulo addressing is controlled by registers specifying the start and end addresses of each modulo buffer in memory. Application-specific hardware can be added here as well to make special addressing modes possible, such as accessing individual bits, nibbles or bytes in memory. The complex data path is controlled by the PCU (program control unit). This unit fetches and decodes instructions from memory, and handles program flow instructions. These are performed with a minimum use of cycles and include subroutine calls and relative, absolute, and calculated jumps. Nested looping is controlled by a loop controller based on an internal stack, with a depth specified by the customer. The contents of the loop stack (loop count, start and end address) can be saved to memory if an interrupt occurs. A repeat controller for single instruction repetition is also available and is used e.g. to move blocks of data from program memory to data memory.

For interrupt-driven applications, one non-maskable and up to 16 maskable interrupt inputs can be used, with possibilities to prevent nested interrupts. Each maskable interrupt line can be switched into a fast interrupt mode. This makes it possible to fetch data samples from an I/O port and store them in memory with a minimum of overhead. Automatic, implicit context switching of pointers is provided to enable full use of fast interrupts.

22.4 THE R.E.A.L. DSP INSTRUCTION SET

The R.E.A.L. DSP uses two different instruction widths. Because code size has to be as small as possible, the most commonly used instructions are compressed into 16 bits. Less common instructions, and those instructions containing immediate data or absolute addresses, use a second 16-bit word. Several classes of instructions are available to control the multiplier, ALUs and parallel moves. The instruction class determines whether the data-path operates with 16-bit or 40-bit operands in the ALUs. The instruction compaction does limit flexibility in operand selection, but this is compensated by the presence of an additional, and more powerful, orthogonal double-word instruction class (32 bits).

To exploit the full parallelism provided by the dual-multiplier quadruple-ALU and AXUs, a special set of instructions, called application specific instructions (ASIs), can be applied, see Fig. 22.3. ASI instructions may be regarded as VLIW instructions (very long instruction words), giving control over all the ALUs, multipliers and AXUs in parallel. Moreover, in ASIs, two fields are reserved to control the ACUs for parallel moves. 16-bit as well as 40-bit arithmetic is supported. ASIs can be conditional. ASIs are stored in a table in the decoder, limiting the total number of ASIs.

The R.E.A.L DSP assembler is able to detect an ASI in the application source code. An entry in the ASI table is reserved automatically, and the contents of this entry are generated from the kind of ASI being selected. No additional entries are generated in case of duplicate ASIs used at different locations in the program. Before starting up an application on the R.E.A.L. DSP, the applied ASIs must be down-loaded into the ASI table. This can be done under program control using two special registers. For production platforms, to save area as well as energy, the ASI table may be fixed in either ROM or special hardware.

As an example, we will explain how to program a block FIR filter taking advantage of the dual multiplier structure. The main difficulty when implementing the block FIR filter using two multipliers is the memory bottleneck: in each cycle, only one sample and one coefficient can be fetched from memory.

A straight-forward FIR filter can thus only calculate one product term per cycle. However, when doing block processing, two output samples can be calculated simultaneously, reusing sample and coefficient values by keeping the previous input and sample values in registers. For example, the two terms in Example 1 can be calculated at the same time, including the update of registers to set up for the next iteration.

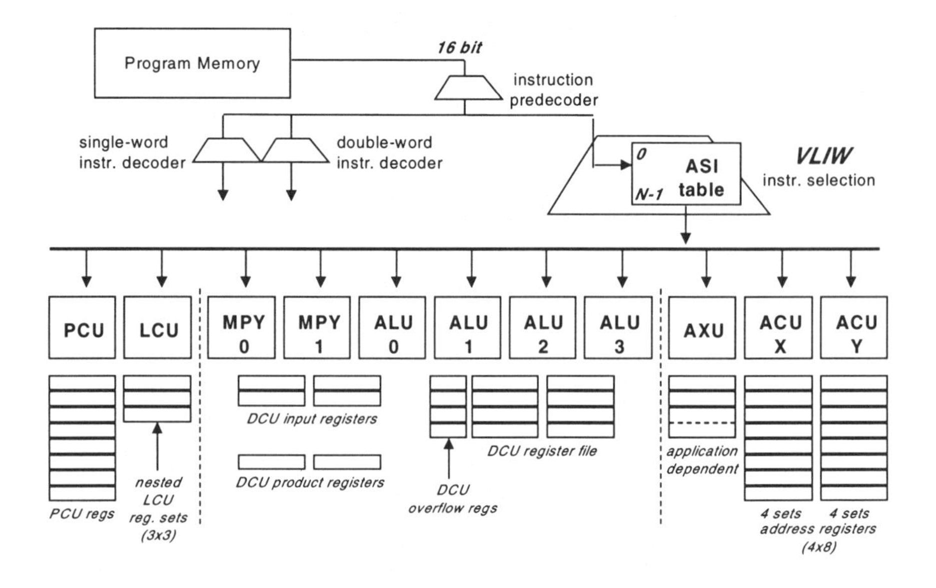

Figure 22.3 R.E.A.L. DSP instruction decoding.

```
Out_n += In_n * Coeff_n
Out_nplus1 += In_nplus1 * Coeff_n # calc terms
In_n = In_nplus1, In_nplus1==XMEM[n+2], # shift data values
Coeff_n = YMEM[n+1]; # next coeff
```

Example 1: FIR filter algorithm (pseudo-code).

A filter kernel would contain two single-cycle instructions, as in Example 2, each calculating one term of sample $[n]$ and one of sample $[n+1]$ (ignoring pipe-line set-up). By switching roles for the x0 and x1 register with regard to being the 'old' or 'new' value, no data move is required to shift the sample, as is shown in Example 2 (actual code).

```
do N/2 {
  a1=a1+p1, a0=a0+p0, p1=x0*y0, p0=x1*y0, x1=*pxm0++, y0=*pym0++;
  a1=a1+p1, a0=a0+p0, p1=x1*y0, p0=x0*y0, x0=*pxm0++, y0=*pym0++;
}
```

Example 2: FIR filter algorithm (actual code).

In this way, the memory can keep up with the MAC section. A 64-tap FIR filter with block size N implemented this way takes $8+37*N$ cycles to execute, i.e. for a 40-sample block it takes approx. 37 cycles per sample, including overhead. A single-multiplier DSP would take at least 64 cycles, excluding any overhead. For many algorithms, block-based processing is feasible, and the R.E.A.L. DSP architecture halves execution time for such algorithms.

Another example of a block-based algorithm is the FFT transform. Again, the memory bandwidth is the bottleneck, and attention has to be paid to the effective use of the internal registers. However, due to the large register set (four 40-bit accumulators, each of which also can be used as two independent 16-bit accumulators), a radix-4 FFT implementation can make full use of the dual multiplier architecture. This reduces the time needed for a 1,024-point complex FFT to around 1,400 cycles.

Based on our experience, even a few ASIs may significantly improve cycle-count efficiency. This can be explained by the fact that most DSP algorithms are based on repetitions of only a few time-critical instruction kernels. Therefore the ASI table is restricted to make the optimal balance between efficiency and core size.

For comparison, the new TI TMS C320C62xx processor [4] takes approx. the same number of cycles for the same task.[1] However, by using the ASI compression technique, the R.E.A.L. DSP does not have the overhead of the massive VLIW program memory requirements of the TI chip, nor the associated energy consumption.

22.5 R.E.A.L. DSP DEVELOPMENT TOOLS

To support program development, a tool set containing an assembler, a linker and a simulator has been developed in combination with a hardware board/debugger combination. To ease code development and maintenance, a C-style syntax is used to support readability. Obviously not having the freedom to write actual ANSI-C', the programmer still has to take into account the actual instruction set and to know which source and destination registers are available for a given operation.

A highly procedural style of programming is supported, with block- and file-based scope for functions and variables. This scoping is fully supported, also by the simulator and debugger tools. An optimising DSP C-compiler is currently under development, using the CoSy™ compiler development platform [5]. An ANSI-C compiler for the R.E.A.L. DSP is already available to support high-level entry of control code.

22.6 R.E.A.L. DSP ASIC IMPLEMENTATION

A number of applications are being served by embedded R.E.A.L. DSP cores, with an instruction rate in the range of 25 to 85 MHz, depending on the technology being used and on the bandwidth performance of the on-chip memories.

To ease the integration of a R.E.A.L. DSP, the core comes in two flavours, *soft* and *firm*. A soft core is distributed at VHDL RTL level, leaving the customer to do the technology mapping. In case of a firm core the design is transferred as a netlist optimised for a given ASIC technology. By exploiting an automatic clock-gating technique on the net-list, reductions in energy consumption of up to 20–25 % have been achieved.

The R.E.A.L. DSP features several hooks for integration on a chip. The DSP chip integrator is free to define its own memory architecture. Using the stall facility, which

---

[1] Based on TI stating 70 μs for a 1K FFT on 200 MHz C62,
TI web page `http://www.ti.com/sc/docs/news/1997/97001a.htm`

allows all DSP registers to be frozen within one cycle, all kind of memory wait-state controllers, cache controllers and bus arbiters can be connected without affecting the contents of the core. Communication between the DSP and its peripherals is done through up to 64 IO registers. An interrupt mechanism is available to avoid the need for software polling. By a special flag, the core signals to the outside world that it is ready to go into sleep mode, allowing the DSP clock to be completely switched off. Any external interrupt reactivates the core.

A large number of peripherals is available. Using the debug support block, an application program can be debugged on the chip. Register and memory manipulation, hardware break-pointing on any DSP bus, and single stepping are supported. The debug support block is controlled by a JTAG port. Repair of embedded DSP application software can be done by a DSP firmware repair block, allowing up to four firmware bugs to be fixed without affecting the DSP as such. Furthermore, interrupt controllers, parallel I/O ports, (a-)synchronous serial I/O ports, programmable timers, and power management units exist.

In order to meet critical ASIC design constraints, special attention is paid to the development and careful selection of target library elements such as an efficient cell library, optimised data-path modules, and embedded memories.

22.7 R.E.A.L. DSP FACTS AND FIGURES

In this section, some facts and figures regarding the implementation of and processing on the R.E.A.L. DSP RD1602x series is presented. The first implementation started a few years ago on a Philips Semiconductors 0.5μm in-house ASIC technology. Since then, the R.E.A.L. DSP has been improved continuously and mapped on several CMOS technologies. Results can be found in Table 22.1.

Table 22.1 Performance of R.E.A.L. DSP in different technologies.

	RD16020	*RD16021*	*RD16022*	*RD16023*
Technology	0.5μm	0.35μm	0.25μm	0.25μm
Performance [MHz] (worst-case)	24 (2.7 V)	39 (2.7 V)	60 (2.2 V)	85 (2.2 V)
Performance [MHz] (typical)	24 (3.0 V)	55 (3.0 V)	90 (2.5 V)	126 (2.5 V)

As an example, Table 22.2 lists the memory usage for GSM phone firmware programmed on the R.E.A.L. DSP RD16022, in addition to worst-case clock-speed performance, and energy consumption. The firmware comprises amongst others equalisation, channel coding, speech coding, and program control. Additional features like tone generation, voice recognition and hands-free are not taken into account. A distinction is made between full-rate, enhanced full-rate, and half-rate speech coding. The triple codec case is also included. To determine the energy consumption, the core is assumed to run at the minimal frequency needed to process the worst-case part in the

GSM phone firmware. Furthermore, the power figures are based on a supply voltage of 1.2 V. All figures include the DSP memories.

Table 22.2 Parameters for GSM phone Firmware.

	full-rate	*enhanced*	*half-rate*	*triple codec*
MiPS (worst-case)	8.1	21.4	20.0	21.4
Power (mW)	2.4	6.4	6.0	6.4
Program memory (kW)	11.6	22.0	25.8	34.7
Data memory (kW)	5.9	13.8	17.0	22.2

Since the R.E.A.L. DSP discussed here is a soft core, performance and power figures must be treated as indicative.

Acknowledgments

The R.E.A.L. DSP architecture, its ASIC implementation, the development tools, and product applications are the result of a close co-operation between various Philips departments. The authors wish to acknowledge the highly valuable and indispensable contributions of many Philips colleagues.

References

[1] R. Woudsma et al. "EPICS, a Flexible Approach to Embedded DSP Cores", *Proc. 5th Int'l Conf. On Signal Processing Applications and Technology*, Dallas Texas, 1994, Volume I, pp. 506–511.

[2] R. Beltman et al. "EPICS10: Development Platform for Next Generation EPICS DSP Products", *Proc. 6th Int'l Conf. On Signal Processing Applications and Technology*, Boston, 1995, Volume I, pp. 810–814.

[3] D. Weinsziehr et al. "KISS-16V2: A One Chip ASIC DSP Solution for GSM", *Proc. IEEE Custom Integrated Circuits Conf.*, 1992, pp. 10.4.1.–10.4.

[4] *TMS320C62xx Programmer's Guide,* Texas Instruments, July 1997.

[5] ACE, Amsterdam "The CoSy Compilation System", http://www.ace.nl

23 DEDICATED VLSI ARCHITECTURES FOR ADAPTIVE INTERFERENCE SUPPRESSION IN WIRELESS COMMUNICATION SYSTEMS

Bruno Haller

Communication Technology Laboratory
and Integrated Systems Laboratory
Swiss Federal Institute of Technology (ETH) Zurich
ETH-Zentrum, CH – 8092 Zurich, Switzerland

Abstract: In this contribution we discuss adaptive signal processing algorithms for interference suppression based on either spatial filtering using base station antenna arrays (→ *"smart" antennas*) or a special form of transversal filtering (→ *linear equalisers* for advanced DS-CDMA receivers). It is demonstrated that for reliable operation in non-stationary environments with different levels of interference, such as encountered in mobile radio, *RLS-based techniques* are inevitable due to their superior transient behaviour. In order to overcome the associated problems of numerical stability and large computational complexity, we propose to employ the QRD-RLS algorithm which solely relies on numerically robust transformations and can be implemented efficiently in VLSI technology based on a network of simple *CORDIC arithmetic* units which makes extensive use of parallel processing. The hardware implementation of such a 2D *systolic array* as well as an alternative pipelined linear array capable of real-time operation in high data rate systems ($>$ 2 Mbps) is presented together with bit-true performance assessments for applications involving adaptive antennas. Special attention is given to the interactive design process required to successfully develop advanced algorithms and dedicated hardware architectures for tackling demanding signal processing tasks.

23.1 INTRODUCTION

Wireless is currently the fastest growing segment of the globally booming telecom market, and this trend will be further reinforced through the introduction of third-generation mobile systems (UMTS and IMT-2000) enabling universal personal communications, as well as upcoming high speed radio networks (WLL and RLANs) providing tetherless local access to broadband services [1, 2]. Great increases in the capacity of current wireless technology will have to be made if such advanced systems are to become a reality. Since the radio spectrum is a scarce and consequently very precious resource, its economical utilisation calls for the application of bandwidth-efficient modulation schemes together with powerful multiple access and channel assignment techniques. This implies that higher performance and efficiency is achievable by applying enhanced signal processing, typically in digital form at the baseband (→ "digital receivers" [3, 4]). Of course, the entailed increase in complexity must be kept within reasonable limits to retain cost-effective solutions, whereby the latter factor is largely influenced by the available technology. Luckily, the continual enormous advancements of microelectronics in terms of circuit density and switching speed [5, 6] have been able to provide system designers with the signal processing horsepower they require to implement their ever more sophisticated algorithms, besides cutting down on the size, cost and current consumption of the components.

Principally, digital VLSI circuits for baseband processing in wireless equipment can be divided into three main categories [3]:

- programmable general-purpose devices (digital signal processors, DSPs),
- accelerator-assisted DSP chips (application-specific signal processors, ASSPs),
- custom-tailored components (application-specific integrated circuits, ASICs).

Today's fixed-point DSPs deliver on the order of 50 MIPS. This amount of processing power is sufficient to realise the baseband parts of the physical layer as well as the speech coding in second-generation TDMA-based cellular terminals (GSM, IS-54/136) [7, 8]. Specialised DSP chips for wireless systems, enhanced with accelerator hardware, e.g. an add-compare-select unit for Viterbi decoding, and equipped with multiple arithmetic-logic units (ALUs), i.e. multiplier-accumulators (MACs), have been introduced on the market lately to help tackle high-data-rate applications [9–11]. While these leading-edge devices offer a great deal of flexibility, they are often unable to satisfy the computational requirements of future broadband systems [12] demanding dedicated hardware solutions based on massively parallel processing and pipelining, which is the exclusive domain of ASICs [13, 15]. Therefore, in this contribution, we will be looking at the major engineering challenges imposed on the designers of dedicated VLSI circuits for the advanced digital signal processing needed in the next generation of high-capacity, high-data-rate wireless communication networks. This will be illustrated based on the case study of special-purpose processors for digital beamforming, i.e. adaptive spatial filtering [16–19].

The rest of this discourse is organised as follows. Section 23.2 briefly outlines some of the main problems involved with the design of dedicated VLSI signal processors. In Section 23.3 we give a short overview on "smart" antennas, followed in Section 23.4 by a description of algorithms and systolic architectures for adaptively controlling (i.e. "steering") the beam pattern of an antenna array, including a few representative results from (bit-level) computer simulations of the proposed structures. Subsequently, a quick look is taken at an alternative application of the described techniques regarding the implementation of advanced DS-CDMA receivers in Section 23.5. Finally, Section 23.6 concludes this presentation with a summary and discussion of the most relevant points.

The aim of this exposé is fourfold: (i) to advocate the use of *adaptive antennas* as an efficient and effective means for *interference mitigation* in radio systems; (ii) to motivate the use of *RLS-based techniques* for high-performance adaptive filtering in wireless communications; (iii) to explain the advantages of employing *CORDIC arithmetic*, instead of the traditional MAC operation, in certain classes of DSP algorithms; and above all (iv) to demonstrate that it is essential to *develop algorithms and hardware architectures concurrently* in order to obtain practically viable solutions to computationally demanding signal processing problems.

23.2 THE ART OF VLSI SIGNAL PROCESSING

The current trend towards hardware intensive signal processing has uncovered a relative lack of understanding of dedicated VLSI architectures. Many hardware-efficient algorithms exist, but these are generally not well known due to the predominance of software solutions employing general-purpose DSP chips which have been in the spotlight of interest over the past two decades. Unfortunately, algorithms optimised for these programmable devices do not always map nicely onto specialised hardware. Therefore, current work in the field of VLSI signal processing is focusing on the *joint study of both algorithms and architectures* for the custom implementation of DSP (sub-)systems, exploiting the interactions between the two to derive efficient solutions and thereby *bridging the gap between system and circuit level design* [20–24]. As a consequence, the common approach presently pursued by many companies and research institutions, which is to seperate the theory and system groups developing algorithms from DSP and VLSI groups implementing them, should be abandoned in favour of a more interdisciplinary design procedure, which means exploiting in a synergistic manner the interrelations between the application, the required algorithms and the underlying hardware architectures, as well as the technology based upon which the system is to be realised. This task calls for the "tall thin" engineer with a broad knowledge in many different fields rather than a specialist deeply involved in only one of the above-mentioned areas [3, 25].

In order to solve the tough problems encountered in real-time signal processing for wireless communications, it is necessary to both devise "good" algorithms and find "good" circuit architectures which fit together well. A "good" DSP algorithm works reliably and is numerically stable, i.e. it must perform satisfactorily under diverse operating conditions and be insensitive to finite-precision effects due to limited operand word lengths and inaccurate computations.[1] Furthermore, algorithms which possess a high degree of parallelism and permit pipelined operation are more directly amenable to high-throughput implementations. On the other hand, "good" circuit architectures must exploit the full potential of advanced VLSI technology as well as be suitable for design automation ($\rightarrow$ "silicon compilation") [27], which enables speedy development cycles to meet stringent time to market requirements. Key features of such VLSI architectures are *regularity, modularity and extendability*. All these properties support a cell-based design approach where complex systems are constructed through repeated use of a small number of (primitive) basic building blocks ("tiles") with only local interconnections and simple control. This methodology forms the basis for building so-called "systolic arrays" [28, 29], which consist of a synchronously clocked regular network of (preferably identical) processing elements—a concept originally proposed by H. T. Kung and C. E. Leiserson in 1978, targeted at performing different matrix computations fundamental to many modern signal processing schemes. For the purposes of our further discussion, we will now concentrate on the application of such VLSI systolic array architectures to an important, representative problem within the realm of future wireless communications, namely the implementation of *digital adaptive beamformers*, in the hope that this example will help to practically illustrate the points made in this section.

23.3 OVERVIEW ON SMART ANTENNAS

The quest for ever better usage of the limited radio spectrum to provide a rapidly growing subscriber population with a broad range of high-quality, increased-data-rate services has recently fostered great, world-wide interest in smart antenna technology, making it one of the hottest topics in wireless communications research today. Yet it must be stressed that adaptive antenna arrays have been studied extensively for more than three decades, although work was mainly restricted to military applications, such as radar and sonar systems, chiefly due to the high complexity and ensued large costs of the required equipment. Hereby, antenna arrays were primarily employed to combat intentional jamming, i.e. as an *interference suppression* technique. On the other hand, multiple-antenna systems for mobile communications have also been studied thoroughly in the past, as a means to mitigate fading by exploiting *space diversity*. The fusion of these two ideas, i.e. *interference suppression in combination with diversity combining* is the essence of current R&D efforts on intelligent antennas for high-capacity mobile radio systems, as documented in [30–35] and the references cited therein.

The aim of a smart antenna is to reinforce the signal of interest (SOI) while suppressing the undesired intersymbol and co-channel interference (ISI, CCI), which are due to multipath propagation and concurrent transmissions from other users, respectively. Thus a smart antenna maximises the signal-to-interference-plus-noise ratio (SINR) at its output.

---

[1] Using floating-point arithmetic for dedicated VLSI chips is most often too costly in terms of chip area and power consumption [26].

This can be interpreted as a process of *adaptive beamforming* where the radiation pattern (spatial response) of the antenna array is automatically adjusted according to the prevailing signal environment, i.e. by steering the main lobe towards the SOI and simultaneously placing nulls in the directions of strong sources of disturbance. Additionally, with M antenna elements, an array generally provides an SNR enhancement by a factor of M (for spatially white noise) plus a diversity gain that depends on the correlation of the fading amongst the antennas (max. M-fold). To take best possible advantage of spatial diversity, the antennas have to be placed far enough apart to guarantee low fading correlation, whereby the required separation depends on the angular spread of the desired signal, i.e. the angle over which it is dispersed as seen by the receiver. Such an array can theoretically completely cancel $K_{\mathrm{I}} < M$ interferers and achieve an $(M - K_{\mathrm{I}})$-fold diversity gain [37]. Significant performance improvements are even attainable when the number of interfering signals exceeds M.

There exists a vast number of different techniques to govern the operation of the adaptive processor that adjusts the weighting for each of the antenna element outputs. These can be categorised according to the chosen weight optimisation criterion—e.g. minimum- mean-square-error (MMSE), maximum SINR or minimum variance / output power (→ MVDR, LCMV) [16, 17]—which consequently determines the way the signal components of interest are discriminated from the interference to protect them against unintentional suppression. This issue can be solved either by incorporating knowledge regarding "source location", i.e. the direction of arrival (DOA) of the desired impinging waves (→ spatial reference beamforming, SRB), or by supplying some form of reference signal which is highly correlated with the desired signal and uncorrelated with the interference (→ *temporal reference beamforming*, TRB) [16, 36]. Alternatively, so-called "blind" methods have also been advocated which rely on special signal properties such as constant modulus, cyclostationarity or finite-alphabet structure of the symbols [16, 36]. For practical reasons, we suggest to use the reference signal approach, because of its lower computational complexity[2] as well as its reduced sensitivity with respect to array imperfections, and because training and synchronisation sequences to support channel estimation, or provide some form of source identification, or both, are present in all operational and foreseeable systems. Additionally, temporal reference antenna control techniques can theoretically approach optimum performance for any given antenna array topology (→ no restrictions regarding element spacing or positioning,[3] which are critical factors both for attaining high space diversity gains as well as for the installation of an array) and hence offer better performance and more reliable behaviour than the spatial and blind methods under most practical circumstances.

To obtain rapid weight adaptation in quickly time-varying situations with high levels of disturbance, we propose to adopt open-loop (direct-solution) techniques. The coefficients of the adaptive combiner are then found through recursive least squares (RLS) minimisation. The corresponding set of equations can be solved in a numerically robust manner by means of orthogonal matrix triangularisation (i.e. $\boldsymbol{QR}$ decom-

[2]DOA estimation algorithms (e.g. MUSIC or ESPRIT [36]) are computationally very demanding.

[3]For SRB schemes to function correctly, the interelement spacing must be small so that the signal fading is fully correlated across the array, which excludes the possibility of exploiting space diversity in conjunction with these methods!

position, QRD). The recursive realisation of this scheme using Givens rotations is known as the QRD-RLS algorithm. It can be mapped very elegantly onto efficient systolic array architectures for high speed real-time operation. Considering the present level of circuit integration, antenna array weight controllers based on such dedicated hardware structures are on the verge of becoming economically feasible for use in commercial wireless communication systems. In the next section, we will present in some detail the algorithms required for temporal reference beamforming, a bit-true description of the associated VLSI architectures, and exemplary performance assessments for two simple operating scenarios.

23.4 QRD-RLS ALGORITHM AND SYSTOLIC ARCHITECTURES FOR ADAPTIVE WEIGHT CONTROL

A temporal reference-based digital adaptive beamformer is illustrated in Fig. 23.1.

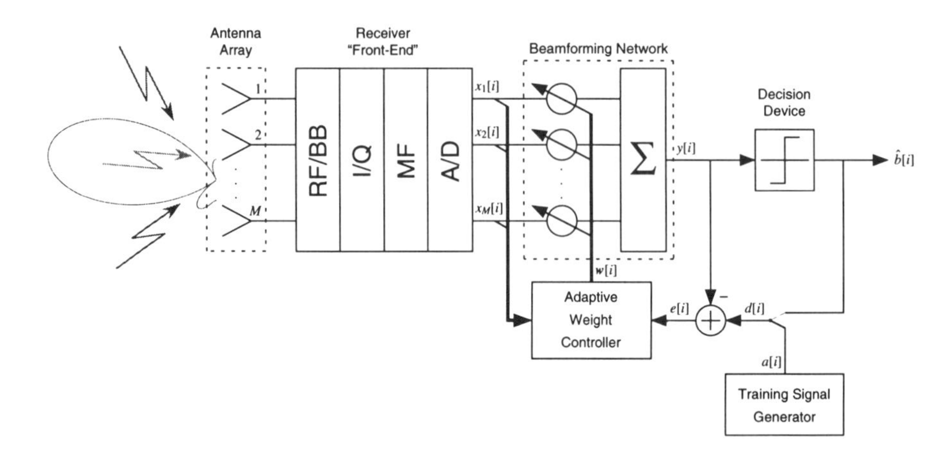

Figure 23.1 Block diagram of a digital temporal reference beamformer.

The array consists of M identical, omnidirectional antenna elements. After quadrature downconversion to baseband, the I and Q components of each array element's signal are separately processed by a matched filter (MF) whose output is sampled once per symbol ($f_s = 1/T_b$) and subsequently digitised.[4] Following this, adaptive combining is carried out by applying complex (amplitude and phase) weights $w_m[i]$ ($m = 1 \ldots M$) to the individual element signals $x_m[i]$ in the beamforming network. In TRB systems, where *a priori* knowledge about the transmitted training sequences or the periodically inserted pilot/reference symbols is available, optimum combining is achieved by selecting the complex weight vector $\boldsymbol{w}[i]$ according to the minimum-mean-square-error (MMSE) criterion. Hereby, the adaptive weight controller adjusts

[4]Oversampling can be employed to improve performance, especially when no feedback exists to control the sampling clock or if several asynchronous SOIs are to be handled concurrently.

$\boldsymbol{w}[i]$ so as to minimise the average power of the residual signal $\varepsilon[i]$, which is the difference between the known desired signal $d[i]$ and the combined array output signal $y[i] = \sum_{m=1}^{M} x_m[i]w_m[i-1]$. The optimum Wiener solution to this problem is given by the matrix form of the Wiener-Hopf (or normal) equations [38]:

$$\boldsymbol{R}_{\boldsymbol{xx}} \boldsymbol{w}_{\text{opt}} = \boldsymbol{r}_{\boldsymbol{x}d} \,, \tag{23.1}$$

where $\boldsymbol{R}_{\boldsymbol{xx}} = \mathrm{E}(\boldsymbol{x}[i]\boldsymbol{x}^{\mathrm{H}}[i])$ represents the input data correlation matrix and $\boldsymbol{r}_{\boldsymbol{x}d} = \mathrm{E}(\boldsymbol{x}[i]d^*[i])$ is the cross-correlation vector between the array element signals $x_m[i]$ and the desired response $d[i]$. Eq. (23.1) forms the basis of the well-known direct/sample matrix inversion (DMI/SMI) algorithm (of complexity $\mathrm{O}(M^3)$) [39]:

$$\boldsymbol{w}[i] = \hat{\boldsymbol{R}}_{\boldsymbol{xx}}^{-1}[i]\hat{\boldsymbol{r}}_{\boldsymbol{x}d}[i] \,, \tag{23.2}$$

which involves forming an estimate (time average) of the correlation matrix $\hat{\boldsymbol{R}}_{\boldsymbol{xx}}[i]$ as well as of the cross-correlation vector $\hat{\boldsymbol{r}}_{\boldsymbol{x}d}[i]$ and subsequently computing the solution to a set of linear equations. This type of algorithm is well suited for a programmable floating-point DSP but extremely difficult to map onto a dataflow processor. Alternatively, several adaptive algorithms can be employed to perform the above-mentioned minimisation task directly. Despite its computational simplicity (it has complexity $\mathrm{O}(M)$) and its favourable numerical behaviour, the popular least mean square (LMS) stochastic gradient technique is often unsuitable in nonstationary environments with variable levels of interference and distortion,[5] as is very common in mobile radio communications. This is due to its poor convergence properties and large misadjustment, resulting in a considerable loss of output SINR compared to the theoretically optimum Wiener solution [38]. Fast convergence is especially essential for rapid weight adaptation when short training sequences are used for transmission efficiency reasons, as well as to guarantee quick retuning of the beamformer coefficients when new sources of disturbance emerge, or present ones disappear. In these situations the recursive least squares (RLS) algorithm is preferred because of its robustness with respect to the signal environment and thanks to its superior speed of convergence. A vast amount of research has been devoted to reducing the demanding computational requirements and overcoming the problematic numerical properties of the RLS algorithm in order to increase its practical applicability. The method of choice for fast and numerically robust RLS filtering is based upon orthogonal triangularisation of the input data matrix via $\boldsymbol{QR}$ decomposition (QRD) [38]. In situations where the data is to be processed on a sample by sample basis (as opposed to block-wise) $\boldsymbol{QR}$-updating through a sequence of Givens rotations is usually employed. The resulting QRD-RLS algorithm requires $\mathrm{O}(M^2)$ operations per updating step. Its numerical robustness stems from the fact that it operates directly on the incoming data matrix, circumventing the need to compute $\hat{\boldsymbol{R}}_{\boldsymbol{xx}}[i]$, whereby the condition number of the data matrix would be squared and the required word length doubled. Additionally, solely numerically stable unitary transformations are used. In the following subsection we briefly review the QRD-RLS algorithm and demonstrate how to compute the *a posteriori* (*a priori*) estimation error $e[i]$ ($\varepsilon[i]$) without explicit computation of the weight vector.

---

[5]The input data correlation matrix then exhibits a large eigenvalue spread: $\chi(\boldsymbol{R}_{\boldsymbol{xx}}) = \lambda_{\max}/\lambda_{\min} \gg 1$.

23.4.1 QRD-RLS algorithm

The function of the adaptive combiner in Fig. 23.1 may be defined mathematically in terms of least-squares estimation (LSE). The aim of the RLS algorithm is to minimise the sum of exponentially weighted squared errors,

$$\min_{\boldsymbol{w}[i]} \sum_{n=1}^{i} \beta^{i-n} \left| d[n] - \boldsymbol{x}^{\mathrm{T}}[n]\boldsymbol{w}[i] \right|^2 . \tag{23.3}$$

The so-called "forgetting factor" $0 \ll \beta \leq 1$ is commonly used to discount old data from the computations (exponential downdating), in order to provide a certain tracking capability when the system operates in a non-stationary environment. This is equivalent to determining the weight vector $\boldsymbol{w}[i]$ which minimises the ℓ_2-norm of the vector of error residuals $\boldsymbol{e}[i]$,

$$\|\boldsymbol{e}[i]\| = \sqrt{\boldsymbol{e}^{\mathrm{H}}[i]\boldsymbol{e}[i]} , \tag{23.4}$$

where

$$\boldsymbol{e}[i] = \mathcal{B}[i](\boldsymbol{d}[i] - \boldsymbol{X}[i]\boldsymbol{w}[i]) \in \mathbb{C}^{i} , \tag{23.5}$$

with the data matrix

$$\boldsymbol{X}[i] \stackrel{\mathrm{def}}{=} \begin{pmatrix} \boldsymbol{x}^{\mathrm{T}}[1] \\ \vdots \\ \boldsymbol{x}^{\mathrm{T}}[i] \end{pmatrix} \in \mathbb{C}^{i \times M} , \tag{23.6}$$

and the weighting matrix

$$\mathcal{B}[i] \stackrel{\mathrm{def}}{=} \mathrm{diag}(\sqrt{\beta}^{\,i-1}, \sqrt{\beta}^{\,i-2}, \cdots, 1) \in \mathbb{R}^{i \times i} . \tag{23.7}$$

Since the Euclidean vector norm is invariant with respect to unitary (orthogonal) transformations $\boldsymbol{Q}$, with $\boldsymbol{Q}^{\mathrm{H}}\boldsymbol{Q} = \mathbf{I}$, we apply the QRD to transform the weighted input data matrix $\mathcal{B}[i]\boldsymbol{X}[i]$ into an upper triangular matrix $\boldsymbol{R}[i] \in \mathbb{C}^{M \times M}$:

$$\boldsymbol{Q}^{\mathrm{H}}[i]\boldsymbol{e}[i] = \begin{pmatrix} \boldsymbol{u}[i] \\ \boldsymbol{v}[i] \end{pmatrix} - \begin{pmatrix} \boldsymbol{R}[i] \\ \mathbf{0} \end{pmatrix} \boldsymbol{w}[i] . \tag{23.8}$$

From this it can be seen that the minimum norm condition for the error residual $\boldsymbol{e}[i]$ is obtained when

$$\boldsymbol{R}[i]\boldsymbol{w}[i] = \boldsymbol{u}[i] . \tag{23.9}$$

This equation defines the least-squares weight solution for the adaptive linear combiner. Since the matrix $\boldsymbol{R}[i]$ is upper triangular, the weight vector $\boldsymbol{w}[i]$ may be obtained via a process of back-substitution.

The quantities $\boldsymbol{R}[i]$ and $\boldsymbol{u}[i]$ in Eq. (23.9) can be calculated recursively on a sample by sample basis according to the QRD-RLS algorithm summarised in Tab. 23.1.[6]

---

[6] A detailed mathematical derivation of the algorithm is contained in [40].

Table 23.1 QRD-RLS algorithm based on complex Givens rotations for TRB.

Initialisation:
$\boldsymbol{R}[0] = \sqrt{\delta}\mathbf{I}_M$ with $0 \leq \delta \ll 1$, $\boldsymbol{u}[0] = \mathbf{0}_M$

for $i = 1, 2, \ldots$ **do:**

$$\begin{pmatrix} \boldsymbol{R}[i] & \varsigma[i] & \boldsymbol{u}[i] \\ \mathbf{0}_M^{\mathrm{T}} & \gamma[i] & \rho[i] \end{pmatrix} \leftarrow \overline{\boldsymbol{Q}}[i] \begin{pmatrix} \sqrt{\beta}\boldsymbol{R}[i-1] & \mathbf{0}_M & \sqrt{\beta}\boldsymbol{u}[i-1] \\ \boldsymbol{x}^{\mathrm{T}}[i] & 1 & d[i] \end{pmatrix}$$

where $\overline{\boldsymbol{Q}}[i] = \boldsymbol{G}_M[i] \cdots \boldsymbol{G}_1[i] \in \mathbb{C}^{(M+1)\times(M+1)}$

$$\text{with } \boldsymbol{G}_m[i] = \begin{pmatrix} \mathbf{I}_{m-1} & \mathbf{0}_{m-1} & \cdots & \mathbf{0}_{m-1} \\ \mathbf{0}_{m-1}^{\mathrm{T}} & \cos\theta_m[i] & \mathbf{0}_{M-m}^{\mathrm{T}} & \sin\theta_m[i]\exp(-\mathrm{j}\phi_m[i]) \\ \vdots & \mathbf{0}_{M-m} & \mathbf{I}_{M-m} & \mathbf{0}_{M-m} \\ \mathbf{0}_{m-1}^{\mathrm{T}} & -\sin\theta_m[i] & \mathbf{0}_{M-m}^{\mathrm{T}} & \cos\theta_m[i]\exp(-\mathrm{j}\phi_m[i]) \end{pmatrix}$$

a priori error
$\varepsilon[i] = d[i] - \boldsymbol{x}^{\mathrm{T}}[i]\boldsymbol{w}[i-1] = \rho[i]/\gamma[i]$

a posteriori error
$e[i] = d[i] - \boldsymbol{x}^{\mathrm{T}}[i]\boldsymbol{w}[i] = \rho[i] \cdot \gamma^*[i]$

with $\gamma[i] = \prod_{m=1}^{M} \cos\theta_m[i]\exp(-\mathrm{j}\phi_m[i]) = \bar{q}_{M+1,M+1}$

The unitary update transformation denoted by $\overline{\boldsymbol{Q}}[i]$ represents a sequence of M complex Givens rotations $\boldsymbol{G}_m[i]$, each one of which operates on two rows of the matrix. The rotation angles ϕ_m and θ_m are chosen such that the elements of the data snapshot $\boldsymbol{x}^{\mathrm{T}}[i]$ are successively zeroed whilst updating $\boldsymbol{R}[i]$ and $\boldsymbol{u}[i]$. Note that the *a posteriori* (*a priori*) error $e[i]$ ($\varepsilon[i]$) can be evaluated without explicitly computing the weight vector and applying the result to Eq. (23.5).

23.4.2 Hardware implementation of the QRD-RLS algorithm

A highly efficient realisation of the QRD-RLS algorithm based on Givens rotations using a parallel and pipelined triangular systolic processor array (tri-array) was first explored by W. M. Gentleman and H. T. Kung [41] and later refined by J. G. McWhirter [42]. As indicated above, the main tasks of the processing elements are the evaluation and execution of plane rotations to annihilate specific matrix elements. With MAC-type (multiply-and-accumulate) arithmetic units the computation of the rotation parameters requires either square-roots and divisions or some trigonometric functions, whereas the vector rotations themselves only need multiplications and additions. This means that the determination of the former quantities takes many more clock/instruction cycles than it does to apply them, leading to an unequal distribution of the work load within the array. An often used alternative method of performing two-dimensional vector rotations relies on the well-known CORDIC (COordinate Rotation DIgital Computer) algorithm introduced by J. E. Volder [43] and extended by J. S. Walther [44]. In the following, we present the CORDIC algorithm together with a bit-level description of a 16-bit fixed-point implementation, and demonstrate how it can be employed to perform all the necessary arithmetic operations in a fully parallel systolic array architecture for adaptive weight control based on the QRD-RLS algorithm.

CORDIC algorithm. The basic idea underlying the CORDIC scheme is to carry out vector ("macro"-)rotations by an arbitrary rotation angle θ via a series of $b_{\mathrm{w}}+1$ "micro-rotations" using a fixed set of predefined elementary angles α_j:

$$\theta = \sum_{j=0}^{b_{\mathrm{w}}} \sigma_j \alpha_j, \quad \sigma_j \in \{-1,+1\}. \tag{23.10}$$

This leads to a representation of the rotation angle θ in terms of the rotation coefficients σ_j. By choosing the elementary angles as

$$\alpha_j \stackrel{\mathrm{def}}{=} \arctan 2^{-j}, \quad j \in \mathcal{J} = \{0,1,2,\ldots,b\}, \tag{23.11}$$

it follows that a μ-rotation can be realised via two simple shift-add operations:

$$\begin{aligned} \begin{pmatrix} \xi_{j+1} \\ \psi_{j+1} \end{pmatrix} &= \frac{1}{\sqrt{1+\tan^2\alpha_j}} \begin{pmatrix} 1 & \sigma_j \tan\alpha_j \\ -\sigma_j \tan\alpha_j & 1 \end{pmatrix} \cdot \begin{pmatrix} \xi_j \\ \psi_j \end{pmatrix} \\ &= \frac{1}{\sqrt{1+2^{-2j}}} \begin{pmatrix} 1 & \sigma_j 2^{-j} \\ -\sigma_j 2^{-j} & 1 \end{pmatrix} \cdot \begin{pmatrix} \xi_j \\ \psi_j \end{pmatrix} . \end{aligned} \tag{23.12}$$

The final result is obtained with a precision of b_{w} bits after the execution of $b_{\mathrm{w}}+1$ μ-rotations and a multiplication with the scaling factor $S_{b_{\mathrm{w}}} = \prod_{j=0}^{b_{\mathrm{w}}} 1/\sqrt{1+2^{-2j}}$. $S_{b_{\mathrm{w}}}$ can also be decomposed into a sequence of simple shift-add operations, which are often performed in a series of additional scaling iterations.

The CORDIC has two modes of operation called "vectoring" (Vec.) to compute the magnitude and phase of a vector

$$\begin{pmatrix} \xi_{\mathrm{out}} \\ \psi_{\mathrm{out}} \end{pmatrix} = \begin{pmatrix} \mathrm{sign}(\xi_{\mathrm{in}}) \cdot \sqrt{\xi_{\mathrm{in}}^2 + \psi_{\mathrm{in}}^2} \\ 0 \end{pmatrix}, \tag{23.13a}$$

$$\theta_{\mathrm{out}} = \arctan \frac{\psi_{\mathrm{in}}}{\xi_{\mathrm{in}}}, \tag{23.13b}$$

whereby the vector $(\xi_{\mathrm{in}}\ \psi_{\mathrm{in}})^{\mathrm{T}}$ is rotated to the ξ-axis, and "rotation" (Rot.)

$$\begin{pmatrix} \xi_{\mathrm{out}} \\ \psi_{\mathrm{out}} \end{pmatrix} = \begin{pmatrix} \cos\theta_{\mathrm{in}} & \sin\theta_{\mathrm{in}} \\ -\sin\theta_{\mathrm{in}} & \cos\theta_{\mathrm{in}} \end{pmatrix} \cdot \begin{pmatrix} \xi_{\mathrm{in}} \\ \psi_{\mathrm{in}} \end{pmatrix}, \tag{23.14a}$$

$$\theta_{\mathrm{out}} = \theta_{\mathrm{in}}, \tag{23.14b}$$

in which case the vector $(\xi_{in}\ \psi_{in})^T$ is rotated by the angle $-\theta_{in}$.

Obviously, the rotation angles ϕ_m and θ_m for a complex Givens rotation can be determined using the CORDIC in vectoring mode, whereas the rotations themselves are carried out using the CORDIC algorithm in rotation mode.

Tab. 23.2 contains the bit-level description of a 16-bit fixed-point CORDIC algorithm as specified in [45]. It forms the basis for the hardware implementations discussed next and was employed for the bit-true simulations of the QRD-RLS-based TRB to be presented in Subsection 23.4.4.

Table 23.2 Bit-level specification of the CORDIC algorithm for plane rotations, i.e. Vec. and Rot., as well as division (Div.) with $b_w = 16$-bit operands.

Input:

Vec./Rot. scaling iteration: $\xi_0 = \xi_{in}/2$ and $\psi_0 = \psi_{in}/2$

Div. : $\xi_0 = 0,\ \psi_0 = \xi_{in}$ and ψ_{in}

add all-zero guard bit extension (6 LSBs)

for $j = 0 \ldots 16$ **do**:

Vec./Rot. μ-rotation: $\xi_{j+1} = \xi_j + \sigma_j \psi_j 2^{-s_j}$ and $\psi_{j+1} = \psi_j - \sigma_j \xi_j 2^{-s_j}$

Div. iteration: $\xi_{j+1} = \xi_j + \sigma_j 2^{-s_j}$ and $\psi_{j+1} = \psi_j - \sigma_j \psi_{in} 2^{-s_j}$

with $\{s_j | j = 0, 1, \ldots, 16\} = \{0, 1, \ldots, 16\}$ and

Vec.: $\sigma_j = \text{sign}(\xi_j) \cdot \text{sign}(\psi_j) = \sigma_{out}$

Rot.: $\sigma_j = \sigma_{in}$

Div.: $\sigma_j = \text{sign}(\psi_{in}) \cdot \text{sign}(\psi_j) = \sigma_{out}$

for $j = 17 \ldots 20$ **do**:

Rot./Vec. scaling iteration:

$\xi_{j+1} = \xi_j + \eta_{j-16} \xi_j 2^{-t_{j-16}}$

and $\psi_{j+1} = \psi_j + \eta_{j-16} \psi_j 2^{-t_{j-16}}$

with $\{t_j | j = 1, 2, 3, 4\} = \{2, 5, 9, 10\}$ and $\{\eta_j | j = 1, 2, 3, 4\} = \{1, -1, 1, 1\}$

Div. dummy iteration: $\xi_{j+1} = \xi_j$ and $\psi_{j+1} = \psi_j$

Output:

erase guard bits

Vec./Rot.: $\xi_{out} = \xi_{21}$ and $\psi_{out} = \psi_{21}$

Div.: $\xi_{out} = \xi_{21}\quad (= \xi_{in}/\psi_{in} < 2)$

In addition to the CORDIC vectoring and rotation modes, Tab. 23.3 also shows how to elegantly implement the division operation (Div.) $\xi_{out} = \xi_{in}/\psi_{in}$ in a bit-serial fashion using the same shift-add operations. As can be seen in Tab. 23.1, a division is needed to calculate the *a priori* estimation error $\varepsilon[i]$.

The actual hardware implementation of the CORDIC equations (23.12) may be done in a number of ways [46] as depicted in Fig. 23.2.

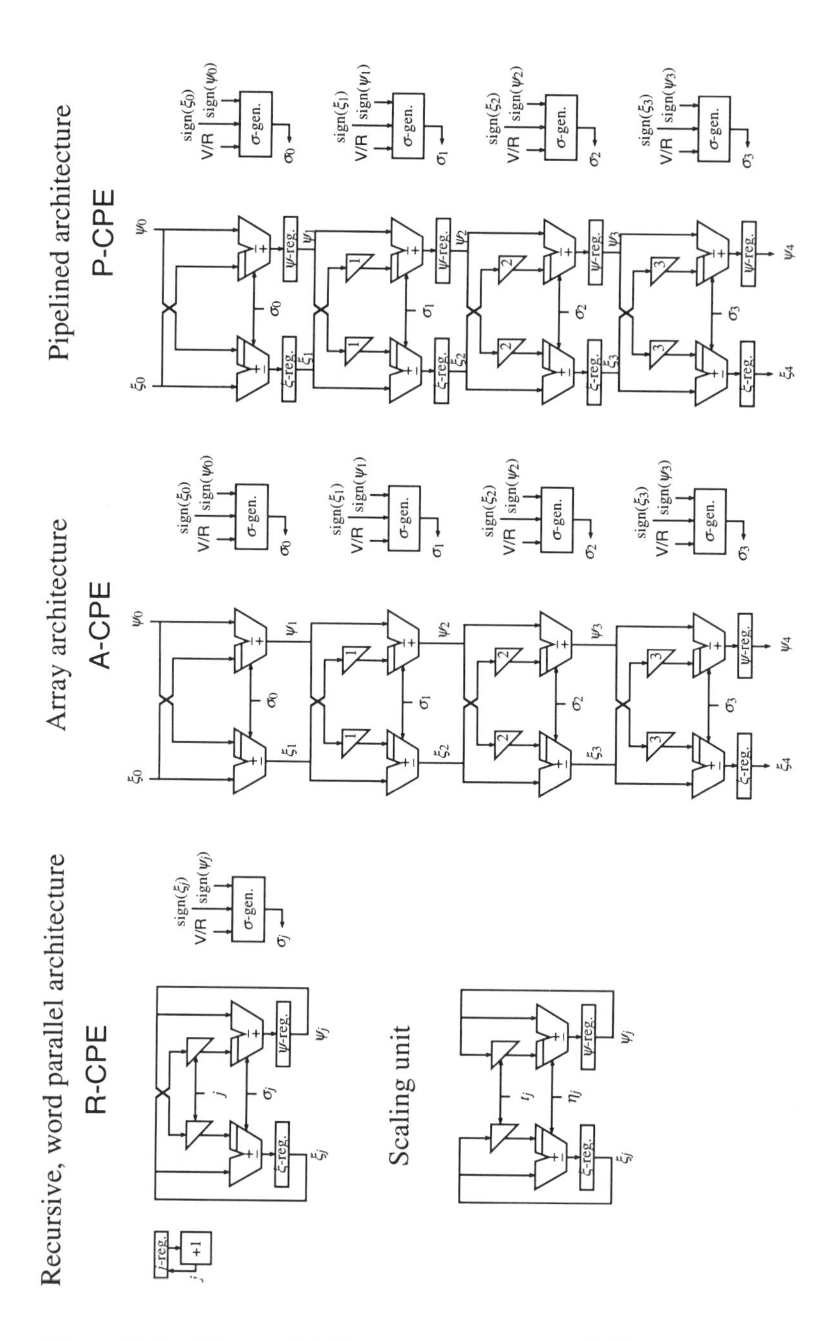

Figure 23.2 Different CORDIC processor element (CPE) architectures.

By choosing a recursive (R-CPE) [47], array (A-CPE) [48] or pipelined (P-CPE) [49] architecture, a trade-off between area, throughput and latency is possible, depending on the application in mind. A comparison of these different CPE realisations, designed in accordance with the specification given in Tab. 23.2, is presented in Tab. 23.3. It is based on the results of VHDL code synthesis using a 5 V, 0.6 µm double-metal CMOS library. In addition to these basic structures, intermediate/hybrid versions of them can also be employed in order to better match application constraints and requirements as will be demonstrated by an example in Subsection 23.4.3 (see Fig. 23.5).

Table 23.3 Comparison of different CPE implementations based on data from VHDL code synthesis using Synopsys tools and the 5 V, 0.6 µm double metal CMOS library from AMS.

Architecture	*Recursive*	*Pipelined*	*Array*
Area A_C [mm^2]	0.6	10.5	9.7
Clock frequency f_{clk} [MHz]	62.5	81.3	4.0
Power consumption P_C [mW]	28.8	634.1	26.0
CORDIC ops [MCops]	2.98	81.3	4.0
Latency τ_C [ns]	336.0	258.3	249.8
AT-product [mm^2/MCops]	0.2	0.13	2.42
PT-product [mW/MCops]	9.68	7.8	6.49

Systolic array architecture. The algorithm summarised in Tab. 23.1 can be straightforwardly mapped onto the systolic array architecture illustrated in Fig. 23.3 for the case of an array with $M = 3$ antenna elements and assuming real-valued input signals to simplify the presentation.[7]

This array entirely consists of CORDIC processor elements (CPEs) which operate synchronously driven by a single master clock. Since all the CPEs need the same amount of time to perform their computations, they can never be flooded with data. The cells in the left-hand triangular array (A-B-C) store the elements of the evolving upper triangular matrix $\boldsymbol{R}$ whilst the right middle column (F-G) holds the updated vector $\boldsymbol{u}$. The input data flow is from top to bottom. Each row of the array (A-F-G-C) performs a Givens rotation, whereby the rotation angles θ are determined by the CPEs designated Vec., configured to the "vectoring" mode of operation, at the beginning of the rows. These rotation angles are then propagated from the left to the right of the array, thereby passing through the CPEs labelled Rot. operating in the "rotation" mode. Due to the fact that the rotation angles traverse the array in a pipelined manner, the elements of the data vector have to be applied to the array in a time-staggered fashion as indicated by the sample indices in Fig. 23.3. Thus data associated with the

[7]For the case of complex-valued input data, each of the boxes representing a CORDIC processor element must be replaced by a "super-cell" comprising three such arithmetic units. For details refer to [50] or [19].

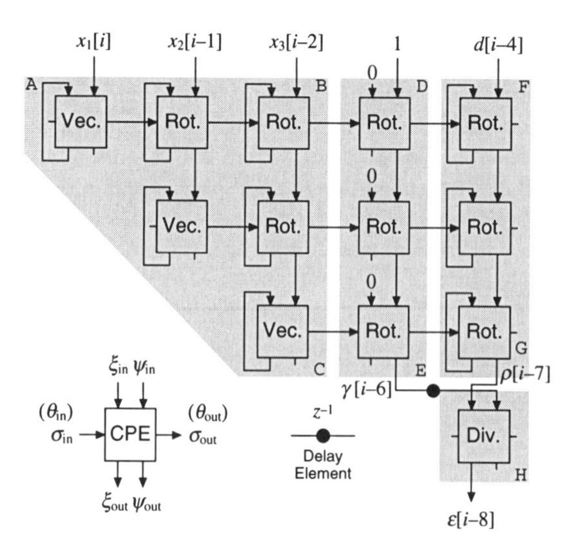

Figure 23.3 Systolic CPE array for the QRD-RLS algorithm.

same index wanders through the array in the form of a "wavefront" which emanates from the top left corner, spreads out and proceeds in the direction of the axis A→H before disappearing again in the bottom right cell. The left middle column of cells (D-E) successively computes the "conversion factor" γ. It is required for computing the *a priori* error ε which is formed by dividing the "rotated error" ρ by γ in the CPE marked H, configured to the Div. mode of operation.

This fully parallel implementation of the array requires $O(M^2)$ processors and achieves a throughput of O(1), i.e. it is capable of processing the input data stream at the sampling rate, thus fulfilling the real-time requirement independent of the number of antennas. Of course, this comes at the price of $O(2M)$ sampling periods latency, due to the intensive pipelining employed. The limiting factor governing the maximum sampling speed is given by the latency τ_C of the individual CORDIC cells, due to the feedback loop (from ξ_{out} to ξ_{in}) present for updating the contents of the array. This problem is attacked in [51] by applying a technique based on "lookahead" computations. Unfortunately, this approach requires a great deal of additional hardware.

Alternatively, this fully parallel 2D-array architecture may be mapped onto a linear array of processors [50], whereby all the operations taking place within each row of cells in the array of Fig. 23.3 are assigned to a single arithmetic unit, leading to a considerable reduction in the total number of required cells. The pipelined CORDIC processor element (P-CPE) is ideally fitted to this structure, whereas the recursive variant (R-CPE) is more suited for the fully parallel implementation. A comparison of these two competing array realisations for a specific application example will be made in the next subsection.

23.4.3 *Application to temporal reference beamforming*

We now describe several ways in which the QRD-RLS algorithm can be used in TRB systems. A simplistic signal flow diagram of such an application is shown in Fig. 23.4.

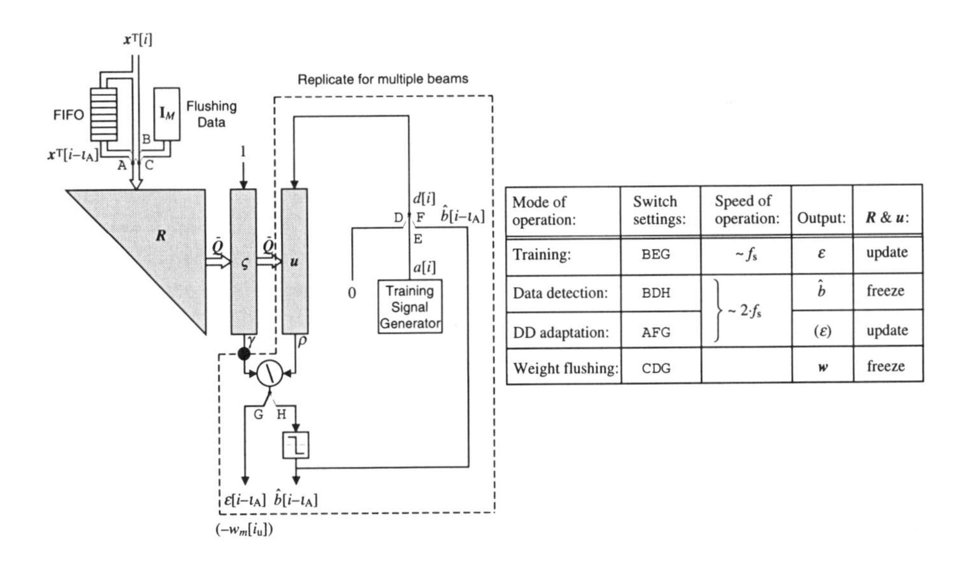

Mode of operation:	Switch settings:	Speed of operation:	Output:	R & u:
Training:	BEG	$\sim f_s$	ε	update
Data detection:	BDH	$\sim 2 \cdot f_s$	$\hat{b}$	freeze
DD adaptation:	AFG		(ε)	update
Weight flushing:	CDG		w	freeze

Figure 23.4 Simplistic signal flow diagram of a TRB system employing the QRD-RLS algorithm.

During the reception of the preamble at the beginning of a new data packet, the switches are set such that the snapshot of antenna element signals $\boldsymbol{x}^T[i]$ and the training sequence $a[i]$ are applied to the array in order to update the upper triangular matrix $\boldsymbol{R}$ and the vector $\boldsymbol{u}$ once every symbol. The corresponding error signal ε appears at the output of the weight controller with ι_A symbol periods delay ($\triangleq$ latency of the array) and can act as a convergence indicator, thus enabling the detection of false time synchronisation of the reference signal. There exists a number of different options to use the array upon completion of the training phase. By freezing the array's contents, i.e. by disconnecting the external feedback loop of all the CPEs, thus preventing any further updating of $\boldsymbol{R}$ and $\boldsymbol{u}$, it acts like a fixed filter with coefficients $\boldsymbol{w}[i_u]$ on the input data. If additionally the reference signal is clamped to zero, the (binary) data can be recovered from the error signal:

$$\hat{b}[i-\iota_A] = \text{sign}(-\varepsilon[i-\iota_A]) = \text{sign}(\boldsymbol{x}^T[i-\iota_A]\boldsymbol{w}[i_u]), \tag{23.15}$$

where i_u is the sample index when the weight vector was last derived. Furthermore, assuming that the so obtained (tentative) decisions $\hat{b}$ are correct, they can be used as a reference signal to update the weights in a decision-directed (DD) mode of operation. In this case, the new values of $\boldsymbol{R}$ and $\boldsymbol{u}$ are computed by applying the past snapshot $\boldsymbol{x}^T[i-\iota_A]$ from a FIFO memory to the array together with the corresponding fed-back data decision $d[i] = \hat{b}[i - \iota_A]$. For continual tracking of the weight coefficients with decision-feedback, the array must be clocked at twice the normal speed, operating alternately in data-detection and DD-adaptation mode within each symbol period. This allows to follow any variations of the environment whose temporal extension exceeds the array latency τ_A, i.e. whose rate of change is less than $(2M+2)f_s$.

Instead of using the array to perform the spatial filtering directly, this operation can be assigned to a separate beamforming network consisting of complex multipliers at the baseband or phase shifters and attenuators/amplifiers at RF or IF (→ analogue beamforming). The combiner coefficients required by the beamforming network can be very elegantly extracted from the array by the following simple "serial weight flushing" procedure. Since the weight vector is equal to the impulse response of the systolic array when $\boldsymbol{R}$ and $\boldsymbol{u}$ are held fixed, we can obtain $\boldsymbol{w}$ by exciting the array with a sequence of M snapshots which together form the identity matrix $\mathbf{I}_M$ along with a vector of M zero reference symbols $\mathbf{0}_M$. The negative weight coefficients then appear at the output ε one after another:

$$\boldsymbol{\varepsilon}_{w} = \mathbf{0}_M - \mathbf{I}_M \boldsymbol{w}[i_u] = \begin{pmatrix} -w_1[i_u] \\ \vdots \\ -w_M[i_u] \end{pmatrix} . \qquad (23.16)$$

Note that some additional inputs (e.g. all zeros) are required to completely flush the pipeline. An alternative method, allowing a continual parallel extraction of the complete weight vector, is treated in [19] and [40].

Clearly, we can easily incorporate several reference signals into our scheme to enable concurrent adaptation of K independent beams. To achieve this, we simply extend the array by replicating the column containing the vector $\boldsymbol{u}$ K times and then applying a K-dimensional vector of reference bits $(d^{(1)}[i]\ d^{(2)}[i]\ \ldots\ d^{(K)}[i])^{\mathrm{T}}$ instead of a single one $d[i]$ with every snapshot. This is equivalent to solving the system of equations (Eq. (23.9)) for different right-side vectors $\boldsymbol{u}^{(k)}$. Such a multibeam antenna can be employed to allow the simultaneous reception of multiple signals, either originating from different angularly distributed users or due to echoes arriving from different directions. Once isolated, the latter multipath components can be aligned in time and appropriately combined to reduce fading effects (→ "2D/space-time rake").

Up to now we have assumed that the local reference is ideally synchronised with the received signal of interest. In practice, sychronisation has to be established before the adaptive array can be trained. We now briefly present a simple and yet highly reliable technique based on the QRD-RLS algorithm to simultaneously establish timing synchronisation and adapt the array using a short, periodically repeated training sequence $a[i]$ of length N_s symbols. To guarantee the best possible performance, we assume that $a[i]$ is orthogonal (uncorrelated) to its delayed replicas, thus making m-sequences a very suitable choice. To achieve initial synchronisation, i.e. to estimate the timing of $a[i]$, we first adapt the array according to the LSE criterion using the QRD-RLS algorithm for each of the N_s possible time offset versions $a[i-n]$ $(n = 0 \ldots N_s - 1)$ of the synchronisation sequence, and then select the one giving rise to the smallest mean square *a priori* error $\mathrm{E}_J\{|\varepsilon[i]|^2\}$ (i.e. highest output SINR), where $\mathrm{E}_J\{\cdot\}$ stands for averaging over J consecutive values. This can be implemented very efficiently using the proposed multibeam weight controller for processing the N_s offset versions of the synchronisation signal in parallel. $\varepsilon^{(n)}[i]$ is generated by the Div.-mode CPE labelled H in Fig. 23.3 and subsequently its squared magnitude is accumulated over J symbol intervals. As stated in [40], array adaptation and timing estimation together require processing of approximately $2M+J$ symbols, where $J=10$ is sufficient to guarantee a high probability of acquisition even at low SNRs.

We complete this subsection by briefly demonstrating how to implement a multi-beam TRB using a linear array of pipelined CORDIC processor elements. Fig. 23.5 illustrates the basic architecture for an example with $M=3$ antennas and $K=3$ independent beams (again assuming real-valued signals), which means that three columns $\boldsymbol{u}^{(1)} \ldots \boldsymbol{u}^{(3)}$ are present on the right of the array in Fig. 23.4. The seven cells in the top row of the corresponding 2D-array are all mapped onto P-CPE$_1$, which comprises seven pipeline stages with three CORDIC iterations performed between each register ($\rightarrow$ mixture of pipelined and array CORDIC implementation). This way no additional dummy data or clock cycles are needed to flush the CORDIC pipeline before applying the next valid inputs. Since the number of cells in each row decreases by one towards the bottom of the 2D-array, the utilisation of the associated P-CPEs is reduced accordingly (to 86% for P-CPE$_2$ and 71% for P-CPE$_3$ in our example).[8] Two control bits, i.e. "feedback" (FB) and "vectoring/rotation" (V/R) determine the operation of each P-CPE and the array as a whole, whereby the frozen mode is assumed when $\mathsf{FB}=0$ and the ξ-inputs are taken from the circular buffers (CBUF) external to the P-CPEs. In Tab. 23.4 a (rough[9]) complexity comparison is made between the fully parallel 2D-array and the linear array solution for an adaptive antenna comprising $M=10$ elements, capable of concurrently processing $K=10$ different beams.

23.4.4 Simulation results

We have conducted a large number of computer experiments to assess the performance of the proposed techniques. A small selection of illustrative examples are presented in the following. In the simulations described here the QRD-RLS algorithm was implemented using the 16-bit fixed-point CORDIC arithmetic specified in Tab. 23.2, in order to take finite-precision effects into account. The antenna element signals x_m are thereby digitised with a resolution of $b_{\mathrm{q}}=12$ bits. All computations for the LMS algorithm are performed in floating-point format with unquantised inputs. The TRB under investigation employs a uniform linear array (ULA) with $M=3$ antennas having an interelement spacing of half a carrier wavelength ($\Delta=\lambda_{\mathrm{c}}/2$). The theoretical array gain offered by this configuration is 4.77 dB.

To illustrate the typical convergence behaviour of the RLS and LMS algorithm, we look at a situation where three signals with different power levels and directions of arrival ($\mathrm{SNR}_{\mathrm{in},1}=30\,\mathrm{dB}$[10], $\mathrm{DOA}_1=5°$; $\mathrm{SNR}_{\mathrm{in},2}=20\,\mathrm{dB}$, $\mathrm{DOA}_2=-35°$; $\mathrm{SNR}_{\mathrm{in},3}=$

[8] C. M. Rader shows how to build a sidelobe canceller using a folded linear array with 100% utilisation in [50].

[9] Wiring for cell interconnection has been omitted, which makes the results for the R-CPE-based realisation overoptimistic. Furthermore, the external CBUFs are not included ($\rightarrow$ frozen op-mode unavailable).

[10] $\mathrm{SNR}_{\mathrm{in}}$ stands for the signal-to-noise ratio at the output of the matched filters (MF) in each antenna branch, i.e. at the input to the adaptive combiner.

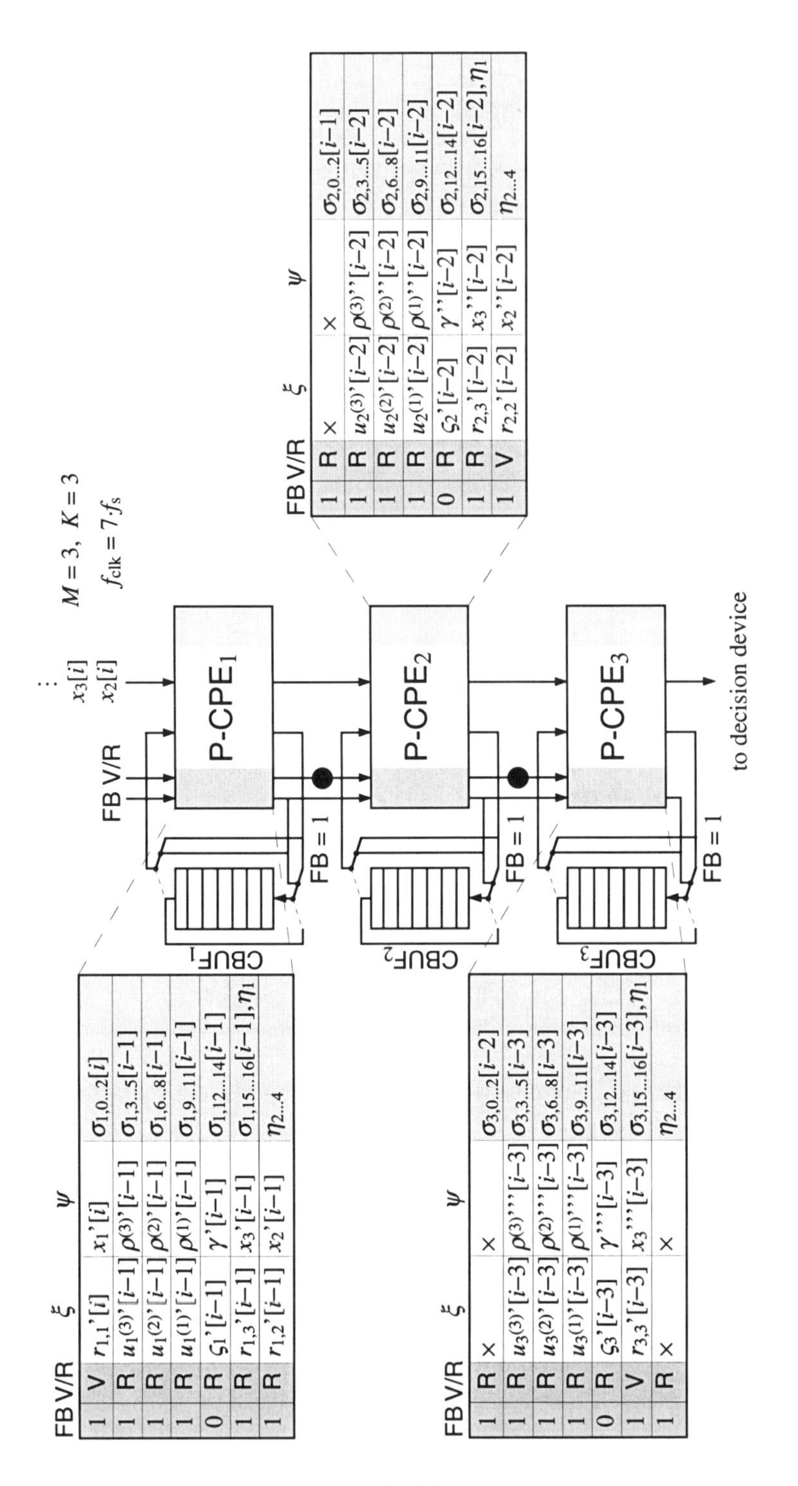

Figure 23.5 Example of a linear systolic P-CPE array for multibeam TRB with $M = 3$ antennas and $K = 3$ beams.

Table 23.4 Complexity comparison (based on the data from Tab. 23.3) of two multi-beam TRB implementations with $M = 10$ antennas and $K = 10$ beams. (†For continual DD updating. ‡Equivalent to approximately 6×10^9 MAC operations per second.)

Architecture	*Fully parallel array*	*Linear array*
Processor element type	R-CPE	P-CPE
No. of CPEs	$(3M^2 + M)/2$ $+3M(K+1) = 485$	$3M = 30$
Area A_{A} [mm^2]	291	315
Clock frequency f_{clk} [MHz]	62.5	81.3
Sampling frequency† f_{s} [MHz]	2.98 (1.49)	3.87 (1.94)
Power consumption P_{A} [W]	14	19
Cops/snapshot (utilisation)	485 (100%)	630 (77%)
Latency τ_{A}	$\approx 21T_{\mathrm{s}}$	$\approx 21T_{\mathrm{s}}$
Total compute power [GCops]	$1.45^{\ddagger}$	2.44

10 dB, $\mathrm{DOA}_3 = 65°$) but sharing the same carrier frequency are to be separated by a receiver through spatial filtering. In this specific case the eigenvalue spread of the input data correlation matrix takes on a large value of $\chi(\boldsymbol{R}_{xx}) = 113$. The simulated learning curves in Fig. 23.6 show the average (over 250 runs) output SINR as a function of the adaptation time.

The QRD-RLS algorithm converges to within -0.5 dB of the maximum SINR attainable with the optimal MMSE weight vector in approximately $10M$ iterations [38]. As can be seen in the top graph of Fig. 23.6, this is true for all three signals. On the other hand, the LMS algorithm only works satisfactorily for the strongest signal. Convergence degrades considerably for the second strongest and becomes unacceptable for the weakest one. This nicely illustrates the algorithm's sensitivity regarding a large eigenvalue spread. If all three received signals have the same SNR ($\rightarrow \chi(\boldsymbol{R}_{xx}) = 2.1$) equal to 10 dB, the dotted learning curve plotted in the bottom of Fig. 23.6 is obtained. It shows improved but still inferior convergence compared to the QRD-RLS implementation. Note also the loss in the maximal theoretically achievable output SINR incurred by the LMS algorithm (dashed lines) relative to the value resulting for the MMSE weight solution. This "misadjustment" is an additional weakness of the LMS technique [38]. As far as the impact of finite-precision CORDIC arithmetic is concerned, the learning curves produced by the chosen QRD-RLS realisation may exhibit a minor droop due to error accumulation. As can be observed in the top curve of Fig. 23.6, this effect is visible at high SNRs, where it leads to a gradual, very slight decay in SINR with time, but gets concealed by noise for levels of SNR below about 20 dB.

The bit error rate (BER) curves for user 1 obtained from simulations of the above scenario for three signals of identical power ($\rightarrow \chi(\boldsymbol{R}_{xx}) = 1.9 \ldots 2.2$ for $E_{\mathrm{b}}/N_0 = 0 \ldots 10$ dB) are plotted in Fig. 23.7.

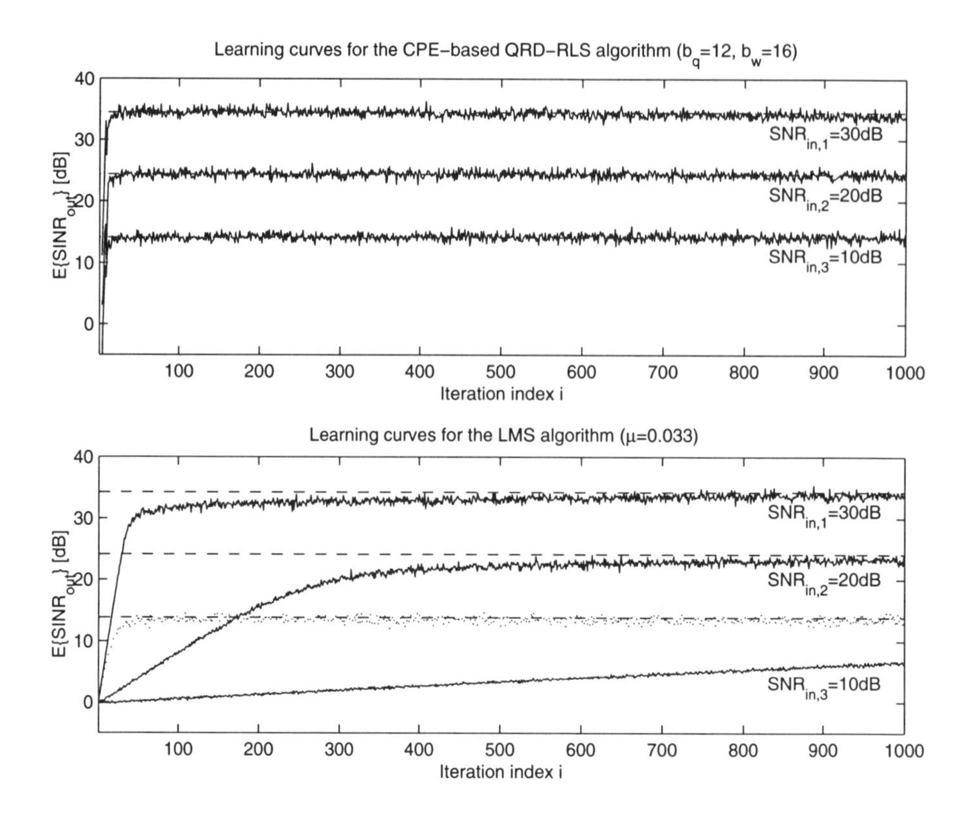

Figure 23.6 Simulated learning curves of the QRD-RLS ($\beta = 1, \delta = 0.004$) and LMS ($\mu = 0.033$) algorithm.

Packetwise transmission is employed with a randomly chosen training sequence of length 30 ($= 10M$) symbols preceding each data burst. No further decision-directed updating is performed after initial weight adaptation. Obviously, the system with a single antenna is useless even at high SNRs. Conversely, the BER of a TRB using the CPE-based implementation of the QRD-RLS algorithm remains close to the BPSK single-user bound. It merely shows a marginal performance degradation relative to the optimal MMSE solution, which in fact is primarily due to the short training interval and only secondarily a result of the signal quantisation and finite-precision calculations. The LMS-controlled beamformer shows a greater loss because of the misadjustment associated with this algorithm. Additionally, the BER would rapidly increase (for the weak users) as the received signal powers become dissimilar.

All the above observations clearly underline the advantages of RLS-based techniques for adaptive antenna control in situations with severe interference where their LMS counterparts fail to deliver satisfactory performance.

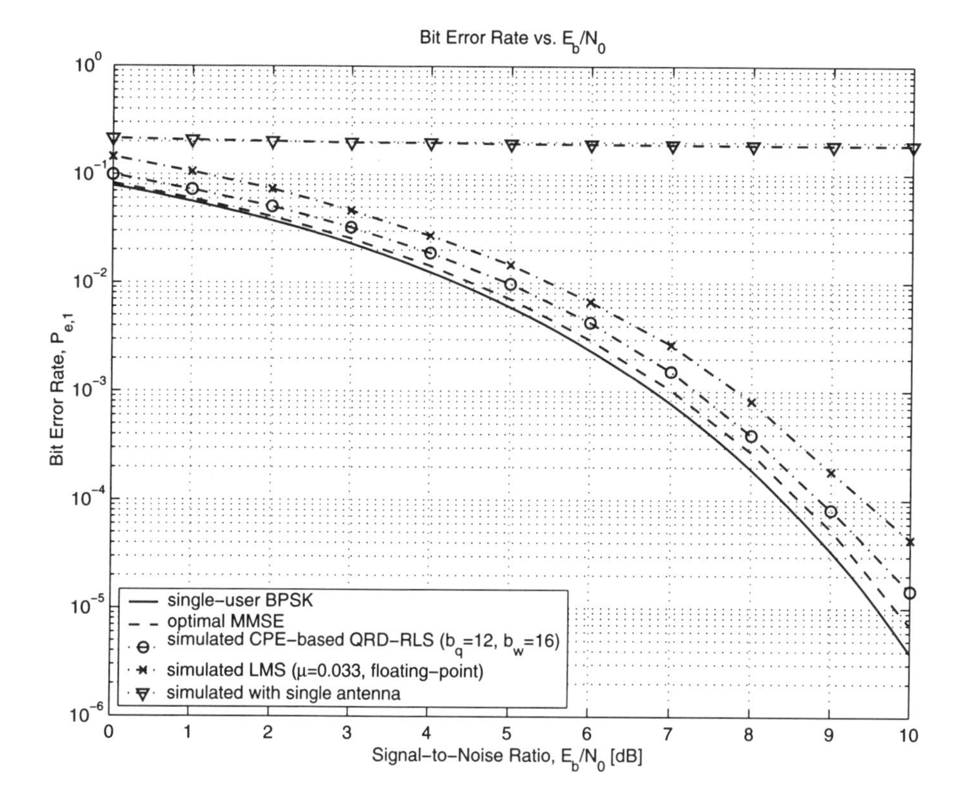

Figure 23.7 Simulated bit error rate curves of user 1 for the QRD-RLS ($\beta = 1$, $\delta = 0.004$) and LMS ($\mu = 0.033$) algorithm. (Same configuration and situation as in Fig. 23.6, except $\text{SNR}_{\text{in},1}$ = $\text{SNR}_{\text{in},2}$ = $\text{SNR}_{\text{in},3}$.)

23.5 APPLICATION TO ADAPTIVE DS-CDMA RECEIVERS

In this section we very briefly touch upon how to apply the QRD-RLS algorithm to a robust single-user DS-CDMA receiver structure [52–56] which jointly acts as an adaptive interference suppressor (for both narrowband interference, NBI, as well as multiple access interference, MAI), multipath combiner (i.e. rake) and despreading correlator. Due to its interference rejection capability, this receiver structure is less sensitive to the near-far problem encountered in spread-spectrum systems, where weak signals from distant transmitters are overwhelmed by strong signals from nearby sources. This relaxes power control requirements considerably and substantially increases network capacity relative to a system employing conventional matched filter or correlator type receivers. The proposed adaptive receiver represents an attractive alternative to the more complex multiuser/joint-detection techniques such as discussed in [57], and is especially suited for decentralised applications (e.g. the mobile terminals in cellular systems), where the receiver has no knowledge about the other users.

Fig. 23.8 shows the block diagram of the considered adaptive receiver. It is essentially a linear equaliser with a tap spacing that equals the chip duration T_c (or some rational fraction of it) and a total length of (at least) one symbol interval $T_b = N_c T_c$, corresponding to a single period of the spreading code.

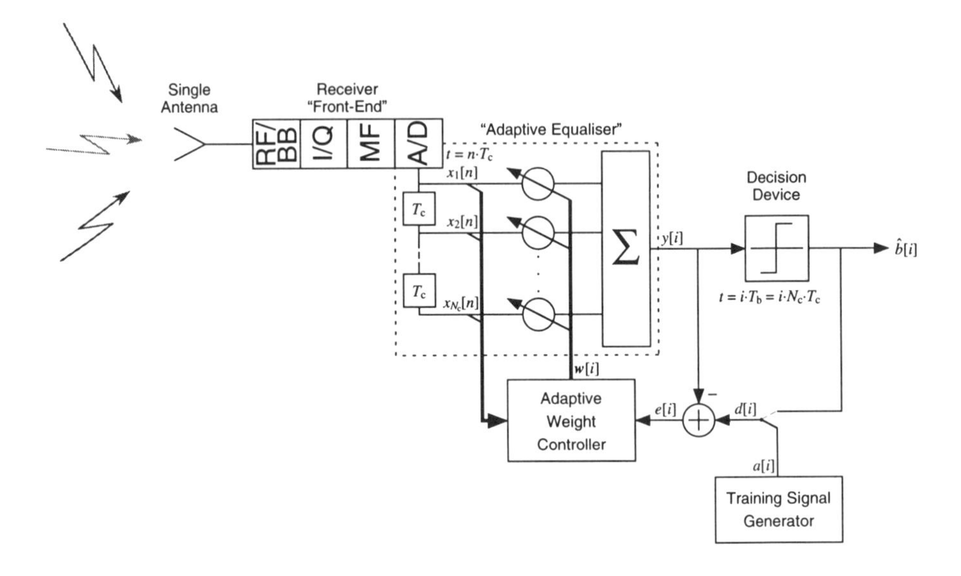

Figure 23.8 Block diagram of the adaptive DS-CDMA receiver.

During the operation of the receiver, the contents of the tapped delay line (TDL) are advanced by a full symbol period before computing each new output value $y[i]$ ($\rightarrow$ downsampling by a factor of N_c) and performing a weight update iteration. As a result, a particular register of the TDL will always correspond to a received signal sample taken at the same time instant within the symbol interval, assuming that the receiver's chip clock has a negligible frequency offset relative to that of the transmitter. In our example, each tap in the TDL contains the samples associated with a particular chip (i.e. a specific element of the spreading sequence). Obviously, the weight controller operates in exactly the same way as for the TRB treated before, therefore the same hardware architectures are applicable. It should be noted that the so-called "fast" RLS algorithms [38] of lower complexity $O(N_c)$ employed in traditional adaptive equalisers for TDMA receivers cannot be applied to this problem since these methods take advantage of the circumstance that the signals used to calculate consecutive outputs $y[i]$ are time-shifted versions of the input sample sequence. This shifting property is not present here due to the fact that the output is not generated at the same rate as the input is clocked through the TDL. For a detailed treatment of algorithmic as well as hardware implementation issues related to this application please refer to [40, 58–62].

23.6 SUMMARY AND CONCLUSIONS

In this contribution we have described algorithms and systolic array architectures for high-speed, high-performance RLS-based digital adaptive beamforming (as well as robust single-user DS-CDMA demodulation) using training sequences for initial adaptation and decision feedback for weight updating during data transmission. The QRD-RLS algorithm has been identified as a very promising candidate for adaptive weight control due to its positive numerical characteristics and the ability to perform the required calculations in a parallel and pipelined manner. CORDIC arithmetic is advocated for the hardware implementation of the individual cells in a possible processor array realisation, since the same number of cycles is required to perform all the needed operations, resulting in a truly systolic data flow. Different possible CORDIC architectures were compared with each other regarding area, throughput and latency, and applied as building blocks to construct both a fully parallel 2D-array, as well as an alternative linear array. Furthermore, a scheme for parallel control of multiple independent beams was briefly discussed together with its possible application for concurrent reception of many signals, either from different angularly distributed users, or multipath components of the SOI with different DOAs, which could be time aligned and combined to reduce the effect of fading. Additionally, it was shown how to employ the QRD-RLS algorithm to simultaneously establish timing synchronisation of the training sequence and adapt the array. Finally, the results of computer simulations were presented to illustrate the performance of the proposed techniques. *It was clearly demonstrated that RLS-based methods greatly outperform their LMS counterparts under severe interference conditions, even when employing finite-precision arithmetic.* The same holds in highly time-varying scenarios. It was pointed out that it is imperative to design algorithms and architectures in a joint process, thereby taking advantage of the interrelation between the two, in order to be able to exploit the merits of advanced signal processing techniques in practice at a reasonable cost.

References

[1] R. Prasad, *Universal Wireless Personal Communications*, Boston, MA: Artech House, 1998.

[2] M. Shafi, A. Hashimoto, M. Umehira, S. Ogose, and T. Murase, "Wireless Communications in the Twenty-First Century: A Perspective," *Proc. IEEE*, Vol. 85, No. 10, pp. 1622–1638, October 1997.

[3] H. Meyr and R. Subramanian, "Advanced Digital Receiver Principles and Technologies for PCS," *IEEE Communications Magazine*, Vol. 33, No. 1, pp. 68–78, January 1995.

[4] H. Meyr, M. Moeneclaey, and S. A. Fechtel, *Digital Communication Receivers: Synchronization, Channel Estimation, and Signal Processing*, New York: John Wiley & Sons, 1998.

[5] *Bell Labs Technical Journal Commemorative Issue: 50th Anniversary of the Transistor*, Vol. 2, No. 4, Autumn 1997.

[6] *Proceedings of the IEEE Special Issue on the Fiftieth Anniversary of the Transistor*, Vol. 86, No. 1, January 1998.

[7] Z. Kostic and S. Seetharaman, "Digital Signal Processors in Cellular Radio Communications," *IEEE Communications Magazine*, Vol. 35, No. 12, pp. 22–35, December 1997.

[8] J. Sanchez and G. Miet, "An Overview of Digital Signal Processing and Digital Signal Processors for Mobile Communications," in this volume.

[9] N. Seshan, "High VelociTI Processing," *IEEE Signal Processing Magazine*, Vol. 15, No. 2, pp. 86–101 & p. 117, March 1998. (`http://www.ti.com/sc/docs/dsps/products/c6000/index.htm`)

[10] M. Alidina, G. Burns, C. Holmqvist, E. Morgan, D. Rhodes, S. Simanapalli, and M. Thierbach, "DSP16000: A High Performance, Low Power Dual-MAC DSP Core for Communications Applications," in *Proc. CICC '98*, Santa Clara, CA, May 11–14, 1998, pp. 119–122. (`http://www.lucent.com/micro/dsp16000/`)

[11] E. Lambers, C. Moerman, P. Kievits, J. Walkier, and R. Woudsma, "R.E.A.L. DSP: Reconfigurable Embedded DSP Architecture for Low-Power/Low-Cost Telecom Baseband Processing," in this volume.

[12] R. Baines, "The DSP Bottleneck," *IEEE Communications Magazine*, Vol. 33, No. 5, pp. 46–54, May 1995.

[13] J. K. Hinderling, T. Rueth, K. Easton, D. Eagleson, D. Kindred, R. Kerr, and J. Levin, "CDMA Mobile Station Modem ASIC," *IEEE J. Solid-State Circuits*, Vol. SSC-28, No. 3, pp. 253–260, March 1993. (`http://www.qualcomm.com/ProdTech/asic/msm.htm`)

[14] K. H. Cho, J. Putnam, E. Berg, B. Daneshrad, and H. Samueli, "30 Mbps Wireless Data Transmission Using an Equalized 5-MBaud M-QAM Testbed," in *Proc. ICUPC '97*, San Diego, CA, October 12–16, 1997, Vol. 1, pp. 104–108.

[15] B. Daneshrad, "VLSI ASICs and Testbeds for High Speed Wireless Data Communications," presentation given at Bell Labs, Crawford Hill, Holmdel, August 5, 1997. (RealPlayer audio recording and slides available at URL `http://www.bell-labs.com/topic/seminars/hoh.wcof/tree/97-08-05/index.htm`)

[16] J. Litva and T. K.-Y. Lo, *Digital Beamforming in Wireless Communications*, Boston, MA: Artech House, 1996.

[17] B. D. Van Veen and K. M. Buckley, "Beamforming: A Versatile Approach to Spatial Filtering," *IEEE Acoust., Speech, Signal Processing Magazine*, Vol. 5, No. 2, pp. 4–24, April 1988.

[18] J.-Y. Lee, H.-C. Liu, and H. Samueli, "A Digital Adaptive Beamforming QAM Demodulator IC for High Bit-Rate Wireless Communications," *IEEE J. Solid-State Circuits*, Vol. SSC-33, No. 3, pp. 367–377, March 1998.

[19] B. Haller, "Algorithms and VLSI Architectures for RLS-Based Time Reference Beamforming in Mobile Communications," in *Proc. Int. Zurich Seminar on Broadband Communications IZS '98*, Zurich, Switzerland, February 17–19, 1998, pp. 29–36.

[20] K. K. Parhi, *VLSI Digital Signal Processing Systems: Design and Implementation*, New York: John Wiley & Sons, 1999.

[21] K. K. Parhi, "Impact of Architecture on DSP Circuits," in *Proc. IEEE Region 10 Int. Conf. TENCON '92*, Melbourne, Australia, November 11–13, 1992, Vol. 2, pp. 784–788.

[22] G. Fettweis and H. Meyr, "On the Interaction Between DSP-Algorithms and VLSI-Architecture," in *Proc. 1990 Int. Zurich Seminar on Digital Communications IZS '90*, Zurich, Switzerland, March 5–9, 1990, pp. 219–230.

[23] H. Meyr, "VLSI in Communications: Interactive Design of Algorithm and Architecture," in *Proc. 3rd Int. Workshop on Digital Signal Processing Techniques Applied to Space Communications*, ESTEC, Noordwijk, NL, September 23–25, 1992, pp. 1.1–1.8.

[24] G. Fettweis, "On Implementing Digital Signal Processing Tasks of Communications Systems in VLSI," *Digital Signal Processing*, Vol. 3, No. 3, pp. 210–219, July 1993.

[25] M. Ismail and N. Tan (Eds.), "Chip Design for Wireless Communication (An Interview with Ericsson's Tord Wingren)," *IEEE Circuits & Devices Magazine*, Vol. 14, No. 1, pp. 8–9, January 1998.

[26] H. Keding, M. Coors, and H. Meyr, "Efficient Design Flow for Fixed Point Systems," in this volume.

[27] A. E. Dunlop, W. J. Evans, and L. A. Rigge, "Managing Complexity in IC Design – Past, Present, and Future," *Bell Labs Technical Journal*, Vol. 2, No. 4, pp. 103–125, Autumn 1997.

[28] S. Y. Kung, "VLSI Array Processors," *IEEE Acoust., Speech, Signal Processing Magazine*, Vol. 2, No. 3, pp. 4–22, July 1985.

[29] S. Y. Kung, *VLSI Array Processors*, Englewood Cliffs, NJ: Prentice Hall, 1988.

[30] T. Ohgane, "Spectral Efficiency Improvement by Base-Station Antenna Pattern Control for Land Mobile Cellular Systems," *IEICE Trans. Communications*, Vol. E77-B, No. 5, pp. 598–605, May 1994.

[31] A. F. Naguib, A. Paulraj, and T. Kailath, "Capacity Improvement with Base-Station Antenna Arrays in Cellular CDMA," *IEEE Trans. Vehicular Technology*, Vol. VT-43, No. 3, Pt. II, pp. 691–698, August 1994.

[32] G. V. Tsoulos, M. A. Beach, and J. P. McGeehan, "Wireless Personal Communications for the 21st Century: European Technological Advances in Adaptive Antennas," *IEEE Communications Magazine*, Vol. 35, No. 9, pp. 102–109, September 1997.

[33] *IEEE Signal Processing Magazine Special Issue on Space-Time Processing*, Vol. 14, No. 6, November 1997.

[34] *IEEE Personal Communications Special Issue on Smart Antennas*, Vol. 5, No. 1, February 1998.

[35] L. C. Godara, "Applications of Antenna Arrays to Mobile Communications, Part I: Performance Improvement, Feasibility, and System Considerations," *Proc. IEEE*, Vol. 85, No. 7, pp. 1031–1060, July 1997.

[36] L. C. Godara, "Applications of Antenna Arrays to Mobile Communications, Part II: Beam-Forming and Direction-of-Arrival Considerations," *Proc. IEEE*, Vol. 85, No. 8, pp. 1195–1245, August 1997.

[37] J. H. Winters, J. Salz, and R. D. Gitlin, "The Impact of Antenna Diversity on the Capacity of Wireless Communication Systems," *IEEE Trans. Communications*, Vol. COM-42, No. 2/3/4, Pt. III, pp. 1740–1751, Feb./Mar./Apr. 1994.

[38] S. Haykin, *Adaptive Filter Theory*, Upper Saddle River, NJ: Prentice Hall, 3rd Ed., 1996.

[39] I. S. Reed, J. D. Mallett, and L. E. Brennan, "Rapid Convergence Rate in Adaptive Arrays," *IEEE Trans. Aerosp. Electron. Syst.*, Vol. AES-10, No. 6, pp. 853–863, November 1974.

[40] B. Haller, *Algorithms and VLSI Architectures for Diversity Combining and Interference Suppression in Wireless Communication Systems*, Doctoral Dissertation, Swiss Federal Institute of Technology, Zurich, Switzerland, 1999.

[41] W. M. Gentleman and H. T. Kung, "Matrix Triangularization by Systolic Arrays," in *Real-Time Signal Processing IV* (T. F. Tao, Ed.), Proc. SPIE Vol. 298, San Diego, CA, August 1981, pp. 19–26.

[42] J. G. McWhirter, "Recursive Least-Squares Minimization Using a Systolic Array," in *Real Time Signal Processing VI* (K. Bromley, Ed.), Proc. SPIE Vol. 431, San Diego, CA, August 1983, pp. 105–112.

[43] J. E. Volder, "The CORDIC Trigonometric Computing Technique," *IRE Trans. Electronic Computers*, Vol. EC-8, No. 3, pp. 330–334, September 1959.

[44] J. S. Walther, "A Unified Algorithm for Elementary Functions," in *Proc. Spring Joint Computer Conf.*, Atlantic City, NJ, May 18–20, 1971, pp. 379–385.

[45] D. König and J. F. Böhme, "Optimizing the CORDIC Algorithm for Processors with Pipeline Architecture," in *Proc. EUSIPCO '90*, Barcelona, Spain, September 18–21, 1990, Vol. 3, pp. 1391–1394.

[46] Y. H. Hu, "CORDIC-Based VLSI Architectures for Digital Signal Processing," *IEEE Signal Processing Magazine*, Vol. 9, No. 3, pp. 16–35, July 1992.

[47] D. Timmermann, H. Hahn, B. J. Hosticka, and G. Schmidt, "A Programmable CORDIC Chip for Digital Signal Processing Applications," *IEEE J. Solid-State Circuits*, Vol. SSC-26, No. 9, pp. 1317–1321, September 1991.

[48] H. M. Ahmed and K.-H. Fu, "A VLSI Array CORDIC Architecture," in *Proc. ICASSP '89*, Glasgow, Scotland, May 23–26, 1989, Vol. 4, pp. 2385–2388.

[49] D. Timmermann, B. Rix, H. Hahn, and B. J. Hosticka, "A CMOS Floating-Point Vector-Arithmetic Unit," *IEEE J. Solid-State Circuits*, Vol. SSC-29, No. 5, pp. 634–639, May 1994.

[50] C. M. Rader, "VLSI Systolic Arrays for Adaptive Nulling," *IEEE Signal Processing Magazine*, Vol. 13, No. 4, pp. 29–49, July 1996.

[51] J. Ma, E. F. Deprettere, and K. K. Parhi, "Pipelined CORDIC Based QRD-RLS Adaptive Filtering Using Matrix Lookahead," in *Proc. 1997 IEEE Workshop on Signal Processing Systems Design and Implementation SiPS '97* (M. K. Ibrahim, P. Pirsch, and J. McCanny, Eds.), Leicester, UK, November 3–5, 1997, pp. 131–140.

[52] C. N. Pateros and G. J. Saulnier, "An Adaptive Correlator Receiver for Direct-Sequence Spread-Spectrum Communication," *IEEE Trans. Communications*, Vol. COM-44, No. 11, pp. 1543–1552, November 1996.

[53] P. B. Rapajic and B. S. Vucetic, "Adaptive Receiver Structures for Asynchronous CDMA Systems," *IEEE J. Select. Areas Commun.*, Vol. SAC-12, No. 4, pp. 685–697, May 1994.

[54] S. L. Miller, "An Adaptive Direct-Sequence Code-Division Multiple-Access Receiver for Multiuser Interference Rejection," *IEEE Trans. Communications*, Vol. COM-43, No. 2/3/4, Pt. III, pp. 1746–1755, Feb./Mar./Apr. 1995.

[55] U. Madhow and M. L. Honig, "MMSE Interference Suppression for Direct-Sequence Spread-Spectrum CDMA," *IEEE Trans. Communications*, Vol. COM-42, No. 12, pp. 3178–3188, December 1994.

[56] H. V. Poor, "Adaptive Interference Suppression for Wireless Multiple-Access Communication Systems," in this volume.

[57] S. Verdú, *Multiuser Detection*, New York: Cambridge University Press, 1998.

[58] L. He and U. Madhow, "Pipelined MMSE Equalizers for Direct-Sequence Spread-Spectrum CDMA Channels," in *Proc. ICUPC '96*, Cambridge, MA, Sep. 29 - Oct. 2, 1996, pp. 120–124.

[59] L. Lucke, L. Nelson, and H. Oie, "Adaptive CDMA Receiver Implementation for Multipath and Multiuser Environments," in *VLSI Signal Processing IX* (W. Burleson, K. Konstantinides, and T. Meng, Eds.), Piscataway, NJ: IEEE Press, 1996, pp. 159–168.

[60] C. Teuscher, *Low Power Receiver Design for Portable RF Applications: Design and Implementation of an Adaptive Multiuser Detector for an Indoor, Wideband CDMA Application*, Ph.D. Thesis, U.C. Berkeley, Berkeley, CA, 1998.

[61] C. Teuscher, D. Yee, N. Zhang, and R. Brodersen, "Design of a Wideband Spread Spectrum Radio Using Adaptive Multiuser Detection," in *Proc. ISCAS '98*, Monterey, CA, May 31 – June 3, Vol. 4, pp. 593–599, 1998.

[62] H. V. Poor and X. Wang, "Code-Aided Interference Suppression for DS/CDMA Communications – Part II: Parallel Blind Adaptive Implementations," *IEEE Trans. Communications*, Vol. COM-45, No. 9, pp. 1112–1122, September 1997.

24 EVOLUTION OF SPEECH CODING FOR WIRELESS COMMUNICATIONS

A system standard: the AMR (Adaptive Multi Rate codec)

Gilles Miet

Philips Consumer Communications
Le Mans, France

Abstract: The world-wide coverage of GSM-like networks led to an explosion of the number of subscribers, causing the saturation of some networks. Besides, the old technology used in GSM systems provided the customer with an unsatisfactory audio quality in many common situations. In parallel, DSP computing capabilities and the state of the art in speech and channel coding have improved. Therefore, to increase both the capacity and the quality of GSM networks, ETSI decided to standardise a new speech transmission system: the AMR. This Adaptive Multi Rate system is a combination of speech and channel codecs controlled by signalling means aimed at providing the best audio quality under background noise and transmission errors. The first part of this chapter explains this concept in comparison with current GSM coders. Another promising technology to achieve high quality speech transmission is wide-band coding. Therefore ETSI is investigating a wide-band AMR mode. The second part of this chapter presents the quality gain that can be provided by wide-band AMR.

24.1 OVERVIEW

With the increase of processing power of current DSPs, ETSI (European Telecommunication Standard Institute) decided to develop a new speech compression/protection system for GSM called AMR (Adaptive Multi Rate codec). The original aim of the AMR project was to develop a system with a continuously varying bit rate depending on the audio signal to transmit. There are indeed many different categories of audio signals. For instance, for a given bandwidth, encoded speech does not require the same bit rate as encoded music. Similarly, narrow-band speech (300 Hz–3400 Hz) does not need the same bit rate as wide-band speech (80 Hz–7000 Hz). Narrow-band speech is commonly used for telephony both because it requires little transmission capacity and because it enables to recognise and understand the correspondent in a relatively quiet environment. However, to increase naturalness and improve the understanding in a noisy environment (which is necessary for wireless communications), wide-band speech transmission is definitively necessary. Therefore, AMR should enable to transmit both narrow-band and wide-band speech, which will be referred to as *narrow-band mode* and *wide-band mode*, respectively.

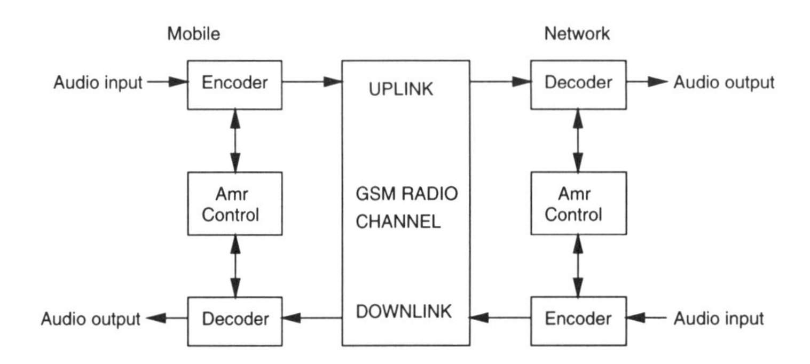

Figure 24.1 Block diagram of the AMR System.

In order to facilitate the penetration of this new standard, it has been decided to make it compatible with the existing GSM physical radio channels. The GSM system actually provides two speech traffic channels, TCH/FS and TCH/HS. The Full Rate Speech Traffic Channel (TCH/FS) provides a gross bit rate of 22.8 kbit/s and the Half Rate Speech Traffic Channel (TCH/HS) half of this rate. If a higher bit rate is needed, a multi-slot solution (the allocation of several traffic channels) is also possible. With the limited time scale due to the high interest of US operators to have the narrow-band mode of this system standardised in December 98, it was decided to first develop AMR for narrow-band [1]. However, the narrow-band AMR will incorporate signalling means to prepare the control of other AMR modes. As described in Fig. 24.1, AMR includes control blocks on the GSM signal processing system. These blocks exchange information with the source/channel codecs and control them. Their role is to command the selection of the encoders and their corresponding decoders such that they match for both links.

24.2 NARROW-BAND AMR

24.2.1 Need for a new standard

Currently, TCH/FS is used to transmit encoded speech with two possible coders, the FR and the EFR. Since the FR (Full Rate codec) offers a mediocre quality at 13 kbit/s, the EFR (Enhanced Full Rate codec) was standardised later. EFR provides a good quality at a bit rate of 12.2 kbit/s. However, quality goes down rapidly as transmission errors increase. Similarly, TCH/HS can be used for speech transmission with the HR (Half Rate codec) at 5.6 kbit/s. This codec gives a very poor quality for both noisy speech and bad radio conditions. For this reason, many operators are reluctant to deploy it on their networks. The narrow-band AMR codec was therefore defined to improve the weaknesses of the existing GSM codecs. In TCH/FS, the target of AMR in Full-Rate mode is to offer more robustness to transmission errors than EFR, the goal being toll quality in all radio conditions. This is a great advantage, but this was not the main reason to drive the AMR project. The motivation to build it up came essentially from the need to have an acceptable codec in TCH/HS. Both European and US operators need AMR to increase the capacity of their networks seamlessly for their subscribers.

24.2.2 Variable bit rates for speech and channel codecs

For a given channel, the narrow-band AMR is a combination of multi-bit rate speech and channel coding systems. This technique is necessary to reach an optimal speech quality for a certain gross bit rate. Indeed, low bit rate speech codecs are very sensitive to the speech nature of the encoded signal because they model speech. When the background noise level increases, the speech codec has to use a higher bit rate to keep up a given speech quality. Additional bits are then necessary to maintain it. Similarly, under bad channel conditions, the bit error rate increases if the protection of encoded speech parameters is fixed. To reduce this error rate, a better protection has to be provided. This is achieved by augmenting the redundancy of the transmitted bit stream, hence the channel codec produces a higher bit rate. Fig. 24.2 shows the effect of the source/channel partitioning under radio error conditions.

The highest source rate gives the best results under no error (speech is more accurately modelled) and the worst results under high error conditions (encoded speech is less protected). Therefore the source/channel bit rate partitioning should vary in function of the environment to provide the best quality. The source or channel bit rate should be increased in the presence of background noise or bad radio transmission, respectively. The curve called "AMR" in Fig. 24.2 represents the quality the AMR codec would provide under transmission errors if it was based on codec A, B and C. With this system using several source/channel codecs, some signalling bits have to be transmitted to control a perfect matching between the coders and the decoders at any time. The conditions in which speech is encoded and transmitted vary very quickly. Consequently, the signalling of the codec mode is transmitted in-band (it uses some bits instead of the source coder) for a given channel. This is shown in Fig. 24.3, where the channel encoder is fed by both encoded speech and signalling bits coming from

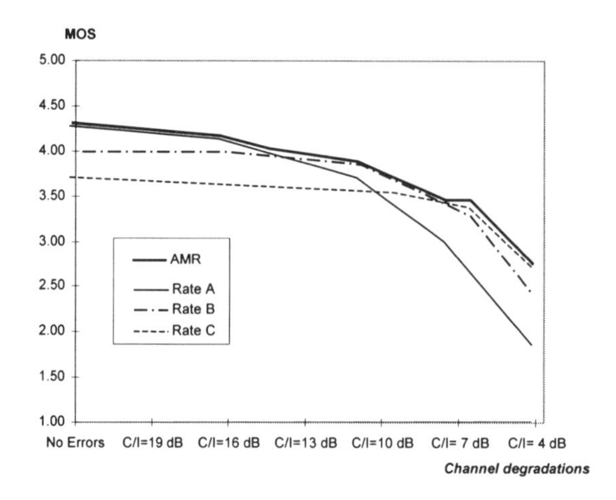

Figure 24.2 Influence of the channel codec bit rate on the quality of speech. The vertical axis represents the perceived speech quality in MOS scale (Mean Opinion Score) provided by ACR (Absolute Category Rating) tests. The horizontal axis shows the radio channel degradations in C/I (Carrier over Interference ratio). For speech codecs: rate $A >$ rate $B >$ rate C, for channel codecs: rate $A <$ rate $B <$ rate C.

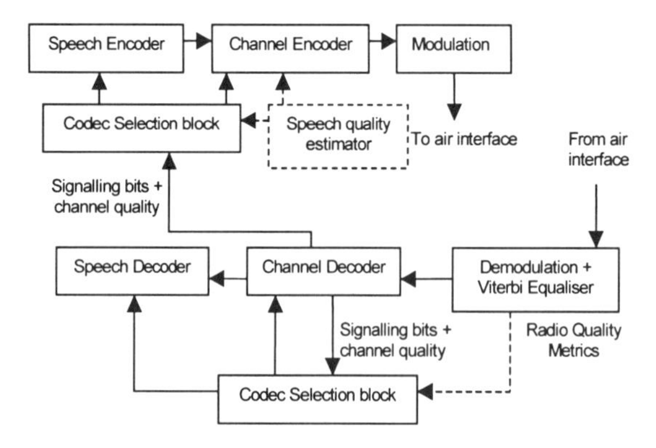

Figure 24.3 Overview of the Narrow-band AMR System processing blocks.

the codec selection block and the speech quality estimator[1]. Similarly, the channel decoder outputs both signalling bits and encoded speech. Fig. 24.3 represents the processing carried out either in the network or in a mobile phone.

Current GSM systems consist of a speech codec, a channel codec, a modulator and a demodulator. Although AMR follows this scheme, it differentiates significantly

[1]The speech quality estimator is a generic term that may represent, for instance, the speech to noise ratio or the background noise level.

by the provision of several codec modes. AMR incorporates also codec selection blocks fed by signalling and quality metrics to select the right codec modes. The narrow-band AMR system being in standardisation process, some uncertainties remain concerning its final structure. They are represented in Fig. 24.3 by dotted lines. For instance, the radio quality metrics coming from the Viterbi equaliser and the speech quality estimator are optional. Effectively, radio quality metrics can be computed in the channel decoder. The speech quality estimator is likely not to be used[2] because the radio quality metrics may be sufficient to decide the codec mode to be selected. These metrics provide the information about the channel protection that should be applied to encoded speech. Since speech sounds better when encoded with a high bit rate, the highest source bit rate is selected by default and decreased in function of the radio quality metrics.

24.2.3 Complexity/performance compromise of the AMR

The complexity requirement of the AMR codec was initially defined globally. Then it was decided to split this requirement between the source codec and the channel codec. This demonstrates very clearly the compromise between the compatibility with the current GSM system and the expected improvements thanks to the AMR. Actually, the main driver for the compatibility is the network. The channel codec being in the BTS (Base Transceiver Station), the AMR system has been defined to enable the use of existing BTS without any change in the hardware. This high constraint prevents the use of Turbo Codes which are the state of the art in channel coding. The complexity of the speech codec was less limited because the speech codec is located in the TRAU (Transcoder Unit) outside the BTS. The overall complexity is still of importance, especially because of the power consumption and price of mobile phones.

The FR AMR performance is targeted to a quality augmentation equivalent to up to 6 dB C/I increase in the radio channel with regard to the EFR. Similarly, HR AMR should provide the same quality as today's GSM FR, which is the most widely spread GSM speech codec.

By doing this, the consumers are presented with a quality either similar (AMR HR) or better (AMR FR) than they are used to with the current GSM FR.

24.3 MULTI-MODE AMR

The current narrow-band system includes hooks in the bit rate control to be able to add wide-band speech and other codecs in a second step. Multi-mode AMR (see Fig. 24.4) can be defined as an audio processing tool box suited to as many audio services as possible. Actually, very low bit rates can be sufficient to transmit non-conversational speech services like voice recognition or voice mails. Higher bit rates are necessary to transmit music or wide-band speech in full duplex. The definition of additional modes for narrow-band AMR is currently quite vague. However, the wide-band AMR agenda has recently been set because of the high interest of this service.

[2]The AMR version recently selected has actually no speech quality estimator.

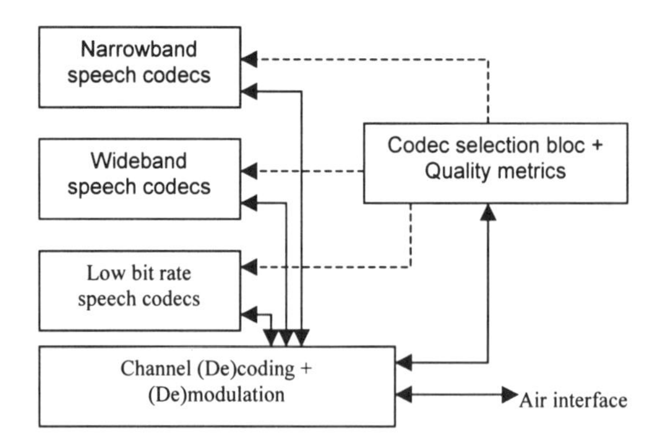

Figure 24.4 Principle of the AMR Multi modes system.

24.4 WIDE-BAND AMR

Wide-band speech (80 Hz–7 kHz) was emphasised at ETSI because it enables to obtain a far better quality than narrow-band speech (300–3400Hz). In fact, the higher bandwidth (3.4 kHz–7 kHz) increases the speech intelligibility (especially for fricatives such as "s" or "f") and the naturalness of speech. The lower bandwidth (80–300 Hz) provides a higher listening comfort and naturalness [2, 5]. A situation that demonstrates the quality difference between narrow-band and wide-band speech is the broadcast of a phone call through FM radio. The listeners notice the difference of voice quality between the voices coming from the studio and the voice over the phone.

24.4.1 Wide-band versus narrow-band quality

The quality gain brought by wide-band speech with regard to narrow-band speech is straightforward. However, the available bit rate on TCH/FS is very low to encode wide-band speech which might lead to strong degradations. Therefore, it is interesting to evaluate the compromise speech bandwidth/degradations. It was achieved by informal listening tests. In this respect, the preference of wider band speech over narrow-band speech was determined by comparing narrow-band speech signals with wider band speech signals generated as described in Fig. 24.5.

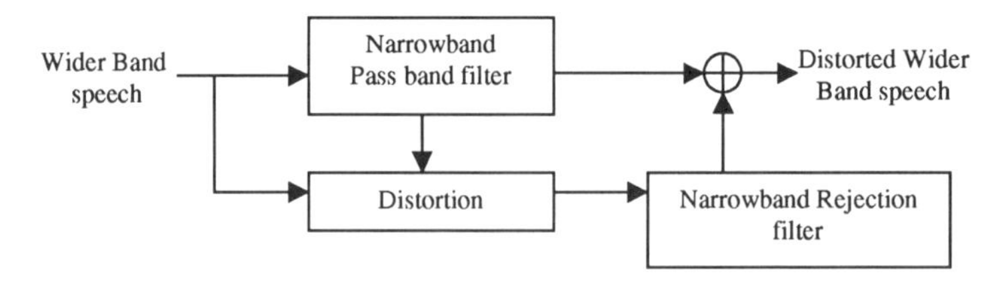

Figure 24.5 Distortion of the wide-band signal envelope.

The wider band speech signals consist of narrow-band speech signals combined with a distorted added band. Three experiments were carried out with three different added distorted bands. The first experiment consisted of the addition of the low band (100–300 Hz). Here the wider band signals were preferred even with a distortion of 3.1 dB. In the second experiment, the high band (3.4–7 kHz]) was added. In this case, the wider speech signals were also preferred with a distortion up to 5.9 dB. The third experiment, consisted of the combination of the two preceding experiments. The wide-band signals were also preferred even with the same distortions as above.

These experiments demonstrate the benefits of widening the bandwidth even if this process introduces significant but not annoying distortions.

24.4.2 Minimum bandwidth for speech to sound wide-band

As mentioned above, wide-band and narrow-band speech bandwidths correspond to 80 Hz–7 kHz and 300–3400 Hz, respectively. However, to provide single slot solutions (compatible with TCH/FS), it might become necessary to lower this bandwidth. Indeed, the speech quality might suffer if the source coder bit rate is too low with regard to the bandwidth of the speech signal to transmit. Consequently, there is an interest in identifying the minimum bandwidth of speech for it to "sound" wide-band. In this respect, there was an experiment undertaken in [4] and recently reported in [5]. This experiment concludes that the lowest frequency should be below 100 Hz and the highest frequency should be above 5 kHz. Besides, the transmission of the lower band 100-300 Hz was found particularly advantageous for male speakers. The higher part (3.4–7 kHz) was especially favourable to female speakers. It was also concluded that the band 20 Hz–100 Hz does not bring any improvement. The contribution [5] thus concluded that wide-band AMR should be targeted to transmit the band 100 Hz–7 kHz. The transmission of 100 Hz–5 kHz was proposed as a fall-back solution.

24.5 CONCLUSION

AMR reveals the recent trend to develop telecommunication signal processing algorithms as a whole. Effectively, old standards have defined several processing blocks independently. In contrast, the different blocks are now developed together, and interact with each other. This enables the designers to provide optimal solutions. Besides, AMR is a huge system that supplies a multiplicity of codecs to enable the GSM standard to adapt to the numerous conditions and applications of wireless communications. Narrow-band AMR increases the robustness under channel errors and limits the degradation with background noise (as compared to the current GSM system). It also provides TCH/HS with an acceptable codec to increase network capacity. Wide-band AMR will be deployed for high quality speech transmission. Other modes (e.g. a mode including noise suppression) are currently being discussed.

References

[1] P. Gaskell, "AMR feasibility study report", V. 1.0, November 1997.

[2] W.B. Kleijn, K.K. Paliwal, "Speech coding and synthesis", Elsevier, 1995

[3] Y. Linde, A. Buzo, R. M. Gray, "An algorithm for vector quantizer design", *IEEE Transactions on Communications*, Vol. COM-28, No. 1, January 1980.

[4] Dominique Pascal, "Comparative performances of two subjective methods for assessing the fidelity of speech transmission systems", *AES* 84th *Convention*, March 1–4, Paris.

[5] ETSI TD SMG11 66/98, "Subjective effect of widening the telephone band from 0 to 7 kHz for speech signal".

25 DIGITAL SIGNAL PROCESSING AND DIGITAL SIGNAL PROCESSORS FOR MOBILE COMMUNICATIONS: BASIC CONCEPTS AND TRENDS

Javier Sanchez

Philips Consumer Communications
Advanced Development Group
Route d'Angers
72081 Le Mans Cedex 9, France

Abstract: The boom of wireless telecommunications arisen in this decade represents a fantastic engine for the technical evolution of Digital Signal Processors (DSP). This constant progress is motivated by the complexity increase of signal processing algorithms, and the product properties required by the market. Indeed, both the rapid evolution of telecommunication standards (to achieve better quality of communications, higher capacity of networks, and new services) and the new features appearing on terminals are increasingly demanding digital signal processing algorithms. In an attempt to implement these algorithms in real-time applications, more elaborate DSP architectures had to be developed. Furthermore, especially in mobile telephony, there is a constant requirement to decrease power consumption and cost of DSPs. This presentation outlines the evolution of such DSPs and tries to draw a trend for the near future.

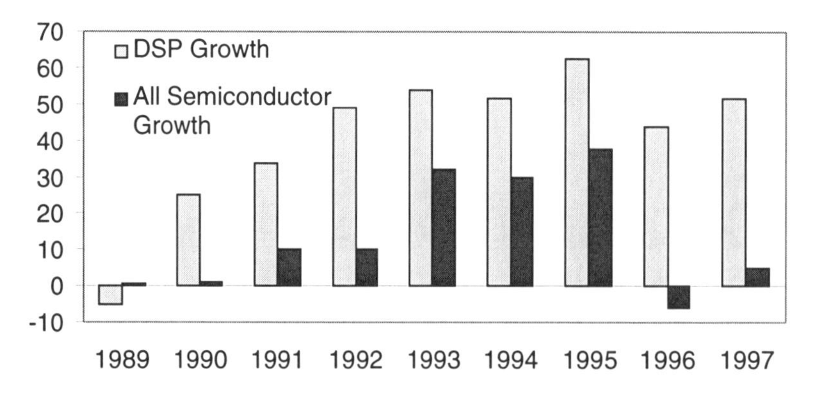

Figure 25.1 Historic worldwide digital signal processor revenue rate. Source: Dataquest, June 1998.

25.1 INTRODUCTION

The successful introduction in the consumer market of the first Digital Signal Processor (DSP) over the last decade has motivated semiconductor manufacturers to concentrate considerable development efforts on this field (see Fig. 25.1). According to *Forward Concepts*, the DSP market represented $3 billion in 1997 and is expected to be on the order of $14 billions in 2002. This fantastic growth can be explained by the popularity of digital signal processing techniques applied to consumer mass products. In this context, cellular telephony has been playing an important role due mainly to its constant market growth as shown in Fig. 25.2. This market is characterised by low-cost products as well as a severe price erosion of up to 30 % per year. Achieving the desired requirement for price/performance translates into a powerful and low-current DSP at smallest possible silicon size: a trade-off continuously researched by DSP manufacturers and appreciated by mobile phone makers. The goal of this chapter is to outline a global panorama of the main challenges and trends encountered when implementing digital signal processing applications in mobile communications. We stress the major issues to face by DSP manufacturers in developing third generation mobile communications systems. Some basic concepts commonly used in evaluating the performance of a DSP are also described. Several figures are also included to lend support to the assertions made.

25.1.1 DSP functions and applications

The digitised-signal-based systems have proved to be very advantageous compared to analog systems because of their functional flexibility, robustness, reduced size, and very often a possibility for superior performance. This fact can be inferred from Fig. 25.3, where some numbers on the digital and analog semiconductor market are given.

Digital signal processing has facilitated the large deployment of high-performance, affordable, and easy-to-use consumer products. Table 25.1 depicts some typical digital signal processing algorithms and some applications where these techniques are com-

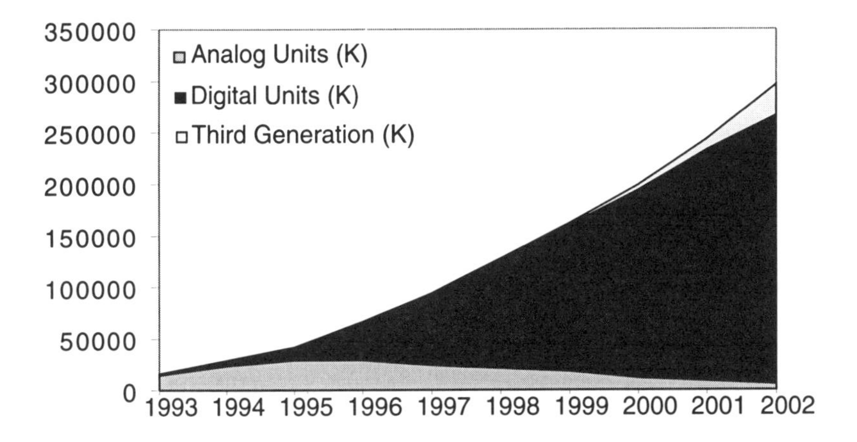

Figure 25.2 Worldwide cellular market. Source: Dataquest, December 1997.

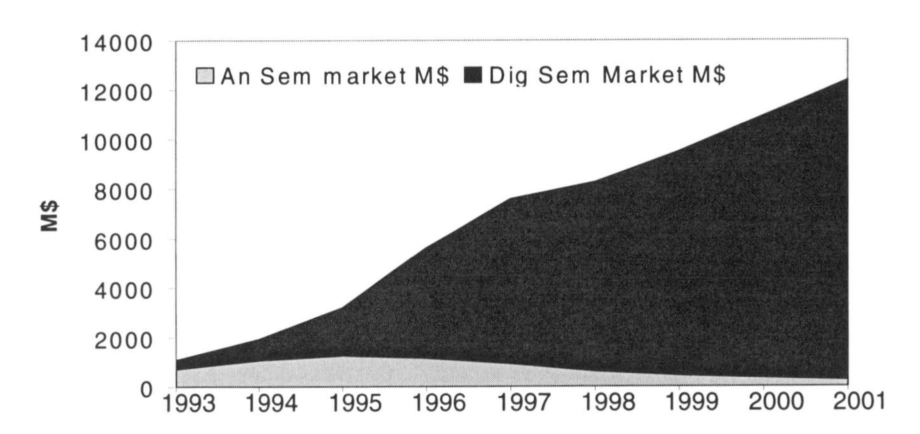

Figure 25.3 Digital and analog semiconductor market. Source: Dataquest, December 1997.

monly used. Most of these techniques are frequently found as part of wireless communication products (e.g. paging, cellular telephony, wireless data networks). This fact has motivated DSP manufacturers to propose processors specifically adapted for wireless applications.

Note that, along with the increasing performance and size reduction of DSPs, multidimensional digital signal processing techniques typically used in military applications have recently aroused considerable interest in the wireless communication sector.

25.1.2 Characteristics of digital signal processing

Mobile phone consumers are constantly expressing demands for new digital signal processing features for whose development marketing, engineering, and manufactur-

Table 25.1 Typical digital signal processing algorithms and common applications.

DSP algorithms	*DSP functions*	*Applications*
FIR filters	Speech codecs (ADPCM)	Digital audio (Car audio, PC audio, TV audio, CD) Wireless communications
IIR filters	Audio (de)compression (MPEG, AC-3)	Professional audio Wireless communications
Adaptive filters	Echo cancellation	Video conferencing Wired communications Wireless communications
Fast Fourier Transform	Noise cancellation Audio equalisation	Consumer audio Music recording Wireless communications
Viterbi (de)coders	Channel equalisation	Wireless communications
Signal (re)construction	Speech recognition Speech synthesis	Advanced interfaces Wireless communications
Array signal processing	Source separation Source finding	Radar Sonar Voice communications (wired, wireless)

ing departments must work together. It is worth mentioning that these demands essentially come from the base-band interface of the mobile phone handset. This tendency is illustrated in Figs. 25.4 and 25.5.

A new digital signal processing feature is first proposed as an "abstract idea" which the engineering department has to represent and simulate from an unambiguous mathematical model according to a specific algorithm. Up to this stage, neither real-time implementation constraints nor bit-level accuracy are considered, since the central point is to evaluate the validity and performance of the algorithm itself. After this assessment step, the DSP platform where this new feature will be operating needs to be specified. It is precisely at this stage that the gap between signal processing research and signal processing development has to be closed. In some cases, it will be necessary to realize some modification to the proposed signal processing algorithm to adapt it to the available resources of the DSP despite a decrease of its performance.

Several parameters have then to be considered to evaluate the efficiency of the DSP architecture in order to achieve the best trade-off between performance and implementation complexity. These parameters are summarised in the following sub-section.

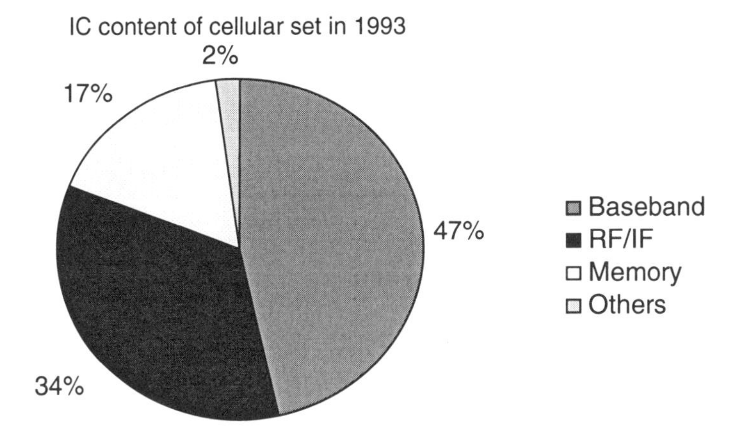

Figure 25.4 Digital cellular set value build-up. Source: Dataquest, December 1997.

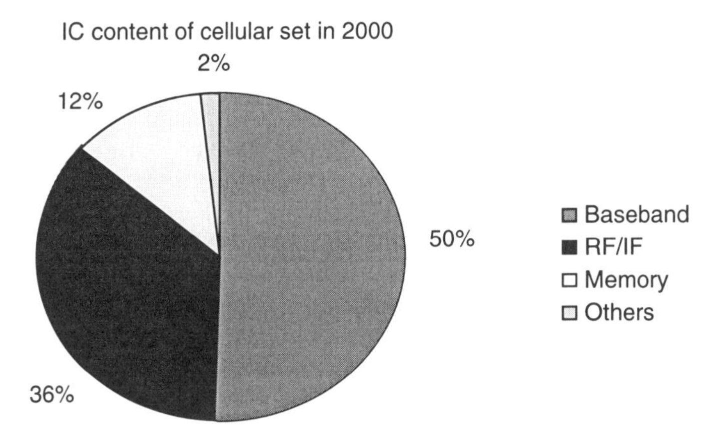

Figure 25.5 Digital cellular set value build-up (cont'd.). Source: Dataquest, December 1997.

25.1.3 Characteristics of digital signal processors

Developing DSPs with better performance at lower cost is a multifaceted challenge for manufacturers. They have to respond to concrete demands of end customers imposing some key driving factors that the DSP manufacturer has to deal with. These factors are illustrated in Fig. 25.6.

When selecting a digital signal processor for implementing a specific digital signal processing algorithm, the first aspect to be considered is the end product for which the application was conceived. For example, in the mobile phone handset sector, the "best" DSP will be the one fitting the requirements of the application at the lowest cost. With this in mind, the developer is confronted with the problem of evaluating the

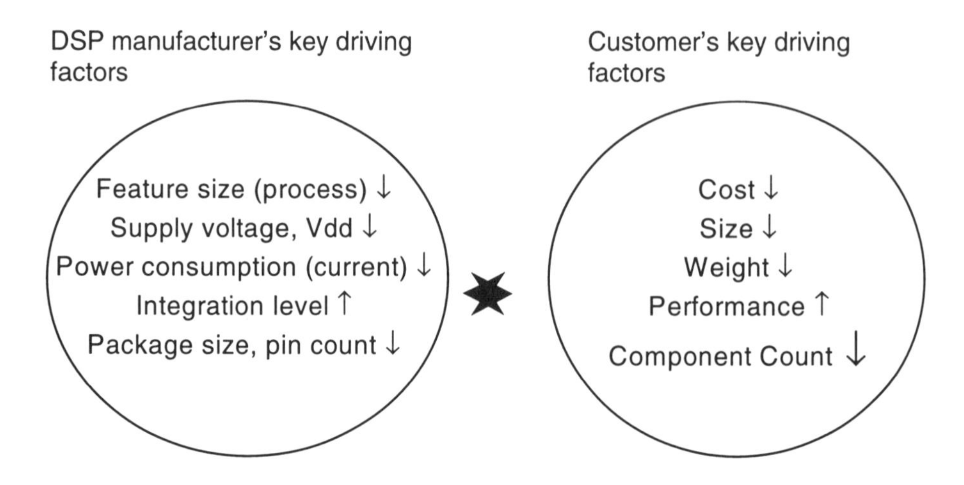

Figure 25.6 Key driving factors between DSP manufacturers and customers

impact of the performance versus cost. Table 25.2 summarises some aspects that can be considered in order to balance this trade-off.

Most of the features described in Table 25.2 are involved in the evaluation of the *execution speed* of the DSP. Commonly, this parameter is measured by the number of instructions in millions per second (MIPS) that the DSP can execute. Another important concern of most cellular telephone manufacturers is the *power consumption* of the DSP. This fact has motivated semiconductor vendors to propose DSPs operating at reduced voltages and integrating programmable clock dividers and *idle modes*. These DSP features have an impact on its final *cost*, which is an important factor for the choice of a DSP to be used in high volume consumer products. Ease of programming can also be an important aspect for the DSP-code developer, especially when algorithms are complex and the time for delivering is reduced.

Finally, the mobile phone manufacturers tend to keep some "extra" resources of the DSP in order to integrate "extra" digital signal processing functions into the end product. We can mention, for instance, the addition of easier-to-use user interfaces in the mobile phone handset (speech recognition) or by improving its audio qualities (background noise suppression, audio equalisation).

25.2 BENCHMARKS OF DIGITAL SIGNAL PROCESSING ROUTINES AND APPLICATIONS

25.2.1 Standard benchmark routines

In the previous section, we have seen the difficulties encountered in selecting the "right" DSP, and several of the considerations which have to be made. There is, however, a more general approach to comparing the performance of a DSP architecture from the execution speed point of view. It consists of a standard set of digital signal processing routines defined by Berkeley Design Technology (BDT), Inc. These rou-

Table 25.2 Features and characteristics of digital signal processors.

DSP feature	*DSP characteristic*
sum-of-products	multiply-accumulate (MAC) in hardware
parallelism	Harvard architecture
bit precision	floating-/fixed-point, fractional/integer Overflow detection/protection, saturation double-precision support
structured memory access	modulo addressing bit-reversed addressing specialised registers
regularity (inner loops)	hardware loop control
fast I/O	low-latency interrupt, DMA dedicated peripherals and system buses
instruction coding	Compacted instructions vs. VLIW encoded vs. orthogonal

tines are listed in Table 25.3. Although some of the routines presented in this table are functionally redundant, they form a basis for benchmarking more complex applications such as speech and channel coders.

BDT also publishes benchmarking results of the performance of commercial DSPs when used to implement these algorithms [1]. Obviously, it is the task of the programmer to find an algorithm that is optimal for each benchmark. According to this, the benchmark results depend not only on the processor's efficiency, but also on the skill of the developer. Nevertheless, these benchmark results give information on the performance and instruction set of the DSP.

25.2.2 DSP architectural features

In order to boost the overall performance of DSPs, several semiconductor manufacturers propose additional architectural features presented as special instructions or independent co-processing units. This is the case in speech and channel coding and decoding which represent two major digital signal processing applications in a mobile phone handset. The Application-Specific Instruction (ASI) set of the Philips R.E.A.L. DSP is a good example of this trend [2, 3]. Table 25.4 shows the number of MIPS required by the R.E.A.L. DSP 16020 to implement typical speech coding standards used in cellular telephony with the aid of ASIs.

Table 25.3 Benchmark routines commonly used in digital signal processing applications.

Benchmark routines	
real vector addition	real vector dot-product
real block FIR filter	Single-sample FIR
Complex block FIR	LMS adaptive FIR
real IIR filter	real vector maximum search
convolutional encoder	finite state machine
256 point Fast Fourier Transform	

Table 25.4 Number of MIPS required by the R.E.A.L. DSP 16020 to implement typical speech coders using ASIs. Source: Philips Semiconductors TCMC.

Speech coder	*Number of MIPS*	*Comments*
EVRC (IS-95)	19	Including noise suppression
VSELP (IS-136)	15.4	High precision LTP
ACELP (IS-136)	11–14.3	–
GSM-FR	2.5	Including DTX/VAD
GSM-HR	12–15.8	Including DTX/VAD
GSM-EFR	15.7	Including DTX/VAD

25.2.3 Evolution of features for DSP implementation

Although mobile phone terminals are still being used for transmitting speech, cellular telephony operators and manufacturers are striving to propose a "mobile multimedia" handset for the mass market. As part of this trend, most telecommunication standardisation institutions are concentrating efforts on this matter either to upgrade current standards or to specify new ones. Table 25.5 illustrates the main standardisation activities in the cellular sector for the near future. Third generation mobile terminals will have to be multi-mode, multimedia and multi-service. They will need to maintain compatibility with second generation networks. The DSP will need to be able to face processing rates beyond 1 GFlops!

Table 25.5 Emerging Cellular Features.

Cluster	*Standard*	*Standard fixed*
GSM upgrades	HSCSD (multi-slot data)	1997
	GPRS (packet data)	1997
	AMR (speech codec)	end 1998
	CTS (residential)	1999
	EDGE (modif. Modulation)	mid 1999
IS-136 upgrades	High speed data, modif. Modulation	1998
IS-95 upgrades	IS-95B (high speed data)	mid 1998
3rd Generation	W-CDMA	2000
	US (several proposals)	2000
	UMTS (Europe/worldwide)	2001

25.2.4 General-purpose DSPs versus application-specific DSPs for mobile phones

In the mobile phone market arena, some manufacturers are constrained to restrain their choice of a DSP technology to that proposed by external semiconductor vendors. This situation may induce a DSP solution not appropriate for the actual application needs of the customer. Conversely, when the mobile phone manufacturer is directly involved in the designing specifications of *his* DSP architecture, the following advantages can be observed:

- Closely coupled relationship between application, architecture, ASIC implementation, and development tools.
- No dependency on external core vendor (neither technical nor commercial).
- Attention points: In-house tool development, link to external application software libraries.
- Ability to customise the architecture to the application according to the actual needs in terms of cost and performance.

25.3 CONCLUSION

As a conclusion, we can say that the new DSPs will be continuing to integrate more new features for increasing performance and for reducing cost, size and power consumption. Third generation terminals will require new user interfaces and considerable *intelligence*. For the semiconductor manufacturers, these trends imply new technological challenges to face the constantly growing cellular market demanding not only low cost and high performance quality, but also upgradable and scalable processing platforms.

References

[1] Berkeley Design Technology, Inc., *Buyer's Guide to DSP Processors*, 1997.

[2] P. Kievits, E. Lambers, C. Moerman, and R. Woudsma, "R.E.A.L.: Philips Embedded Digital Signal Processor", Philips Research, Centre for Technical Training, Eindhoven, The Netherlands, Oct. 1997.

[3] E. Lambers, C. Moerman, P. Kievits, and R. Woudsma, "R.E.A.L. DSP: Reconfigurable Embedded DSP Architecture for Low-Power/Low-Cost Telecom Baseband Processing", *1st IEEE CAS Workshop on Wireless-Communications, Circuits & Systems*, Lucerne, Switzerland, June 1998.

VI Blind Channel Equalization

26 BLIND CHANNEL EQUALIZATION

Ruey-wen Liu

Department of Electrical Engineering
University of Notre Dame
Notre Dame, IN 46556, USA

26.1 INTRODUCTION TO THE FOLLOWING PAPERS

There has been an increasing interest in the problems and the techniques related to the so-called "blind channel identification and equalization" tasks in recent years. While channel identification and equalization have been studied extensively, their "blind" counterparts are relatively new.

A typical communication system consists of an information-bearing signal, a transmission channel, and the received signal. When the signal is known, then the system function of the channel can be identified by using the received signal. This is known as the problem of channel (system) identification. When the channel is known, then a channel equalizer can be designed in such a way that the complete system is an ideal delayer or the best estimator for the information-bearing signal. This process is known as the problem of channel equalization. Both problems of channel identification and channel equalization are well understood and well developed.

A typical communication system, however, is usually inherently blind, i.e., neither the information-bearing signal nor the channel is known. As such, the techniques of channel identification and channel equalization can not be immediately applied. The current technology relies heavily on so-called *training sequences*, which are used solely for the purpose of the identification and equalization of the channel. These training sequences are known to the receiver. With them, the channel can be identified and can then be equalized at the receiver's end with the known techniques. Almost all current communication systems are designed based on the use of training sequences. Since they do not bear information, they add to the overhead cost.

The ever-increasing demand for more speed and more kinds of functional communication systems will sooner or later push the training sequence technology to or beyond its limits. This time may arrive rather sooner than later. In wireless communication, the channel varies rapidly, and hence training sequences have to be sent repeatedly in order to track the channel. Using more training sequences decreases the information throughput, and hence increases the cost of information transmission. For example, in wireless ATM systems, the use of training sequence technology may impose substantial overheads.

As another example, in digital HDTV, digital signals are sent frame by frame. This limits how frequently the training signals can be sent, which in turn limits the performance of the HDTV in a fast-varying environment. A recent simulation study with airplane flutter [1] showed the need for blind equalization in dynamic channels that could not be handled adequately by the current standard, i.e. an equalizer training on the training sequence alone. Blind equalization can improve the performance significantly.

In other applications, the source signal may not even be accessible, such as in biomedical systems or the passive detection of signals in a multipath environment if the signal waveforms are unknown. In these cases, training-sequence technology can not be applied.

All these examples show the need for blind channel equalization, which is mandated by applications.

By blind channel identification and equalization we mean that the channel is identified and equalized without the use of training sequences. Mathematically, it becomes a problem to find an "inverse" of the channel when knowing neither the channel itself nor the input. At first glance, the problem seems to be ill-posed. It turns out that this apparently intractable problem can have elegant solutions with efficient algorithms by exploiting some qualitative and statistical information on the channel and its input signals, together with a multiple-channel-transmission technology.

For more information about the subject, the recent special issue on Blind System Identification and Estimation [2] may be a good place to start with.

This part of the book consists of three papers on current topics by prominent authors.

The first paper is on the subject of wireless multiple-access communication systems. The author, Vincent Poor, is a leader in the field of adaptive interference suppression. His work set standards in the field.

The second contribution is on wireless ATM systems. The author, Lang Tong, is a pioneer in the important field of multiple-channel technology. The application of blind methods to ATM systems is relative new, and he again is a pioneer in the field.

The third paper is on the design of a blind demodulator. The author, John Treichler, is at the forefront of the research into the constant modulus algorithm, which is the first algorithm ever used for blind equalization as well as the most popular one in industry. In his presentation, many practical issues are presented which are known only to experienced designers like him.

References

[1] M. Ghosh, "Blind decision feedback equalization for terrestrial television receivers", *Special issue on Blind Systems Identification and Estimation, Proceedings of the IEEE* , pp. 2070–2081, Oct. 1998.

[2] *Special Issue on Systems Identification and Estimation, Proceedings of the IEEE*, edited by R.w. Liu and L. Tong, Oct. 1998.

27 ADAPTIVE INTERFERENCE SUPPRESSION FOR WIRELESS MULTIPLE-ACCESS COMMUNICATION SYSTEMS*

H. Vincent Poor

Department of Electrical Engineering
Princeton University, Princeton, NJ 08544, U.S.A.

Abstract: Future wireless communication systems will involve significant adaptive signal processing capabilities to enhance the performance of high-rate transmission through limited-bandwidth, dynamic multiple-access channels. Structured interference is a dominant, performance-limiting characteristic of such channels, and this paper provides a brief overview of basic adaptive signal processing algorithms that are useful for combating this problem. We will focus primarily on adaptive linear methods for the suppression of multiple-access interference, although we will also discuss briefly the mitigation of such impairments as dispersion impulsive noise, and the exploitation of receiver diversity arising either from natural multipath or from the use of multiple antennas. Algorithms based on traditional techniques such as recursive-least-squares and on more recently developed methods involving subspace tracking will both be described.

*This work was supported by the the U. S. National Science Foundation under Grant NCR-96-13108 and by the U. S. Office of Naval Research under Grant N00014-94-1-0115.

27.1 INTRODUCTION

Systems for wireless communications are currently undergoing very rapid development due to a large number of emerging applications, such as third-generation cellular telephony, personal communications, indoor wireless, nomadic computing, wireless local subscriber loops, and peer–to–peer tactical communications. Many of these new systems will incorporate considerable signal-processing intelligence in order to provide advanced services such as multimedia transmission, see [10]. In order to make optimal use of available bandwidth for these services and to provide maximal flexibility, many such systems will operate as multiple-access systems, in which channel bandwidth is shared by many users on a random-access basis using protocols such as code-division multiple-access (CDMA) signaling. Moreover, in order to support the high data rates inherent in such services, ratios of bit rates to bandwidths will be pushed to their limits. These stringent signaling requirements are made more difficult to attain by the fact that wireless channels can be very hostile media through which to communicate. Physical impairments such as multiple-access/co-channel interference, multipath transmission, amplitude fading, impulsive noise, and dispersion due to limited bandwidth, all contribute to make it difficult to transmit data reliably and quickly through wireless channels. Moreover, the dynamism resulting from user mobility and the above-noted random-access nature of mobile channels, amplify the effects of these impairments, and make them much more difficult to ameliorate. Solutions to these difficulties lie in the use of adaptive algorithms to perform advanced signal processing functions, and this paper addresses one important issue in this context — namely, interference suppression.

In particular, we will discuss primarily the problem of suppressing multiple-access interference (MAI), known as *multiuser detection.* In this context, although adaptive nonlinear algorithms are discussed briefly, we will focus on adaptive linear methods of multiuser detection, and in particular on two techniques: adaptive minimum output energy (MOE) detection, and linear detection based on subspace tracking. Our discussion will necessarily be brief, and we refer the reader to the recent survey [3] for a general discussion of adaptive multiuser detection and, in particular, of the MOE approach, and to [14] for a detailed description of the subspace approach. Similarly, the cited literature is representative rather than exhaustive, and the reader is referred to [3], [11] and [14] for more extensive bibliographies. In addition to the basic problem of MAI mitigation, we will also review very briefly algorithm enhancements which deal with multipath, receiver antenna diversity, intersymbol interference, and impulsive noise.

27.2 SIGNAL MODEL

As a model, we consider a K-user binary communication system employing the direct-sequence, spread-spectrum CDMA (DS/CDMA) protocol. In the absence of other channel impairments, the signal received at a receiver in such a system is a superposition of K data signals embedded in noise:

$$r(t) = \sum_{k=1}^{K} x_k(t) + n(t)\,, \tag{27.1}$$

where $x_k(t)$ is the data signal of the k-th user and $n(t)$ represents additive white Gaussian noise (AWGN). The signal $x_k(t)$ is given by

$$x_k(t) = A_k \sum_{i=-\infty}^{\infty} b_k(i) s_k(t - iT - \tau_k), \qquad (27.2)$$

where T denotes the information symbol interval, and where A_k, $\{b_k(i)\}$, s_k and τ_k denote respectively the received amplitude, symbol stream, normalized signaling waveform, and delay of the k-th user. The symbols are assumed to be binary and antipodal (± 1). In the DS/CDMA format, the signaling waveforms are of the form

$$s_k(t) = \sum_{j=0}^{N-1} c_k(j) \psi(t - jT_c), \qquad 0 \leq t \leq T, \qquad (27.3)$$

where N is the processing gain; $\{c_k(j)\}_{j=0}^{N-1}$ is a signature sequence of ± 1s assigned to the k-th user, and ψ is a normalized chip waveform of duration $T_c = \frac{T}{N}$. We assume throughout that $K < N$.

The basic problem of interest is to extract the data symbols of one or more users from the received multiplex (27.1)–(27.2) when the signaling waveforms are non-orthogonal, as is typically the case in wireless CDMA applications. (It should be noted that this problem is not limited to CDMA systems, as phenomena such as multipath and co-channel interference also give rise to nonorthogonal signaling constellations. Thus, the methods discussed in this paper may also be applicable in systems using formats that are nominally orthogonal.) Optimal methods for such data extraction, assuming complete knowledge of the signaling information, are discussed in [11]. The best performance in such systems comes from the use of maximum-likelihood detection, which can be implemented via a dynamic program having $\mathcal{O}(2^K)$ time complexity per binary decision. The resulting performance improvement over conventional methods attained by using this method can be dramatic. Lower complexity versions, including iterative structures and methods that are primarily linear, can also result in significant performance gains, albeit somewhat less than for fully optimal multiuser detectors. The techniques to be considered here are primarily of the linear variety, as will be described below.

The performance gains of multiuser detection come at two costs — computational complexity and informational complexity. Reduction of the first of these costs has been the focus of considerable research, and has largely been solved through the use of linear and iterative multiuser detectors, as noted above. The second cost, i.e. informational complexity, arises because multiuser detection generally requires that the receiver make use of features of the received signal not exploited by conventional detectors, and so more information about the signaling environment is needed in multiuser detection than is needed in the conventional case. This motivates the considerable recent research on the development of adaptive techniques for multiuser detection to mitigate this problem. Adaptivity not only allows multiuser detection to be applied without additional protocol overhead with respect to that required by conventional methods, but it also allows multiuser detectors to operate in the dynamic environments found in many wireless applications, as noted in the Introduction. A broad spectrum of methodologies has been brought to bear on this problem, including adaptive linear filtering, neural networks, stochastic approximation, and the EM algorithm. Here, we will focus on adaptive linear filtering. A review of some of the nonlinear adaptive methods can be found in [6].

27.3 ADAPTIVE MOE DETECTION

Until further notice, we particularize the model of (27.1)–(27.2) to the synchronous case, in which all signals arrive at the receiver with the same delay $\tau_1 = \ldots = \tau_K = 0$ (The asynchronous case will be discussed further below.) In this case, we can reduce the continuous-time model of (27.1)–(27.2) to a discrete-time model by chip-matched filtering and chip-rate sampling. The resulting samples, in the i-th symbol interval, can be collected into an N-long vector

$$\mathbf{r}(i) = \sum_{k=1}^{K} A_k b_k(i) \mathbf{c}_k + \mathbf{n}(i), \tag{27.4}$$

where $\mathbf{c}_k$ is a vector containing the signature sequence of user k, and $\mathbf{n}(i)$ is a vector of independent and identically distributed (i.i.d.) Gaussian noise samples.

Again, the basic problem of interest is to detect the symbol streams of one or more of the users from the received signal. Due to the synchrony of transmission, in detecting symbols in the i-th signaling interval, we can restrict attention to the vector $\mathbf{r}(i)$ of samples received in that interval. *Linear multiuser detection* detects the i-th bit of the k-th user by quantizing a real-valued linear mapping of the received vector:

$$\hat{b}_k(i) = \text{sgn}\left\{\mathbf{w}_k^T \mathbf{r}(i)\right\}, \qquad \mathbf{w}_k \in \mathbb{R}^N, \tag{27.5}$$

where the weight vector $\mathbf{w}_k$ can be chosen according to one of several criteria. The most useful of such detectors are the zero-forcing or *decorrelating* detector, which chooses these weights to zero out the signals from interfering users, and the *minimum-mean-square-error* (MMSE) detector, which chooses the weights to minimize the mean-square error (MSE) between the filter output and the desired bit [4, 7]:

$$\text{MSE} = E\left\{\left|b_k(i) - \mathbf{w}_k^T \mathbf{r}(i)\right|^2\right\}. \tag{27.6}$$

Each of these approaches can achieve significant performance gains over the conventional, or matched filter, detector, which corresponds to (27.5) with $\mathbf{w}_k = \mathbf{c}_k$. This latter detector is not designed to be resistant to multiple-access interference and can become MAI limited.

Both the decorrelator and the MMSE detector can be adapted in a straightforward way through the use of training sequences. For example, the minimization of the MSE of (27.6) when the sequence $b_k(i)$ is known is a classical problem in adaptive linear filtering (e.g. [4]). However, the use of training sequences can be limiting in wireless applications (although some proposed systems involve the insertion of pilot symbols at regular intervals). Thus, it is of interest to consider techniques which can blindly adapt the weights of (27.6) without prior knowledge of the data stream of any user.

Two basic techniques for blind adaptive multiuser detection are the *minimum output energy* (MOE) approach (e.g. [2, 8, 9]) and the *subspace* approach (e.g. [13, 14]). In this section, we discuss the MOE approach; the subspace approach is discussed in the following section.

In the MOE approach, it is recognized that minimizing the output energy of a linear multiuser detector, within a projection constraint on the weights, is equivalent (for the purposes of detection) to minimizing the MSE of (27.6). In particular, on using (27.4) and assuming that the symbols from different users are uncorrelated, we can write

$$\text{MSE} = 1 - 2\mathbf{w}_k^T \mathbf{c}_k + E\left\{\left|\mathbf{w}_k^T \mathbf{r}(i)\right|^2\right\} . \tag{27.7}$$

From this expression, it is easily seen that MMSE detector weights can be found (up to a constant multiple) by minimizing the output energy

$$E\left\{\left|\mathbf{w}_k^T \mathbf{r}(i)\right|^2\right\} \tag{27.8}$$

subject to the constraint $\mathbf{w}_k^T \mathbf{c}_k = 1$.

Since the output energy can be measured directly without knowledge of the symbol streams of any user, this property allows the use of standard methods to adapt the linear detector blindly provided we know the signature sequence of the user of interest, i.e. $\mathbf{c}_k$. In particular, the MOE problem can be solved using an anchored form (to enforce the constraint) of either least-mean-squares (LMS) [2] or recursive least-squares (RLS) [8]. Of course, RLS converges more rapidly than LMS, at the expense of greater complexity ($\mathcal{O}(N^2)$ for RLS versus $\mathcal{O}(N)$ for LMS). This complexity, however, can be managed efficiently for the multiuser detection problem through the use of a systolic array implementation described in [9]. Thus, in dynamic environments, or those with narrowband interference (in which case LMS converges very slowly), RLS may be the better choice.

27.4 ADAPTIVE SUBSPACE DETECTION

An alternative to the MOE method of blind adaptive multiuser detection is provided by the use of subspace methods. These methods exploit the fact that the multiple-access interference lies in a finite-dimensional subspace spanned by the interfering signals' waveforms. Subspace estimation and tracking techniques (e.g. [16]) can be used to track this subspace, and then appropriate linear multiuser detectors can be constructed from information about this subspace. In principle, the subspace approach can outperform the MOE approach, since the former adapts in lower dimensions than the latter, thereby producing gain against the noise.

To understand this approach, consider the covariance matrix of $\mathbf{r}(i)$:

$$\mathbf{C} = \mathbf{C}_s + \sigma^2 \mathbf{I}_N , \tag{27.9}$$

where $\mathbf{C}_s = \sum_{k=1}^K A_k^2 \mathbf{c}_k \mathbf{c}_k^T$, and where $\mathbf{I}_N$ denotes the $N \times N$ identity matrix. (Here σ^2 is the noise variance.) Since this matrix consists of the sum of the rank-K matrix $\mathbf{C}_s$ and the rank-N identity, its eigendecomposition can be written in the form

$$\mathbf{C} = \mathbf{U}_s \boldsymbol{\Lambda}_s \mathbf{U}_s^T + \mathbf{U}_n \boldsymbol{\Lambda}_n \mathbf{U}_n^T , \tag{27.10}$$

where the columns of $\mathbf{U}_s$ are the K principal orthonormal eigenvectors of $\mathbf{C}$, and where the columns of $\mathbf{U}_n$ are the remaining $N-K$ orthonormal eigenvectors of $\mathbf{C}$. The $K \times K$ diagonal matrix $\mathbf{\Lambda}_s$ has the K largest eigenvalues of $\mathbf{C}$ on its diagonal, and $\mathbf{\Lambda}_n = \sigma^2 \mathbf{I}_{N-K}$. Since the columns of $\mathbf{U}_s$ form a basis for the span of $\mathbf{c}_1, \ldots, \mathbf{c}_K$, we can interpret $\mathbf{U}_s$ as defining a *signal subspace* and $\mathbf{U}_n$ as defining a *noise subspace*.

If we know the signal subspace, then the weights in (27.5) corresponding to the decorrelator and MMSE detector for user k, can be written respectively as

$$\mathbf{d}_k = \gamma_k^d \, \mathbf{U}_s \left(\mathbf{\Lambda}_s - \sigma^2 \mathbf{I}_K\right)^{-1} \mathbf{U}_s^T \, \mathbf{s}_k \tag{27.11}$$

and

$$\mathbf{m}_k = \gamma_k^m \, \mathbf{U}_s \, \mathbf{\Lambda}_s^{-1} \mathbf{U}_s^T \, \mathbf{s}_k \, , \tag{27.12}$$

where γ_k^d and γ_k^m are positive constants (which are irrelevant to detection). Thus, these linear detectors can be viewed as operating by first projecting $\mathbf{r}(i)$ onto the signal subspace: $\mathbf{y}(i) = \mathbf{U}_s^T \, \mathbf{r}(i) \in \mathbb{R}^K$; projecting $\mathbf{c}_k$ onto the signal subspace and then forming a linear multiuser detector in the signal subspace by choosing weights $\mathbf{a}_k \in \mathbb{R}^K$, such that $\hat{b}_k(i) = \operatorname{sgn}\left\{\mathbf{a}_k^T \mathbf{y}(i)\right\}$. In this framework, the decorrelating detector and the linear MMSE detector are given in terms of the weight vectors

$$\mathbf{a}_k^d = \operatorname{diag}\left\{\frac{1}{\lambda_1 - \sigma^2}, \ldots, \frac{1}{\lambda_K - \sigma^2}\right\} \mathbf{p}_k \tag{27.13}$$

and

$$\mathbf{a}_k^m = \operatorname{diag}\left\{\frac{1}{\lambda_1}, \ldots, \frac{1}{\lambda_K}\right\} \mathbf{p}_k \tag{27.14}$$

where $\mathbf{p}_k$ is the projection of $\mathbf{c}_k$ onto the signal subspace: $\mathbf{p}_k = \mathbf{U}_s^T \, \mathbf{s}_k \in \mathbb{R}^K$; and where $\lambda_1, \ldots \lambda_K$ are the diagonal elements of $\mathbf{\Lambda}_s$.

The quantities needed to specify the above detectors are, essentially, the signal subspace parameters and the signaling waveform ($\mathbf{c}_k$) of the desired user. In principle, the signal subspace can be estimated straightforwardly since $\mathbf{C}$ and its eigencomponents can be estimated directly from the stream $\{\mathbf{r}(i)\}$ of received vectors. This implies that $\mathbf{d}_k$ and $\mathbf{m}_k$ can be obtained *blindly* from $\{\mathbf{r}(i)\}$, together with knowledge of $\mathbf{c}_k$, just as in the case of the MOE detector. However, eigenvalue decomposition or singular-value decomposition are of complexity $\mathcal{O}(N^3)$, and thus are computationally prohibitive for typical system parameters. However, this problem can be solved by using one of several $\mathcal{O}(NK)$-per-update subspace *tracking* algorithms which have been developed over the past few years. Experiments with the PASTd algorithm [16], reported in [14], show that this approach is quite promising. Not only is the complexity lower than that of MOE-RLS (which requires $\mathcal{O}(N^2)$ operations per update), but the converged signal–to–interference-plus-noise ratio (SINR) is larger due to the above-noted gain resulting from adaption of K, rather than N, parameters.

27.5 ENHANCEMENTS

As noted in the Introduction, practical wireless channels are impaired by a number of significant phenomena beyond multiple-access interference. Among these are multipath, fading, dispersion, narrowband interference, and impulsive noise. Also, practical receivers may have multiple antenna elements to take advantage of spatial diversity. All of these effects can be treated jointly with MAI mitigation using enhancements to the techniques described above. Here, we will mention briefly some of these enhancements. Further details can be found in [3, 8, 12–15].

Path diversity. Adaptive subspace detectors are particularly easily modified to account for many channel deviations from the basic model of (27.4). This is true, for example, for channels exhibiting path diversity, such as arises from resolvable nondispersive multipath or from multiple receiving antennas. In these situations, subspace tracking can be applied for each path (i.e. for each multipath delay, or each antenna element) to yield per-path linear multiuser weights. The outputs of these per-path filters give rise to a vector whose covariance matrix has, as its principal eigenvector, the vector of optimal path-combining weights. This phenomenon (discussed in [14] and [15]) allows for blind joint adaptive channel identification and multiuser detection in these environments.

Dispersion. In dispersive channels, intersymbol interference (ISI) arises in addition to MAI. Such channels can be modeled by replacing the non-dispersed data signals $x_1(t), \ldots, x_K(t)$ in (27.1)–(27.2) with versions that have been dispersed by time-invariant linear filtering (corresponding to channel bandlimiting). Assuming that this filtering spans a finite number, say L, of symbol intervals, then this model gives rise to a multi-input/multi-output (MIMO) version of the synchronous channel model (27.4). (This formalism also allows for unknown, dissimilar delays.) As in the non-dispersive, synchronous case, this situation gives rise to a signal subspace structure, which can be exploited to blindly identify the MIMO channel, and then to perform combined equalization (i.e. ISI suppression) and multiuser detection. The details of this approach are described in [13].

Impulsive noise. In many radio channels, the ambient noise (represented by $\mathbf{n}(i)$ in (27.4)) is decidedly non-Gaussian due to natural and man-made impulsive noise. In such channels, linear methods such as those described above often perform poorly in terms of bit-error rate. This performance loss can be mitigated, and in fact performance can be improved over Gaussian-channel performance, if the impulsive noise is properly accounted for in detector design. One way of doing this is to consider the synchronous multiuser detection problem (in the context of (27.4)) as a regression problem of estimating $\theta_k(i) \triangleq A_k b_k(i)$ from the observation

$$\mathbf{r}(i) = \sum_{k=1}^{K} \theta_k(i)\mathbf{c}_k + \mathbf{n}(i)\,. \tag{27.15}$$

In particular, if we let $\hat{\theta}_k(i)$ be an estimate of $\theta_k(i)$, then $\hat{b}_k(i) = \text{sgn}\left\{\hat{\theta}_k(i)\right\}$ is an estimate of the symbol $b_k(i)$. By using *robust regression* in this context, we can guard against impulsive noise while retaining the advantages of multiuser detection. Moreover, this type of detector can be adapted blindly via subspace methods. These techniques are described [12].

"Long-code" systems. As a final comment, we recall that the adaptive methods described in this paper have been derived from the model (27.2) in which the signaling waveform repeats periodically at the symbol rate. In systems such as that described by the IS-95 cellular standard, this model is not accurate because the basic signaling waveform has a period much longer than a single symbol interval. Techniques for dealing with this issue have been the subject of several recent studies. These techniques primarily make use of chip-level structure that is invariant to the period of the spreading code. The reader is referred to [1] and [5] for recent approaches to adaptive signal processing in such situations.

References

[1] C.D. Frank and E. Visotsky, "Adaptive interference suppression for CDMA with long spreading codes", *Proc. 36th Allerton Conf. Commun. Contr. Comput.*, University of Illinois, Monticello, IL, September 23–25, 1998.

[2] M. L. Honig, U. Madhow, and S. Verdú, "Adaptive blind multi-user detection", *IEEE Trans. Inform. Theory*, Vol. 41, No. 4, pp. 944–960, 1995.

[3] M. L. Honig and H. V. Poor, "Adaptive Interference Suppression." Chapter 2 in [10], pp. 64–128.

[4] U. Madhow and M. L. Honig, "MMSE interference suppression for direct-sequence spread-spectrum CDMA", *IEEE Trans. Commun.*, Vol. 42, No. 12, pp. 3178–3188, 1994.

[5] U. Madhow, L. J. Zhu and L. Galup, "Differential MMSE: A new adaptive algorithm for equalization, interference suppression, and beamforming", *Proceedings of the Thirty-third Asilomar Conference on Signals, Systems and Computers*, Pacific Grove, CA, November 1–4, 1998.

[6] H. V. Poor, "Adaptivity in multiple-access communications", *Proc. 34th IEEE Conf. Decision Contr.*, New Orleans, LA, December 13–15, 1995.

[7] H. V. Poor and S. Verdú, "Probability of error in MMSE multiuser detection", *IEEE Trans. Inform. Theory*, Vol. 43, No. 3, pp. 858–871, 1997.

[8] H. V. Poor and X. Wang, "Code-aided interference suppression in DS/CDMA spread spectrum communications - Part I: Interference suppression capability", *IEEE Trans. Commun.*, Vol. 45, No. 9, pp. 1101–1111, 1997.

[9] H. V. Poor and X. Wang, "Code-aided interference suppression in DS/CDMA spread spectrum communications - PartII: Parallel blind adaptive implementation", *IEEE Trans. Commun.*, Vol. 45, No. 9, pp. 1112–1122, 1997.

[10] H. V. Poor and G. W. Wornell, *Wireless Communications: Signal Processing Perspectives.* (Prentice-Hall: Upper Saddle River, NJ, 1998)

[11] S. Verdú, *Multiuser Detection.* (Cambridge University Press: New York, NY, 1998)

[12] X. Wang and H. V. Poor, "Adaptive multiuser detection in non-Gaussian channels", *Proc. 35th Allerton Conf. Commun. Contr. Comput.*, University of Illinois, Monticello, IL, September 29–October 1, 1997, pp. 450–459. [See also: X. Wang and H. V. Poor, "Robust adaptive array for wireless communications", *IEEE J. Selected Areas in Commun.*, Vol. 16, No. 8, pp. 1352–1366, 1998.]

[13] X. Wang and H. V. Poor, "Blind equalization and multiuser detection for CDMA communications in dispersive channels", *IEEE Trans. Commun.*, Vol. 46, No. 1, pp. 91–103, 1998.

[14] X. Wang and H. V. Poor, "Blind multiuser detection: A subspace approach", *IEEE Trans. Inform. Theory*, Vol. 42, No. 2, pp. 677–690, 1998.

[15] X. Wang and H. V. Poor, "Blind adaptive multiuser detection in multipath CDMA channels based on subspace tracking", *IEEE Trans. Signal Processing*, Vol. 46, No. 11, pp. 3030–3044, 1998.

[16] B. Yang, "Projection approximation subspace tracking", *IEEE Trans. Signal Processing*, Vol. 44, No. 1, pp. 95–107, 1995.

28 CHANNEL ESTIMATION AND EQUALIZATION IN WIRELESS ATM*

Lang Tong

School of Electrical Engineering
326 Frank H.T. Rhodes Hall
Cornell University
Ithaca, NY 14853, USA

Abstract: Wireless ATM has attracted considerable research attention in recent years because of its promise of delivering multimedia traffic ubiquitously to mobile users, and its seamless connection to the (land-line) broadband ISDN. However, originally designed for time-invariant and high-quality physical links (such as fiber-optic networks), implementations of ATM in a wireless environment face several challenges. Of particular significance is the ability to provide a reliable physical layer without excessive overhead. In this paper, we consider channel estimation and equalization problems in wireless ATM. Our goal is to eliminate or greatly reduce the transmission of training signals associated with channel equalization. The proposed approach, dubbed Protocol-Aided Channel Equalization (PACE), exploits simultaneously the structure of the ATM cell, the medium access control (MAC), and information-carrying data symbols.

*This work was supported in part by the National Science Foundation under Contract NCR-9321813 CCR-9804019 and by the Office of Naval Research under Contract N00014-96-1-0895.

28.1 INTRODUCTION

Since it was proposed in 1993 as a possible solution for the next-generation multimedia wireless networks, wireless ATM (Asynchronous Transfer Mode) has attracted world-wide research interests. Like ATM itself, wireless ATM remains to be a subject of debate. To its proponents, wireless ATM offers (i) the flexibility of assigning bandwidth on demand for multimedia traffic; (ii) the compatibility with land-line ATM; (iii) the small-size cell structure suitable for implementing ARQ (automatic repeat request), and other error protection measures which are indispensable in a wireless environment. To its critics, wireless ATM suffers from, among others, high overheads associated with both the header of small-size ATM cells and the need for sending training sequences for equalization, potentially for each cell. In contrast to conventional TDMA cellular systems, where users are assigned to periodically spaced time slots, wireless ATM (and many communication networks for multimedia applications) is an asynchronous packet transmission network. By asynchronous we mean that different users transmit information asynchronously in time slots that may or may not be pre-determined. A user may be silent for an extended period of time and, when the need arises, may want to transmit at a high data rate that occupies a large portion of the available bandwidth. Furthermore, the user may also want to have a guaranteed quality of service (QoS).

Adaptive receiver design is critical in asynchronous packet networks regardless of whether wireless ATM is a viable choice for the next-generation wireless multimedia networks. The main challenge lies in the trade-off between performance, complexity and spectral efficiency. The constraints on spectral efficiency are much stronger in a multimedia asynchronous network, where not only channels change in time, but also spatially. Specifically, users may transmit at an unpredictable time, and their performance requirement may be different. In synchronous networks, for example, the base-station can load channel or equalizer coefficients from previous time slots. This becomes unreliable for asynchronous networks because the channel may have changed drastically from the last transmission. The QoS requirement in ATM also raises an interesting problem in receiver design. Real-time communication must have a short time delay, which requires the receiver to have a simple structure. In contrast, the transmission of e-mails and data files must have low error probability. In such a case, a high-performance receiver with reasonable latency may be preferred. Each mobile user can have multiple connections, e.g., a user may perform file transfer and have a phone conversation at the same time. Therefore, even if the physical channel is the same, the receivers for different connections from the same user may be different.

Most existing approaches to equalization in wireless ATM are conventional techniques that rely on the transmission of training signals. This, unfortunately, leads to sometimes unreasonably high overhead. The importance of achieving adaptive receiver design with minimum system overhead was pointed out succinctly by Ayanoglu, Eng, and Karol [2]:

> For a burst-mode modem, the time-varying channel needs to be learned often, potentially each time a packet is to be transmitted, especially if packets are coming from different sources. This generates an overhead, which results in a necessary reduction in the efficiency of the system. Keeping this overhead to its minimum is a research challenge.

In this paper, we discuss several issues in channel estimation and equalization for wireless ATM. Specifically, we investigate the possibility of eliminating the preamble at the beginning of each burst by using the so-called protocol-aided channel equalization (PACE). This is motivated by the fact that, in wireless ATM, cell headers are often transmitted and many MAC protocols [13, 17, 20] are designed, or may be modified, in such a way that headers are known *a priori* to the receiver. The key component of PACE is a blind or semi-blind channel estimation that must have fast convergence rate, possibly achieving equalization within one data cell.

28.2 WIRELESS ATM

Asynchronous Transfer Mode (ATM) is a switching and multiplexing technique for Broadband Integrated Services Digital Networks (B-ISDN). Developed for communications of multimedia traffic including data, voice, video, etc., ATM combines features of telephone networks, which provide guaranteed Quality-of-Service (QoS) once the connection is made, and features of data networks, which offer flexibility to handle a large number of users and have an efficient usage of bandwidth. Literature on ATM is extensive, and readers are referred to [26, 27]. For our purposes, only a few properties of ATM are discussed.

28.2.1 Circuit switching, packet switching, and virtual circuit packet switching

Traditional telecommunication systems such as the original telephone network are circuit-switched networks. As *connection-oriented* services, circuit switching—as illustrated in Fig. 28.1—means that (i) the signal propagates through a fixed route determined at the time when the call is set up; (ii) once the route is determined, there is a fixed allocation of the bandwidth to this particular connection. For example, if a call is requested between A and D, the network, according to its current traffic, assigns a route (e.g. ABHFD) along with a guaranteed bandwidth, say 64 Kbits/s. Before this call is completed, if another call is requested by G to I, and if links HF and DF have already been allocated to their full capacity *at the time of request*, then this new call cannot be established and will be rejected. Note that circuit-switched networks guarantee the QoS once the connection is set up.

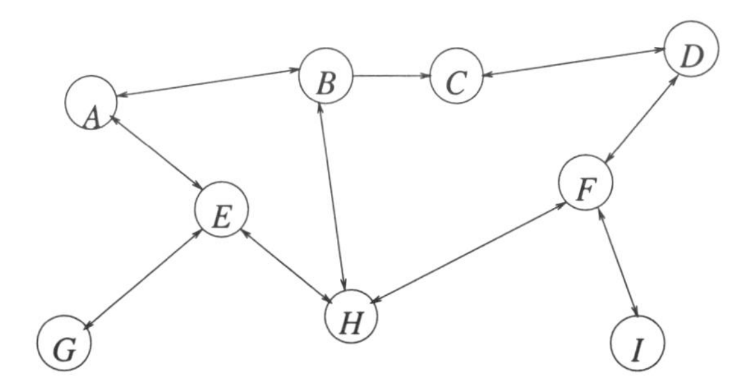

Figure 28.1 Network switching.

One disadvantage of the circuit-switched network is its lack of efficient usage of the network resources. For example, suppose that the call between A and D is set up through route ABHFD, and suppose that there is no active traffic between the two nodes (user A may simply log in on a remote computer but not issue any command), the link HF, although fully allocated to its capacity, is not actually used to the capacity, and it could have been assigned to another connection. Unfortunately, since the resource allocation is determined at the time of call-setup, this link can not be shared.

Packet switching offers *connection-less* services, and it is widely used in data communication networks. The internet is one such example. If A wants to send data to D, the data will be segmented into *packets* of fixed or variable size. These packets may propagate through different paths (AEHFD and AEHBCD), and they may arrive at D at different time due to different traffic conditions on different routes. Buffers are usually needed by the nodes along the routes, and these packets are reassembled at the destination. In contrast to circuit switching networks, packet switching networks determine the route for each packet independently according to the traffic condition of the network. In other words, network resources are allocated dynamically. For the example discussed earlier, connections AD and GI can both be accommodated by multiplexing their packets on various routes. If HF is congested, a packet from A to D can take AEHBCD.

One of the disadvantages of packet-switched networks is that the network does not guarantee that the packet will arrive on time. This is significant in multimedia applications when certain communications such as voice and video require on-time delivery. Another complication of packet switching is that each packet must have an identifier that indicates where it comes from and what its destination is. The transmission of this information imposes a tax to the network. Obviously, the longer the packet is, the lower is this tax.

As a compromise between the two, *virtual circuit packet switching* combines features of both circuit and packet switching. It is connection oriented; all signals from A to D propagate through the same route determined at the time of request. In order to obtain flexibility and efficient use of network resources, the signal is segmented into packets, and the network resource is allocated dynamically to each packet (although all packets travel through the same route). If A has no packet to send, then it will not occupy any link in the route, and all links can be allocated to other users. On the other hand, because there is a fixed route, all packets from A to D will arrive in sequence. Further, packets can be labeled by priorities. Voice-signal packets and data-file packets will be treated differently. The former require minimum latency but possibly allow higher error rates, while the latter require low error rate but allow a certain amount of delay.

ATM is virtual-circuit—often referred to as virtual-channel—packet switching, and it offers the flexibility of supporting multimedia traffic with QoS guarantee. The packet in ATM, the so-called ATM cell, has the fixed 53-byte size that includes a 5-byte header and 48 bytes of payload. The header has a 2-byte virtual channel identifier (VCI) and a one-byte virtual path identifier (VPI). Jointly, VCI and VPI specify the routing of the ATM cell. The rest of the header includes control information, the type of traffic, priority information and error control. Like packet switching, virtual circuit packet switching also imposes a tax on the network. Because the requirement of QoS, the packet size cannot be too large. Indeed, one of the reasons of choosing the

53-byte cell size is to minimize the delay involved in the packetization of voice signals. Consequently, the overhead associated with the packet information can be quite significant. Therefore, while the network is efficiently used by multiplexing different users, it is inefficiently used by transmitting not only information-carrying signals, but also signals that are not useful to the end-user.

28.2.2 Wireless ATM

The merge of communications and computing, the demand of multimedia services, and the perceived future proliferation of tether-less on-demand services such as high-definition television (HDTV) and World-Wide Web (WWW) prompted active research in wireless ATM. Is ATM suitable for wireless applications? Can a technology originally developed for bandwidth-rich applications be used in a bandwidth-constrained environment? What is the impact of noise, interference, and time-varying fading on an ATM network? Although these questions have not been answered satisfactorily, there have been, since 1992, a number of proposals and prototypes that demonstrate feasibility and advantages of wireless ATM. Examples include NEC's WATM [21], the Radio ATM by Olivetti Research Ltd [18], the BAHAMA by the Bell Laboratory [6], the WAND as part of the Advanced Communications Technologies and Services (ACTS) research program [8].

Wireless ATM is an active research field that covers not only the physical layer, but all layers of the network architecture. For our purposes, we mention here only a few aspects of wireless ATM that are closely related to physical layer design. Interested readers may consult the excellent survey articles [1, 2, 19] and the references therein.

Most wireless ATM implementations share several common features. Usually, they have a cellular structure that involves a base-station and mobile users as shown in Fig. 28.2. Through the base-station, each mobile user communicates with others inside and outside the cell. The base-stations are connected with Mobile ATM switches. In most cases, ATM cells are transmitted over the air between the base-station and mobile units.

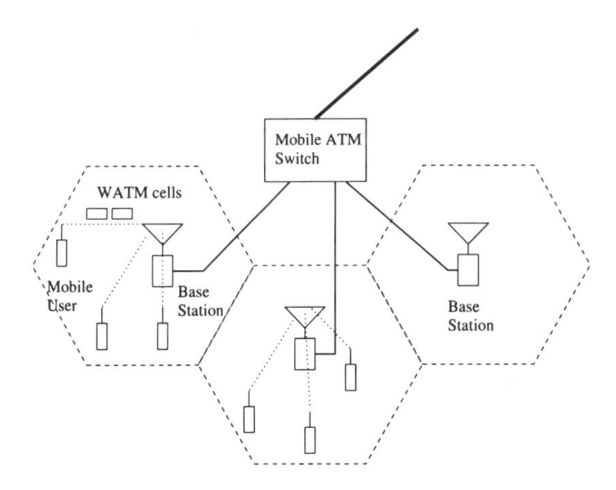

Figure 28.2 A wireless ATM architecture.

Physical layer designs of these systems are different. Roughly, they aim to provide 20–25 Mbps in either the 2.5 GHz or 5 GHz band. The bandwidth efficiency should be around 2–4 bits/s/Hz. Modulation being considered are QPSK, 16QAM, GMSK, OFDM and possibly spread-spectrum signals. The channel usually has a time spread of 200–500 μs. Time-division multiple access (TDMA) and time-division duplex (TDD) are used in many of the systems.

Directly above the physical layer is the medium-access-control (MAC) layer where the MAC protocol determines how each mobile unit accesses the base-station in the up-link and how the base-station delivers information to mobile units in the down-link. Compared to traditional ATM, this part is unique in wireless communication. In designing MAC protocols, one must consider both the QoS requirements and the efficient utilization of the available bandwidth. Again, MAC protocols are different for various implementations. However, most prototypes have reservation-based MAC protocols [4, 13, 15, 16, 20], i.e., the allocation of time slots are based on requests of mobile users.

28.3 CHANNEL ESTIMATION AND EQUALIZATION IN WIRELESS ATM

One of the most important functions of receiver design for communications over interference channels is channel estimation and equalization. Interference can be caused by other users or by the signal itself through multi-path reflection and other channel distortions. The latter is often referred to as inter-symbol interference (ISI). Since most wireless ATM implementations are based on TDMA, ISI is the primary concern, and we shall restrict our discussion to dealing with ISI.

While the need for equalization in wireless ATM is well known, few details about equalization in wireless ATM are reported in the open literature, although many prototypes suggest conventional training based equalization schemes (cf. [2, 9, 10, 17, 20]). Training-based methods rely on the transmission of so-called training signals which the receiver knows *a priori*. Based on the received signal and the known training symbols, estimators can be formed directly, or in an indirect way, by estimating first the channel parameters. These approaches have the advantage of using well-tested equalization techniques and require no interaction between physical and upper layers.

There are several drawbacks of using training-based equalization in wireless ATM. First, bandwidth efficiency may be greatly affected [2]. For example [20], a training preamble of 27 bytes may be needed for a single ATM cell of 53 bytes. Second, for reservation-based medium access control (MAC), the use of a preamble in access contention lengthens the request message. For a fixed contention period, this increases the probability of collision and decreases network efficiency. To improve the overall efficiency in bandwidth utilization, it is necessary to reduce or even eliminate the overhead associated with the training. Potentially, this can be achieved by the use of *blind* equalization methods [7, 24]. Perhaps the first application of blind equalization was suggested by Cox [5], where the constant modulus algorithm (CMA) was considered promising. Its simplicity and its close connection with the MMSE receiver make CMA an attractive solution for the mobile user. On the other hand, CMA lacks the convergence speed necessary for the equalization of up-link channels, especially for certain traffic when there are significant channel variations between the transmission of adjacent cells. For these applications, fast blind equalization techniques [12, 14, 24, 25, 29] that exploit both time and space diversity are preferred for short-data packet transmissions. Perhaps a much more interesting issue is the effect which sending training signals has on the efficiency of bandwidth utilization. When the channel is time invariant, the training signal needs to be transmitted only once and, asymptotically, there is no degradation in transmission throughput. On the other hand, when the channel varies, training signals must be sent repeatedly, which may greatly reduce the transmission throughput. Fig. 28.3 illustrates one such case with a high-frequency (HF) channel. The channel is a two-ray multi-path channel with variable-fading bandwidth under the Watterson model [28]. When the fading is time invariant, training needs to be done only once. Using the symbol error rate of 10^{-2} as the threshold, increasingly frequent transmission of training sequences is required as the fading bandwidth increases in order to maintain the symbol error rate at 10^{-2}. As shown in Fig. 28.3, if the normalized Doppler spread is below 10^{-4}, about 10 % of the transmissions are used for training in order to maintain the desired symbol error rate. Once the normalized Doppler spread exceeds 10^{-4}, the throughput deteriorates rapidly.

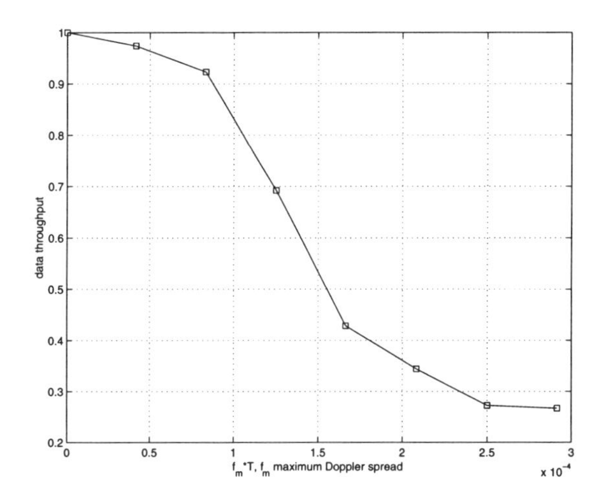

Figure 28.3 Efficiency vs. fading bandwidth for HF channels under the Watterson Model. SNR= 27 dB. Desired raw symbol error rate at 10^{-2}. MMSE block DFE is used in equalization. Symbol rate: 2400 symbol/s.

28.3.1 PACE: Protocol-aided channel equalization

The Protocol-Aided Channel Equalization is motivated by the observation that in ATM, most parts of the cell header for a fixed connection do not change from one cell to the next. For example, both VCI and VPI are fixed once the connection is set up. Other parts of the cell header change according to a predetermined pattern. For example, the ATM sequence number increases by one during regular transmission. Although MAC protocols can be used to aid channel equalization in various ways, we consider here only the use of ATM cell headers.

The main question is: Can the header be exploited for equalization? If yes, how? We examine this question in three separate steps.

Can the ATM header be used? The answer to this question depends on whether the receiver knows *a priori* the header of the particular ATM cell, and therefore largely on specific implementations of MAC protocols.

Consider a generic demand-assignment protocol in TDMA/TDD system as used in the WATMnet by NEC. The frame structure of this MAC is shown in Fig. 28.4. When a mobile unit needs to transmit data, it will send a request in the mobile-base-station control slots. Upon receiving the request, the base-station decides, according to the priority of the request, the traffic of the network, and the available bandwidth, in which time slot in some frame that follows this mobile unit can transmit. This decision is broadcasted in the base-station-mobile time slot. Because of such types of handshaking, the base-station knows, in a particular time slot in the up-link, where the signal comes from, with which communication link the signal is associated, and where the data is to be routed. In other words, most of the ATM header is known to the receiver.

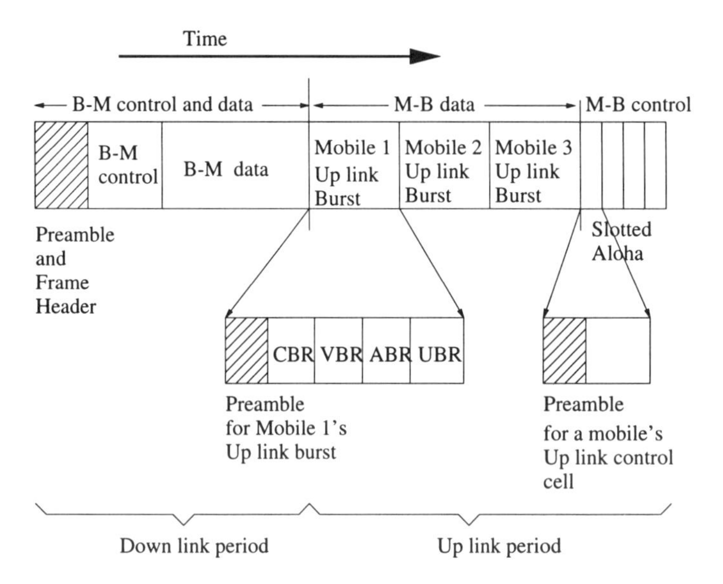

Figure 28.4 NEC WATMnet's MAC frame, B: Base station, M: Mobile users.

Is the ATM header sufficient for equalization? The answer to this question depends on the type of channel and modulation. If the channel does not have large memory, a few symbols would be sufficient. On the other hand, if the channel has considerable delay spread, and high-order modulations such as 16 QAM are used, the header itself would not be sufficient.

To formulate this problem formally, we consider the indirect approach where the channel impulse response is to be estimated first. At issue is whether the header alone is sufficient for the identification of the channel. The concept of identifiability is fundamental in estimation theory [22], and it can be described as follows: let $\mathbf{h}$ be the vector containing channel parameters, $\mathbf{x}$ be the received signal vector, and $f(\mathbf{x};\mathbf{h})$ be the likelihood function. The channel is identifiable if $f(\mathbf{x};\mathbf{h}_1) = f(\mathbf{x};\mathbf{h}_2)$ implies $\mathbf{h}_1 = \mathbf{h}_2$. In other words, identifiability means that there is no arbitrariness in the estimation of the channel parameter. In the up-link of a TDMA system, the so-called guard period is usually inserted between signals from different users so that they will not interfere with each other. In the down-link, on the other hand, the signal is broadcasted from the base-station, and no guard period is necessary. The presence of guard periods does have an effect on channel estimation, because the receiver knows that the transmitter is silent during this period. It can be shown that if the channel has order L (there are $L+1$ parameters), then the minimum length (in terms of number of symbols) for the header must satisfy

$$\text{min. header length} = \begin{cases} 2L+1 & \text{without guard time} \\ L+1 & \text{with guard time} \end{cases} \tag{28.1}$$

It is clear that using a header alone offers only limited capability in estimating the channel impulse response.

Can the payload portion of the cell be used for equalization? The answer to this question is not trivial. Intuitively, although the receiver receives only the distorted version of the transmitted signal, the received signal does contain certain information about the channel. The real issue is whether such information is sufficient for the identification of the channel and how the identification can be performed.

We consider now the estimation of multiple channels as shown in Fig. 28.5, where the data symbols are transmitted through a vector of channels. This model can come from the presence of an array of receivers, or simply from sampling the received signal at a rate higher than the symbol rate. Denoting the vector impulse response by

$$\mathbf{h}_k \triangleq [h_k^{(1)}, \cdots, h_k^{(P)}]^t , \tag{28.2}$$

we have the following system equations in the time domain

$$\mathbf{y}_k = \sum_{i=0}^{L} \mathbf{h}_i s_{k-i} + \mathbf{n}_k , \tag{28.3}$$

where $\mathbf{y}_k$ is the received (noisy) signal. Depending on the type of traffic and the mode of transmission, the channel model takes slightly different forms.

When the payload part of the cell is used for channel estimation, the corresponding symbols are unknown and we have the so-called blind channel identification problem, i.e., given $\mathbf{x}_k$, can $\mathbf{h}_k$ be identified (from the likelihood function) up to a constant factor? Note that since the data symbols s_k are unknown, the constant ambiguity is unavoidable unless the magnitude and phase of certain symbols are known. We then have the following identifiability results:

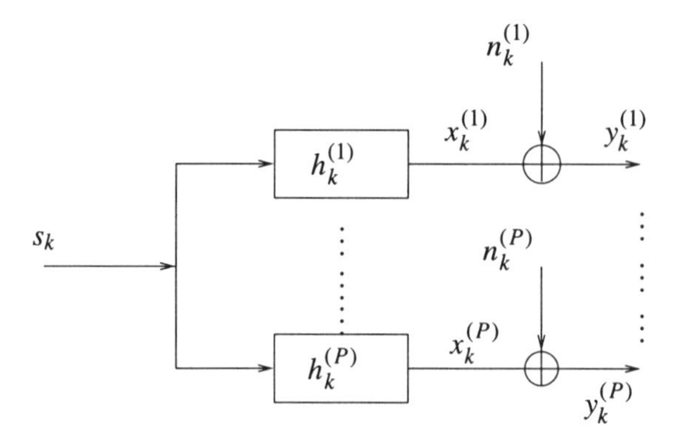

Figure 28.5 A multichannel model.

Theorem 1 *Consider the vector channel model given in* (28.3).

1. *If there is no guard period, then the channel coefficients are identifiable if and only if*
   (a) $\{h_i(z)\}$ *are co-prime;*
   (b) $\{s_k\}$ *has linear complexity greater than* $2L+1$.
2. *If* $\mathbf{h}$ *is to be identified using a single ATM cell with guard periods at the beginning and the end of the cell, then* $\mathbf{h}$ *is identifiable if and only if* $\{h_i(z)\}$ *are co-prime and* $\{s_k\} \neq \{0\}$.
3. *If multiple ATM cells, separated by guard periods, are used for identification, then* $\mathbf{h}$ *is identifiable if and only if* $\{s_n(z)\}$ *are co-prime, where* $s_n(z)$ *is the z-transform of the nth cell.*

Sketch of the Proof: The proof of 1. is given in [23]. To show identifiability, it suffices to examine the noiseless case. To show 2., let $s(z)$ be the z-transform of the input sequence. With guard periods, the observation at the ith channel can be written as

$$x_i(z) = h_i(z)s(z)\,. \tag{28.4}$$

Hence

$$\frac{x_i(z)}{x_j(z)} = \frac{h_i(z)}{h_j(z)}\,. \tag{28.5}$$

When $\{h_i(z)\}$ are co-prime, they are uniquely determined up to a constant by $\{x_i(z)\}$.

To show 3., let $s_n(z)$ be the z-transform of the input sequence for the nth cell, and $x_i^{(n)}(z)$ be the z-transform of the observation. We then have

$$x_i^{(n)}(z) = h_i(z)s_n(z)\,. \tag{28.6}$$

Using the same argument as above, replacing $h_i(z)$ by $s_n(z)$, one concludes that $s_n(z)$, hence $h_i(z)$ can be uniquely determined from $x_i(z)$ up to a scaling factor. □

Tab. 28.1 summarizes the comparison case when only the header is used in the identification and when the payload is used in the identification. Here we compute the maximum channel length for that identification is possible. It is evident that the use of the payload portion of the ATM cell greatly enhances the channel's identifiability.

Table 28.1 Maximum identifiable channel order.

		QPSK	*8-PSK*	*16QAM*	*64 QAM*
Without Guard Time	Blind Identification	53	35	26	17
	Using header only	11	7	5	3
With Guard Time	Blind Identification	any	any	any	any
	Using header only	23	15	11	7

Blind and semi-blind channel estimation for ATM. Having established the benefit of using the header as well as the payload part of the ATM cell for channel estimation and equalization, we now address the question how to exploit header and payload structures for channel estimation and equalization. This problem can be cast as a blind and semi-blind channel-estimation and equalization problem which has been an active research area in recent years [24].

We outline here one of several possible implementations of PACE algorithms illustrated in Fig. 28.6. Details can be found in [3]. This approach is a combination of (blind) channel estimation and block minimum-mean-square-error decision-feedback equalization (MMSE-DFE) [11]. Given the received data from one cell or a block of cells, an initial channel estimate $\tilde{\mathbf{h}}$ is obtained by a maximum-likelihood-based blind-channel estimation (BCE). The initial channel estimate is then used in (MMSE-DFE) to produce a tentative decision $\tilde{\mathbf{s}}_p$ of the payload. Using these detected symbols, the channel is re-estimated by a decision-directed channel estimator (DDCE) that uses standard techniques such as LMS or RLS. The improved channel estimate $\hat{\mathbf{h}}$ is used next to provide the final symbol estimate $\hat{\mathbf{s}}_p$. Obviously, additional DDCE-MMSE can be used to improve performance with the price of increased complexity.

The role of the ATM cell header is twofold. In channel estimation, cell headers and their structure can be used to improve the quality of channel estimation. Although it may be possible that a channel is estimated using only the cell header, the performance may not be satisfactory. The use of blind channel estimation is important because it blocks possible channel estimation error propagation in the conventional block detection schemes. In symbol detection, the knowledge of the ATM cell headers blocks the propagation of detection errors that severely hinder the detection performance.

28.4 CONCLUSION

It is well known that channel estimation and equalization is vital in reliable communications over interference channels, and equalization techniques for single-user

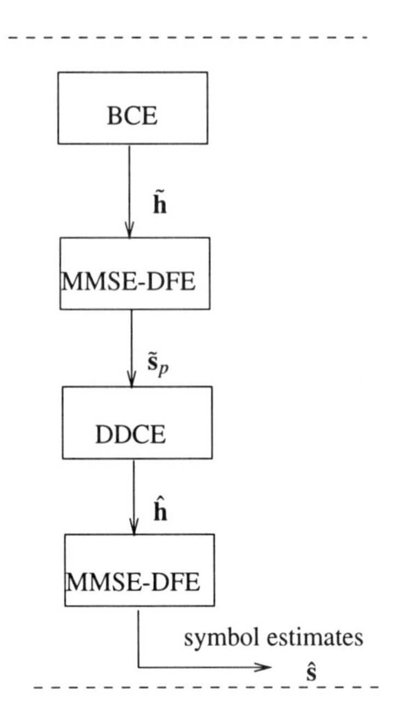

Figure 28.6 Channel estimation and equalization for ATM.

point-to-point communications are mature. However, equalization for wireless ATM or other types of asynchronous networks present new challenges and new opportunities. In this chapter, we addressed several aspects of this problem by examining the role of ATM cell structure in channel estimation and equalization. In order to exploit the ATM cell structure for equalization, the incorporation of the MAC protocol is necessary. To a great extent, how to effectively integrate the physical layer with MAC protocol designs is an open research area.

Acknowledgments

The author wishes to acknowledge and thank the help of Mr. Jeffrey Bao in conducting simulation shown in Fig. 28.3 and many helpful discussions with him.

References

[1] A. Acampora, "Wireless ATM: A Perspective on Issues and Prospects", *IEEE Personal Communications*, pp. 8–17, August 1996.

[2] E. Ayanoglu, Kai Y. Eng, and M. Karol, "Wireless ATM: Limits,Challenges,and Proposals", *IEEE Personal Communications*, pp. 18–34, August 1996.

[3] Q. Bao and L. Tong, " Protocol-Aided Channel Equalization in Wireless ATM", *submitted to IEEE J. Select. Areas. Commun.*.

[4] F. Bauchot et. al., "MASCARA, a MAC Protocol for Wireless ATM", In *Proc. ACTS Mobile Summit*, pp. 647–651, Granada, Spain, Nov. 1996.

[5] D.C. Cox, "Wireless Personal Communications: What is it?", *IEEE Personal Communications*, 2(1):22–35, Feb 1995.

[6] K. Y. Eng et. al., "A Wireless Broadband Ad-Hoc ATM Local-Area Network", *ACM/Blatzer Wireless Networks J.*, 1(2):161–174, 1995.

[7] C.R. Johnson et. al, "Blind equalization using the constant modulus criterion: A review", *IEEE Proceedings*, pp. 1927–1950, Oct 1998.

[8] Dravopoulos et. al, "The Magic WAND", Technical report, ACTS Project AC085, August 1998.

[9] D. Falconer, "A system architecture for broadband milimeter wave access to an ATM Lan", *IEEE Personal Communications*, (8):36–41, August 1996.

[10] L. Fernandes, "Developing a system concept and technologies for mobile broadband communications", *IEEE Personal Communications*, 2(1):54–59, Feb 1995.

[11] G. K. Kaleh, "Channel equalization for block transmission systems", *IEEE J. Sel. Areas Comms.*, 13(1):110–121, January 1995.

[12] Y. Hua, "Fast maximum likelihood for blind identification of multiple FIR channels", *IEEE Trans. Signal Processing*, SP-44(3):661–672, March 1996.

[13] M. Karol, Z. Liu, and K. Eng, "Distributed Queuing Request Update Multiple Access (DQRUMA) for Wireless Packet (ATM) Networks", In *Proc of International Conference on Communications*, pages 1224–1231, Seattle, USA, June 1995.

[14] E. Moulines, P. Duhamel, J.F. Cardoso, and S. Mayrargue, "Subspace-Methods for the Blind Identification of Multichannel FIR Filters", *IEEE Trans. SP*, SP-43(2):516–525, Feb. 1995.

[15] P. Narasimhan and etal, "Design and Performance of Radio Access Protocols in WATMnet, a Prototype Wireless ATM Network", In *Proc. ICUPC97 Conf.*, San Diego, CA, Oct. 1997.

[16] D. Petras, "Medium Access Control protocol for wireless, transparent ATM access", pp. 79–84, Long Island, NY, Nov 1995.

[17] J. Porter and etal, "The ORL Radio ATM system: Architecture and Implementation", Technical report, Olivetti Research Limitted, January 1996.

[18] J. Porter and A. Hopper, "An ATM based protocol for wireless LANs", Technical Report 94.2, Olivetti Research Limitted, April 1994.

[19] D. Raychaudhuri, "Wireless ATM Networks: Architecture,System Design and Prototyping", *IEEE Personal Communications*, pp. 42–49, August 1996.

[20] D. Raychaudhuri and et. al, "WATMnet: A Prototype Wireless ATM System for Multimedia Personal Communication", *IEEE J. Selected Areas in Communications*, 15(1):83–95, January 1997.

[21] D. Raychaudhuri and N.D. Wilson, "ATM-Based Transport Architecture for Multiservices Wireless Personal Communication Networks", *IEEE J. Select. Areas Commun.*, 12:1401–1414, Oct. 1994.

[22] T.J. Rothenberg, "Identification in parametric models", *Econometrica*, 39, 1971.

[23] L. Tong and J. Bao, "Equalizations in Wireless ATM", In *Proc. 1997 Allerton Conference on Communications, Control, and Computing*, pp. 64–73, Urbana, IL, Oct. 1997.

[24] L. Tong and S. Perreau, "Multichannel blind channel estimation: From subspace to maximum likelihood methods", *IEEE Proceedings*, 86(10):pp. 1951–1968, October 1998.

[25] L. Tong, G. Xu, and T. Kailath, "Blind identification and equalization based on second-order statistics: A time domain approach", *IEEE Trans. Information Theory*, 40(2):pp. 340–349, March 1994.

[26] M. Toy, *Development and Applications of ATM: Selected Readings*, IEEE Press, New York, 1996.

[27] J. Walrand and P. Varaiya, *High-Performance Communication Networks*, Morgan Kaufmann, San Francisco, 1996.

[28] C. C. Watterson, J. T. Juroshek, and W. D. Bensema, "Experimental confirmation of an HF channel model", *IEEE Trans. Commun. Technology*, 18(6):pp. 792–803, December 1970.

[29] G. Xu, H. Liu, L. Tong, and T. Kailath, "A Least-Squares Approach to Blind Channel Identification", *IEEE Trans. Signal Processing*, SP-43(12):pp. 2982–2993, December 1995.

29 BLIND SEPARATION AND COMBINATION OF HIGH-RATE DIGITAL QAM RADIO SIGNALS

John Treichler, C. R. Johnson, Jr., and S. L. Wood

Applied Signal Technology, Inc., Sunnyvale, CA 94086, USA
Cornell University, Ithaca, NY 14853, USA
Santa Clara University, Santa Clara, CA 95050, USA

Abstract: Blind signal separation is currently an active research topic, and many techniques for accomplishing it have been discussed in the literature. In his excellent survey article [4], Cardoso describes blind signal separation (BSS) as the recovery of "unobserved signals of 'sources' from several observed mixtures". This paper describes a particular implementation, a recently constructed two-channel processor capable of operating on convolutive mixtures of QAM signals running at symbol rates as fast as 40 MHz. This processor has been applied to several practical problems, including the cancellation of cross-pole interference from a digital radio signal of interest and the diversity combination of signals received from two apertures or antenna feeds.

29.1 INTRODUCTION

The need to transmit growing amounts of information through channels of fixed bandwidth has encouraged the development of spectrally efficient transmission techniques. In digital microwave radio systems, for example, spectral efficiency can be attained by transmitting multiple bits per symbol using quadrature amplitude modulation (QAM), by transmitting simultaneously on orthogonal signal polarizations at the same nominal carrier frequency, or both. When doing both, it is commonly found that even though the signals are transmitted on orthogonal polarizations, refraction and reflection encountered during the signal's propagation to the receiving antennas tends to additively mix the two. Thus, each of the two signals is received in the presence of interference from the other. While some interference can be tolerated, particularly when the modulations employed are simple (such as QPSK), this interference can badly degrade the accurate reception and demodulation of high-order QAM signals. This practical problem is an example of a class of theoretical problems which have come to be known as "blind signal separation (BSS)" [4]. We shall first treat the problem at hand and then determine how it fits into the more general BSS construct.

A number of papers and reports were published in the 1980s, for example [1–3, 6], which addressed solutions to the problem of cross-pole interference. While technically accurate, the recommendations made in those papers could not be applied to the design of digital microwave radio demodulators at the time, because the required high-speed semiconductor technology was not yet available. Its recent arrival, however, has permitted the construction of such a demodulator. This unit, described in more detail in [7], accepts two intermediate frequency (IF) inputs and produces an output consisting of demodulated data symbols. The demodulator is capable of handling QAM signals with constellations of up to 128 at symbol rates of greater than 40 MHz. The demodulator is capable of blind acquisition, implying that it can initialize all of its control loops, including the channel equalizers and cross-pole interference cancellers, without the need for explicit or embedded training signals from the transmitter.

Section 29.2 describes the cross-pole interference problem in mathematical terms and states the analytical solution. There it is shown that two other practical problems can be viewed as subsets of the "cross-pole" problem and that all fall under the rubric of blind signal separation. Section 29.3 briefly describes how a demodulator was built to implement this solution. Some experimental results are shown in Section 29.4.

29.2 THREE RELATED INTERFERENCE PROBLEMS

Fig. 29.1 shows a simplified view of a microwave communications system employing frequency reuse. Two separate input data streams are applied to two modulators operating at exactly or nominally the same center frequency. The outputs of the two modulators are then applied to the orthogonally polarized feeds of the same antenna. The receiving facility uses a similar dual-polarized antenna and is oriented so that the respective feeds at the transmitter and receiver are spatially aligned. In the absence of impairments such as multipath propagation, the signals recovered at the two orthogonal receiver feeds are essentially those transmitted. Even though they share the same bandwidth, their orthogonal polarizations allow them to be demodulated separately and without regard for the other. In fact, however, the two signals do contaminate each other. This "cross-coupling" typically stems from one or two factors, mechanical imperfections or misalignments of the antennas and reflection and scattering encountered in the propagation from the transmitter to the receiver. The combination of these factors can be well modeled using the block diagram shown in Fig. 29.2. Each receiver input is the sum of causally and possibly time-varyingly filtered versions of each of the transmitted signals. In a well designed transmission system the gains of the two "direct" paths are much higher than those of the "cross-pole" paths, and therefore mitigation of this coupling is not necessary, even though equalization of the multipath-induced distortion in each direct path is. Many practical situations do arise, however, in which the cross-pole terms are commensurate or even larger than the direct paths. In these cases, processing of some sort must be performed to permit accurate demodulation of the transmitted signals.

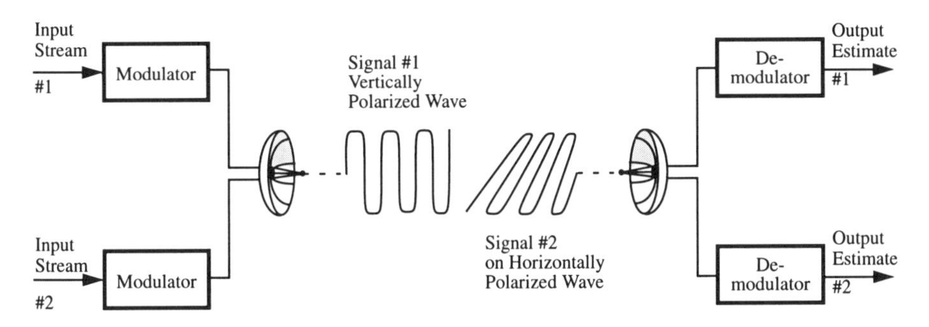

Figure 29.1 A typical configuration for frequency reuse in digital microwave radios using transmission on orthogonal polarizations.

Estimation of either of the two transmitted signals in such a situation can be addressed by posing it as a multichannel Wiener filtering problem. Given the two received signals, what linear combination of them produces the least-mean-squared difference between it and a suitably delayed version of the transmitter input? Fig. 29.3 repeats the model of the cross-pole channel and shows the two filters whose impulse responses must be chosen. If the desired transmitted signal is available to the receiver, or if the demodulator's output decisions are already accurate most of the time, the choice of filter pulse responses can be determined using any number of techniques for minimizing the mean or average squared error. If these conditions are not satisfied, then blind adaptation rules must be applied.

There are two other problems commonly encountered in the reception of digital signals, so-called "flat fading" and interference from narrowband signals. It is sometimes practical to solve these problems in the same way, that is, by providing the receiving processor with two versions of the transmitted signal. Even though the solution and its implementation are the same, the techniques for solving these problems go by different names. The problem of flat fading is usually addressed with "diversity combination", the artful summation of two or more received versions of the same transmitted signal. These different versions are usually attained by using multiple antennas or sometimes by using orthogonal feeds on the same antennas ("polarization diversity"). If the flat fading characteristics encountered are statistically independent of each other, then the quality of the summed signal will be as good as (and often better than) any of the individual received versions. Historically, the circuitry employed for diversity combination was capable of adjusting the gain and sometimes the phase of the diversity inputs, thus limiting the applicability of diversity combination to those cases in which the inputs did not encounter different propagation delays or dispersion. Consider now Fig. 29.4, a model for the signals received in a two-channel diversity receiving system. In principle, the two signals are received in the presence of different delay and

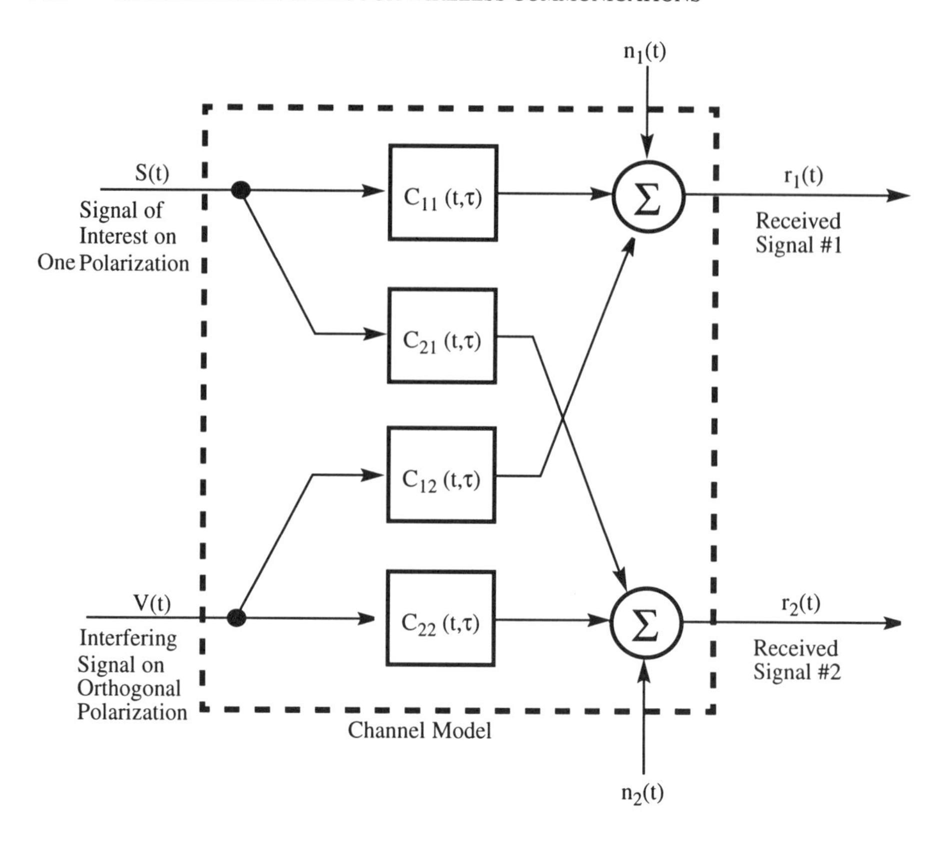

Figure 29.2 A propagation channel model including cross-coupling of the signals transmitted on orthogonal polarizations.

dispersion, as modeled by the filters C_{11} and C_{21}. The optimal Wiener receiver in this case is shown on the right side of Fig. 29.4. It consists of two properly chosen filters, whose outputs are summed to produce the input to the demodulator. A comparison of Fig. 29.3 and 29.4 illustrates the fact that the diversity combination problem can be viewed as a subset of the cross-pole problem. Even though there is no interference present, the optimal receiver is still a two-channel processor with the filter pulse responses chosen to maximize SNR (or any other appropriate quality measure).

The third related problem is commonly called "sidelobe cancelling" and is illustrated in Fig. 29.5. The signal of interest is received in the presence of an interfering signal, often because the receiving antenna is not spatially selective enough to reject it. Also shown in Fig. 29.5 is a technique commonly employed to combat this problem, adaptive subtraction of the interferer using a signal obtained by an antenna aimed as directly as possible at the interferer. As in the diversity combination problem, the use of this "sidelobe cancellation" has traditionally presumed that the versions of interference present at the differencing node have been subjected to the same dispersion and

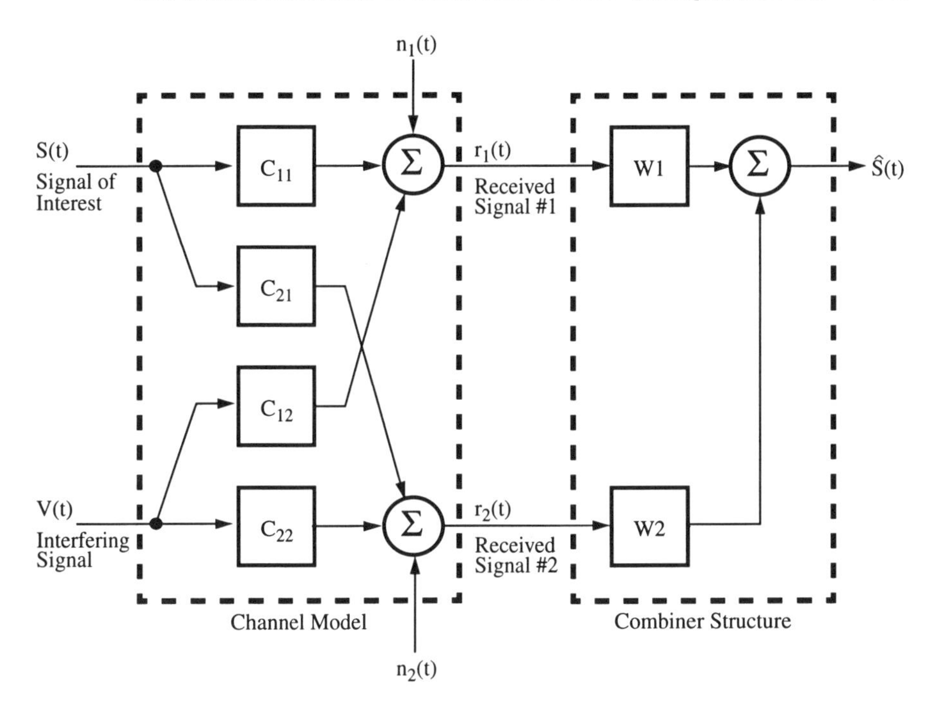

Figure 29.3 The Wiener filter solution for the case of cross-coupled channels.

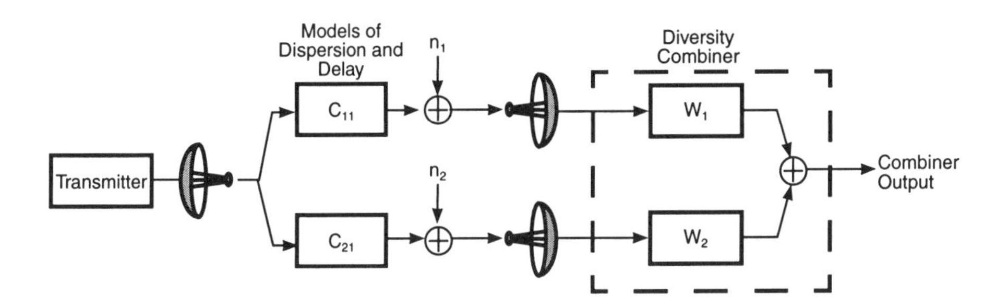

Figure 29.4 The Wiener filter solution for the diversity combination of differentially delayed or dispersed signals.

delay. Since this is rarely true, the utility of this approach has been limited to the case of interference which is very narrowband compared to the signal of interest.

If we compare Fig. 29.3 to Fig. 29.5 we see that the sidelobe canceller can be viewed once again as a special case of the cross-pole canceller. Furthermore we see that the dual-channel receiver architecture shown in Fig. 29.3 would permit delay and dispersion compensation between the two received versions of the interference, thus

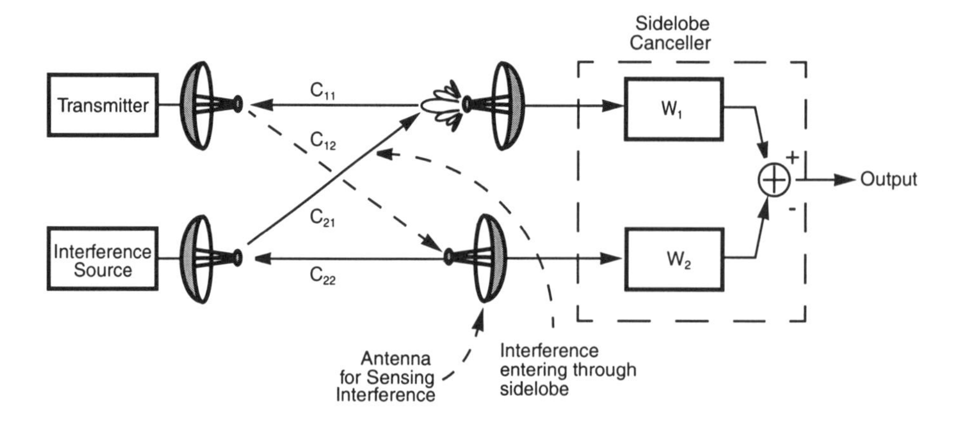

Figure 29.5 The Wiener filter solution for the cancellation of an interferer received in the sidelobe of an antenna.

making the system capable of operating successfully against wideband interferers. Thus we conclude that with the proper selection of filters W_1 and W_2 in Fig. 29.3, an adaptive combiner could address all three of these practical problems, and, in fact, extend the performance usually attainable with diversity combining and sidelobe cancelling systems, because of its ability to deal with differential delay and dispersion.

29.3 A SOLUTION

The problem of blind cross-pole combination was addressed in [3] in 1985 for frequency-modulated (FM) and QPSK signals. It employed the general architecture shown in Fig. 29.6 and used the Constant Modulus Algorithm (CMA) to adapt the filters. Subsequent work, reported in [5], has shown that CMA can be used for "one-channel" equalization of higher-order QAM. In fact, as illustrated in this paper, CMA can also be used to direct the blind cross-pole combining of high-order QAM, and can therefore be viewed as one method of solving the blind signal separation problem [4]. It is interesting to note that the digital implementation implied by Fig. 29.6 was not feasible at that time and only recently has become so for signals of the bandwidths typically employed in digital microwave radios.

Fig. 29.7 shows the block diagram of a cross-pole canceller built from a pair of demodulators which have been linked together to form a single output. This output is applied to an adaptive law which is used to blindly and jointly adapt the two filters. Once adequately adapted, decisions based on the output signal itself can be considered accurate enough to be used as the reference signal for conventional dual-channel LMS adaptation. The general similarity of the architecture shown in Fig. 29.3 and 29.6 and that seen in Fig. 29.7 are evident. The design seen in Fig. 29.7 reveals more detail, however, including the design of the individual demodulators and the linkage between them needed to guarantee that both operate at exactly the same symbol rate. Factors considered in the architecture of the individual demodulators are discussed in [5],

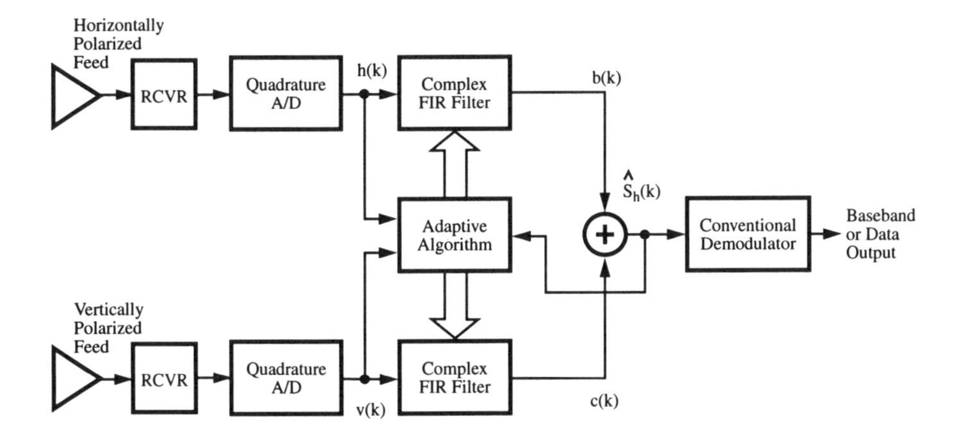

Figure 29.6 A dual-channel processor for recovering one of the two transmitted QAM signals (drawn from [3], circa 1985).

while additional factors associated with dual-channel implementation are discussed in [7].

29.4 PERFORMANCE

The performance of this dual-channel processor can be evaluated in many ways. Here we choose two. Fig. 29.8 shows the output constellation in steady state. The signal of interest has a 32-QAM constellation and is operating somewhat in excess of 31 megasymbols per second. The signal is received in the presence of additive noise 24 dB down from the signal's carrier. Fig. 29.8a shows the convergent constellation in the absence of cross-pole interference. Fig. 29.8b shows the effect of receiving the cross-polarized interference of exactly the same modulation type at a level of 18 dB below this signal's carrier. The cluster variance (CV), a measure of the effective signal–to–nonsignal ratio at the demodulator's decision circuit, degrades to about 17.5 dB with the introduction of this cross-pole interference. Once running, the two-channel combiner produces the constellation shown in Fig. 29.8c. A visual comparison of the constellations shown in Fig. 29.8a and Fig. 29.8c shows that the degradation is due to the additive noise, not the cross-polarized interferer. The cluster variance at convergence is 23.4 dB, only slightly worse than the signal–to–additive noise ratio at the receiver's input.

A more careful evaluation of the cross-pole demodulator's performance can be obtained by examining Fig. 29.9. In this figure the symbol error rate at the demodulator's output is plotted as a function of the cross-pole interferer's power, both with the adaptive canceller engaged and without. The favorable impact of the adaptive canceller is clearly demonstrated.

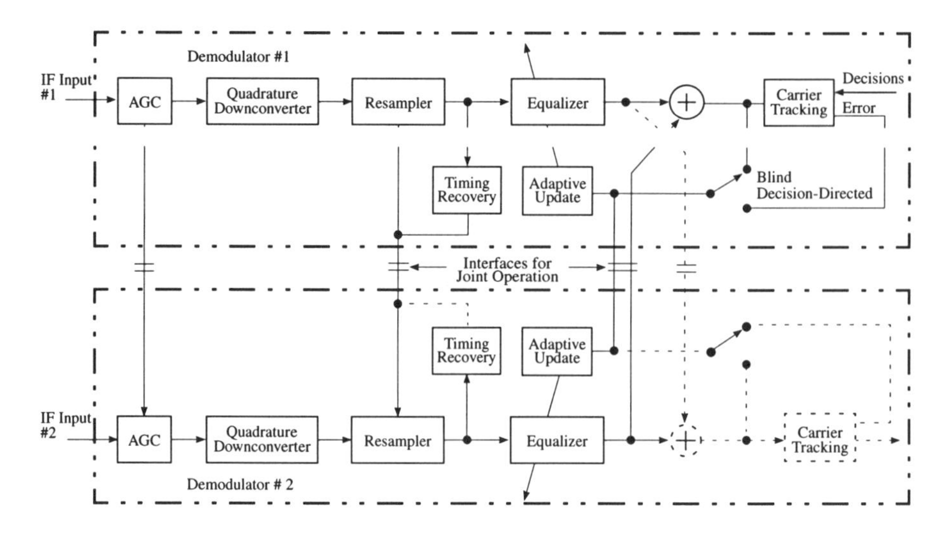

Figure 29.7 Block diagram of a cross-pole canceller composed of two digitally implemented blind QAM demodulators.

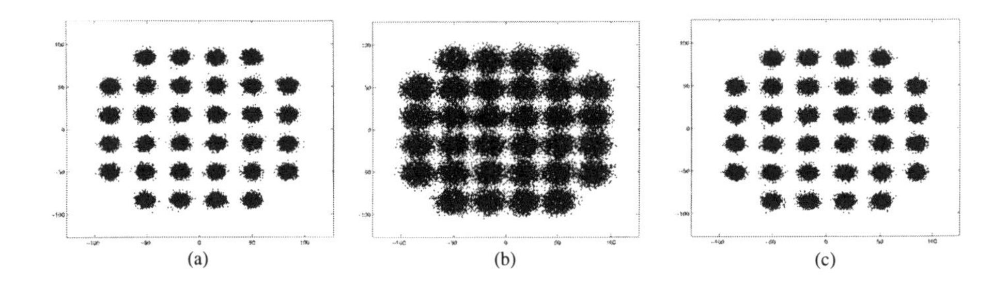

Figure 29.8 Comparison of signal constellations: (a) without interference but with additive noise 24 dB below the signal carrier; (b) with 32-QAM interference at a level of 18 dB below the signal carrier; and (c) after successful operation of the adaptive cross-pole canceller.

29.5 CONCLUSIONS

This paper has described the theoretical framework linking the problems of cross-pole interference cancellation, dual-channel diversity combination, and sidelobe cancellation, and has shown that all three can be optimally addressed using a two-channel adaptive combiner.

This optimal architecture was used as the basis for the hardware discussed in Section 29.3. This processor accepts two inputs with bandwidths of up to 50 MHz and uses the Constant Modulus Algorithm [8] to give it the ability to acquire the signal of interest in the absence of training procedures initiated by the transmitter. The results

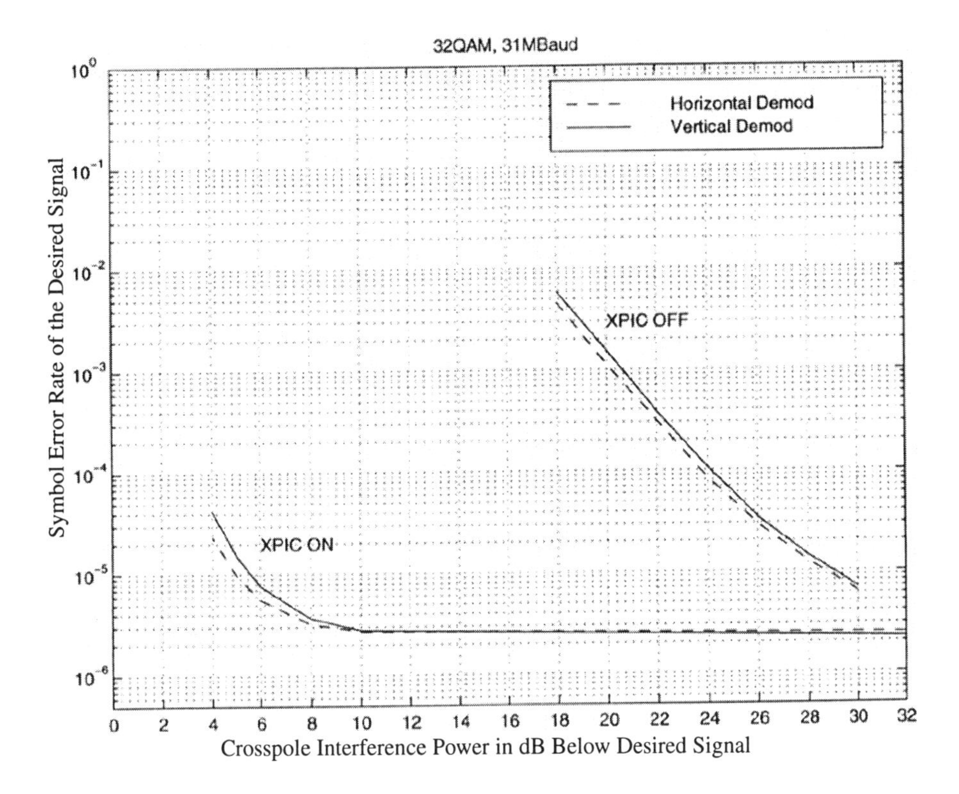

Figure 29.9 Comparison of symbol error rate performance with and without cross-pole canceller in the presence of a cross-pole interferer ranging in power from 0 to 40 dB below the power level of the desired signal.

presented in Section 29.4 show that this implementation is in fact capable of blind signal separation (BSS) in real time for wideband signals of significant complexity.

Acknowledgments

Many engineers have worked hard to design the hardware described in this paper and then to make it work. The study work leading to this architecture was done in the late 1980s [6] with the actual construction and testing being done over the past two years by a team led by Jon Bohanon and Tom Giovannini [7].

References

[1] J. Namiki and S. Takahara, "Adaptive receiver for cross-polarized digital transmission", in *Proc. ICC '81*, June 1982, pp. 46.31–46.35.

[2] J.R. Treichler and M.G. Larimore, "New Processing Techniques Based on the Constant Modulus Adaptive Algorithm", *IEEE Transactions on Acoustics, Speech,*

and Signal Processing, Vol. ASSP-33, No. 2, April 1985, pp. 420–431.

[3] J. Salz, "Data Transmission over Cross-coupled Linear Channels", *AT&T Technical Journal*, Vol. 64, No. 6, July-August 1985, pp. 1147–1160.

[4] J.F. Cardoso, "Blind Signal Separation: Statistical Principles", *Special Edition of the Proceedings of the IEEE on Blind System Identification and Estimation*, Vol. 86, No. 10, October 1998, pp. 2009–2025.

[5] J.R. Treichler, M.G. Larimore, and J.C. Harp, "Practical Blind Demodulators for High-Order QAM Signals", *Special Edition of the Proceedings of the IEEE on Blind System Identification and Estimation*, Vol. 86, No. 10, October 1998, pp. 1907–1926.

[6] "Diversity Study Final Report", *Applied Signal Technology Final Report FR-023-88*, 29 September 1988.

[7] J.R. Treichler and Jon Bohanon, "Blind Demodulation of High-order QAM Signals in the Presence of Cross-pole Interference", *Proceedings of the 1998 MILCOM Conference*, Monterey, CA, June 3, 1998.

[8] J.R. Treichler and B.G. Agee, "A New Approach to the Multipath Correction of Constant Modulus Signals", *IEEE Trans. On ASSP*, Vol. ASSP-31, No. 2, April 1983, pp. 459–471.

30 GLOSSARY

A/D Analog to digital

AC-3 Audio coding standard for surround applications, version 3

ACELP Algebraic-code-excited linear prediction

ACP Adjacent channel power

ACR Absolute category rating

ACTS Advanced Communications Technologies and Services. European project. Information on the ACTS project WAND can be found on the internet at `http://www.tik.ee.ethz.ch/~wand`

ACU Address calculation unit

AC Alternate current

AD6640 Analog Devices' AD6640 (12 b, 65 MHz) analog-to-digital converter

ADC Analog-to-digital converter

AGC Automatic gain control

ALT Alternate channel power

ALU Arithmetic-logic unit

AM Amplitude modulation

AMPS The AMPS family of wireless standards were intended to be just another analog-radio-telephone standard (e.g. Advanced Mobile Phone Service followed IMTS: Improved Mobile Telephone Service).

AMR Adaptive modulation rate

ANSI-C Standardized C programming language

APLAC An object-oriented analog circuit simulator and design tools

ARQ Automatic repeat request

ASIC Application-specific integrated circuit

ASI Application-specific instructions

ASSP Application-specific signal processor

ATI Assignment-time instantiation

ATM Asynchronous transfer mode

AWGN Additive white Gaussian noise

AXU Application-specific eXecution Unit

BAHAMA Broadband Ad Hoc ATM Anywhere, a wireless broadband ad-hoc ATM local-area network

BB Base band

BDT Berkeley Design Technology (BDT), Inc.

BER Bit error rate

BJT Bipolar junction transistor

BPF Bandpass filter

BRIGHT Family of RF chipsets for GSM, BiCMOS Radio IC for GSM by Hitachi and TTPCom

BSS Blind signal separation

BTS Base transceiver station

BW Bandwidth

CBUF Circular buffer

CCI Co-channel interference

CD Compact disk

CDMA Code division multiple access

CLC952 National Semiconductor's CLC952 (12 b, 41 MHz) analog-to-digital converter

CMA Constant modulus algorithm

CMOS Complementary metal oxide semiconductor

CORDIC Coordinate rotation digital computer

COSSAP Synopsys' Digital signal processing design (DSP) tool suite to create, explore, and test algorithms, architectures, and implementations for DSP applications.

CP Compression point

CTS Cordless Telephony System

CV Cluster variance

DAC Digital-to-analog converter

dB Decibel

DC Direct current

DCS 1800 Digital communications system (GSM at 1800 MHz). The DCS 1800 and PCS 1900 systems are two adaptations of the GSM standard.

DCU Data computation unit

DDCE Decision-directed channel estimator

DDS Decision-directed signal

DD Decision directed

DECT Digital Enhanced Cordless Telecommunications standard maintaining a high speech and reception quality.

DFE Decision-feedback equalization

DF Decision feedback

DMA Direct memory access

DMI Direct matrix inversion

DMP Dual-modulus prescaler

DNL Differential nonlinearity

DOA Direction of arrival

DQPSK Differential quadrature phase shift keying

DRAM Dynamic random access memory

DR Dynamic range

DS-CDMA Direct-sequence code-division multiple access

DSP Digital signal processing (or processor)

DTI Declaration-time instantiation

DTX Discontinuous transmission

DVGA Digitally controlled variable-gain amplifier

EDGE Enhanced data rates for global evolution

EM Evolutionary methods

EPICS DSP architecture, by Philips

ESD Electrostatic discharge

ETSI European Telecommunications Standards Institute

EVRC Enhanced variable rate-coder

FDD Frequency division duplexing

FDMA Frequency-division multiple access

FET Field-effect transistor

FFT Fast Fourier transform

FH-QPSK Frequency-hopping QPSK

FH Frequency hopping

FIFO First in first out

FIR Finite impulse response

FM Frequency modulation

FRIDGE Fixed-point pRogramIng DesiGn Environment, a tool suite that permits a design flow from an ANSI-C floating-point algorithm to a fixed-point description.

FSK Frequency shift keying

GMSK Gaussian minimum shift keying

GPRS General Packet Radio Service

GPS Global Positioning System

GSM Global System for Mobile Communications

GSM-EFR GSM with Enhanced Full-Rate codec

GSM-FR GSM with Full-Rate codec

GSM-HR GSM with Half-Rate codec

GSM1800 GSM in the 1800 MHz band

GSM900 GSM in the 900 MHz band

HBT Heterojunction bipolar transistor

HDTV High-definition television

HDL High description language

HEIBiC High-energy implanted BiCMOS

HEMT High-electron-mobility transistor

HF High frequency

HSCSD High-Speed Circuit Switched Data

HW Hardware

IB Bias current

IC Integrated circuit, collector current

IEEE The Institute of Electrical and Electronics Engineers

IF Intermediate frequency

IIP3 Input 3rd order intercept point

IIR Infinite impulse response

IM3 3rd order intermodulation

IMFDR3 3rd order intermodulation free dynamic range

IMT-2000 International mobile telecommunications 2000

INL Integral non-linearity

IO Input/output

IPR Intellectual property rights

IP3 3rd order intercept point

IRSSADC Image-rejection sub-sampling ADC

IS (Interim Standard): A designation of the American National Standards Institute—usually followed by a number—that refers to an accepted industry protocol; e.g. IS-95, IS-136, IS-54.

IS-54 A dual-mode standard for U.S. digital cellular system and Advance Mobile Phone Service based on time-division multiple access.

IS-95 The standard for code & vision multiple access.

IS-136 The latest generation of the digital standard time division

ISCAS International Symposium on Circuits and Systems

ISDN Integrated services digital network

ISI Intersymbol interference

ISM Integrated system management

ISS Institute for Integrated Systems in Signal Processing

ISSCC International Solid-State Circuits Conference

JSSC Journal of Solid-State Circuits

JTAG Standard instructions for performing functional and interconnect tests of DSP chips on circuit boards.

JTIDS Joint tactical information distribution system

KISS A one-chip ASIC DSP solution for GSM

LAN Local area network. A group of client computers connected to a server.

lbp Location of the binary point

LCMV Linearly constrained minimum variance

LC L for inductor, C for capacitor

LDMOS Laterally diffused metal oxide semicoductor

LMS Least mean square

LNA Low-noise amplifier

LO Local oscillator

LPCVD Low-pressure chemical vapor deposition

LQFP-48 Low-profile quad flat pack, 48-pin

LSB Least significant bit

LSE Least-squares estimation

LTP Long-term prediction

MAC Multiplier-accumulator

MAI Multiple-access interference

MBE Molecular-beam epitaxy

MBiC Modular BiCMOS

MCM Multi-chip module

MDAC Multiplying digital-to-analog converter

MEMS Micro-electro-mechanical systems

MESFET Metal-semiconductor FET

MF Matched filter

MIMO Multiple input multiple output

MIM Metal-insulator-metal

MIPS Million instructions per second

MLSE Maximum-likelihood sequence estimation

MMSE Minimum mean-square error

MOE Minimum output energy

MOSFET Metal-oxide-semiconductor field-effect transistor

MPEG Moving Picture Experts Group. The international consortium of hardware, software and publishing interests dedicated to developing international standards for the compression of moving video images in digital systems.

MPI Maximum-precision interpolation

MSB Most significant bit

MSE Mean squared error

MSK Minimum shift keying

MS Mobile station

MVDR Minimum-variance distortionless response

NBI Narrowband interference

NB Narrowband

NF Noise figure

NMOS *n*-type metal oxide semiconductor

NPN *n*-type bipolar transistor

OFDM Orthogonal frequency division multiplex

OPLL Offset phase-locked loop

PA Power amplifier

PACE Protocol-aided channel equalization

PAE Power added efficiency

PCB Printed circuit board

PCS1900 Personal Communications Service. A two-way, 1900 MHz digital communications service.

PCU Program control unit

PC Personal computer

PDA Personal Digital Appliance

PDC Personal Digital Cellular. PDC is one of the world's three main digital wireless standards. Currently used in Japan.

PDF Probability density function

PFD Phase frequency detector

PHEMT *p*-type high-electron-mobility transistor

PLL Phase-locked loop

PMOS *p*-type metal oxide semiconductor

PM Phase modulation

PSD Power spectral density

QAM Quadrature amplitude modulation

QFP Quad flat pack

QPSK Quadrature phase shift keying

QRD QR (orthogonal-triangular) decomposition algorithm

RB Temporal-reference beamforming (or beamformer)

RD1602x DSP family by Philips Semiconductors

R.E.A.L. Reconfigurable DSP archictecture by Philips Semiconductors

RF Radio frequency

RLAN Radio local area network

RLS Recursive least squares

RMS Root mean square

ROM Read-only memory

RTA Rapid thermal anneal

RTE Rapid thermal epitaxy

RTL Resistor-transistor logic

RX Receiver

R&D Research and development

SAW Surface acoustic wave

SBM Single-balanced mixer

SB Sideband

SC Switched-capacitor

SEM Scanning electron microscope

SFDR Spurious-free dynamic range

SIMS Secondary ion mass spectrometry

SINR Signal-to-interference-plus-noise ratio

SMD Surface-mounted device

SMI Sample matrix inversion

SNR Signal-to-noise ratio

SOI Signal of interest

SOS Silicon-on-sapphire

SPICE Simulation program with integrated circuit emphasis

SQR Signal-to-quantization-noise ratio

SSB Single sideband
SSOP Shrink small outline package (package type)
SW Software
TACS Total-Access Telecommunication System
TCH/FS Full-rate speech traffic channel
TCH/HS Half-rate speech traffic channel
TDD Time division duplexing
TDL Tapped delay line
TDMA Time-division multiple access
TEM Transmission electron microscope
TI Texas Instruments
TQ9222 TriQuint Semiconductor's TQ9222 datasheet
TV Television
TX Transmitter
UHF Ultra-high frequency
ULA Uniform linear array
UMTS Universal mobile telecommunication system
UPC Universal Personal Communications
US United States
UV Ultraviolet
VAD Voice adaptive detection
VCI Virtual channel identifier
VCO Voltage-controlled oscillator
VHDL Very high description language
VLIW Very long instruction words
VLSI Very-large-scale Integration
VPI Virtual path identifier
VSELP Vector-Sum-Excited Linear Predictive
WAND Wireless ATM Network Demonstrator
WATM Wireless asynchronous transfer mode
WB Wideband
WCDMA Wideband code-division multiple access
WLAN Wireless local area network
WLL Wireless local loop
WWW World-wide web

Index